D1087748

Ecology and Behavior of Food-Enhanced Primate Groups

MONOGRAPHS IN PRIMATOLOGY

Volume 1: Child Abuse: The Nonhuman Primate Data
Martin Reite and Nancy G. Caine, *Editors*

Volume 2: Viral and Immunological Diseases in Nonhuman Primates
S.S. Kalter, *Editor*

Volume 3: Blood Groups of Primates: Theory, Practice, Evolutionary Meaning
Wladyslaw W. Socha and Jacques Ruffié

Volume 4: Female Primates: Studies by Women Primatologists
Meredith F. Small, *Editor*

Volume 5: Clinical Management of Infant Great Apes
Charles E. Graham and James A. Bowen, *Editors*

Volume 6: Nonhuman Primate Models for Human Growth and Development
Elizabeth S. Watts, *Editor*

Volume 7: The Lion-Tailed Macaque: Status and Conservation
Paul G. Heltne, *Editor*

Volume 8: Behavior and Pathology of Aging in Rhesus Monkeys
Roger T. Davis and Charles W. Leathers, *Editors*

Volume 9: Primate Conservation in Tropical Rain Forest
Clive W. Marsh and Russell A. Mittermeier, *Editors*

Volume 10: Comparative Behavior of African Monkeys
Evan L. Zucker, *Editor*

Volume 11: Ecology and Behavior of Food-Enhanced Primate Groups
John E. Fa and Charles H. Southwick, *Editors*

Volume 12: Nonhuman Primate Studies on Diabetes, Carbohydrate Intolerance, and Obesity
Charles F. Howard, Jr., *Editor*

Ecology and Behavior of Food-Enhanced Primate Groups

Editors

John E. Fa
Departamento de Ecologia
Instituto de Biologia
Universidad Nacional Autonoma de Mexico
Mexico City, Mexico

Charles H. Southwick
Department of Environmental, Population, and Organismic Biology
University of Colorado
Boulder, Colorado

ALAN R. LISS, INC., NEW YORK

Address all Inquiries to the Publisher
Alan R. Liss, Inc., 41 East 11th Street, New York, NY 10003

Library of Congress Cataloging-in-Publication Data

Ecology and behavior of food-enhanced primate groups.

(Monographs in primatology ; 11)
Includes bibliographies and indexes.
1. Primates—Food. 2. Primates—Feeding and feeds.
3. Primates—Behavior. 4. Primates—Ecology.
5. Mammals—Food. 6. Mammals—Feeding and feeds.
7. Mammals—Behavior. 8. Mammals—Ecology. I. Fa, John E. II. Southwick, Charles H. III. Series.
QL737.P9E26 1988 599.8'0453 88-654
ISBN 0-8451-3410-8

Contents

SECTION II: DEMOGRAPHY AND LIFE HISTORY PATTERNS

SECTION III: BEHAVIOR AND SOCIAL ORGANIZATION

Contributors

Jeanne Altmann, Department of Biology, University of Chicago, and Department of Conservation Biology, Chicago Zoological Society, Chicago, IL 60637 **[ix]**

James R. Anderson, Laboratoire de Psychophysiologie, Universite Louis Pasteur, 67000 Strasbourg, France **[231]**

Michael W. Andrews, Department of Psychiatry, State University of New York, Health Science Center at Brooklyn, Brooklyn, NY 11203 **[247]**

Carol M. Berman, Caribbean Primate Research Center, Punta Santiago, Puerto Rico 00749; present address: Department of Anthropology, State University of New York at Buffalo, Buffalo, NY 14261 **[269]**

Arnold S. Chamove, Department of Psychology, University of Stirling, Stirling FK9 4LA, UK **[231]**

Montague W. Demment, Department of Agronomy and Range Science, University of California, Davis, CA 95616 **[25]**

Andrew P. Dobson, Department of Biology, Princeton University, Princeton, NJ 08544; present address: Department of Biology, University of Rochester, Rochester, NY 14627 **[167]**

John E. Fa, Departamento de Ecologia, Instituto de Biologia, Universidad Nacional Autonoma de Mexico, Mexico City, Mexico 04510 D.F.; present address: Apartado Postal 22-027, Mexico 22 D.F. **[xv,53]**

Debra L. Forthman Quick, Center for Environmental Studies, Arizona State University, Tempe, AZ 85287; present address: 9986 Braddock Drive, Culver City, CA 90232 **[25]**

Toshitaka Iwamoto, Department of Biology, Miyazaki University, Miyazaki-shi 880, Japan **[79]**

Jutta Kuester, Affenberg Salem, 7777 Salem, Federal Republic of Germany **[199]**

Phyllis C. Lee, Sub-Department of Animal Behaviour, University of Cambridge, Cambridge CB3 8AA, UK **[297]**

James Loy, Department of Sociology and Anthropology, University of Rhode Island, Kingston, RI 02881 **[153]**

Anna Marie Lyles, Department of Biology, Princeton University, Princeton, NJ 08544 **[167]**

Iqbal Malik, Department of Biology, Institute of Home Economics, University of Delhi, NDSE, New Delhi 110049, India **[95]**

The numbers in brackets are the opening page numbers of the contributors' articles.

Bernadette M. Marriott, Caribbean Primate Research Center and Department of Obstetrics and Gynecology, University of Puerto Rico, Medical Sciences Campus, San Juan, Puerto Rico 00936; present address: 2725 North Calvert St., Baltimore, MD 21218 **[125]**

Andreas Paul, Affenberg Salem, 7777 Salem, Federal Republic of Germany **[199]**

Dennis R. Rasmussen, Animal Behavior Research Institute, Madison, WI 53715; present address: Department of Behavioral Endocrinology, Wisconsin Regional Primate Research Center, Madison, WI 53715 **[313]**

Leonard A. Rosenblum, Department of Psychiatry, State University of New York, Health Science Center at Brooklyn, Brooklyn, NY 11203 **[247]**

M. Farooq Siddiqi, Department of Geography, Aligarh Muslim University, Aligarh, Uttar Pradesh, India **[113]**

Charles H. Southwick, Department of Environmental, Population, and Organismic Biology, University of Colorado, Boulder, CO 80309 **[xv, 95, 113]**

Paul Winkler, Institut für Anthropologie, Universität Göttingen, 3400 Göttingen, Federal Republic of Germany **[3]**

Foreword

The stomach's empty: there it all begins.
—Bertoldt Brecht, *Three-Penny Opera*

How do individuals of various primate species fill their stomachs? What is the impact of differences in food resources on social structure, demography, ranging patterns, and daily activities? What biological and social processes are food-limited in the ecological sense of increasing when food availability increases? The answers to these questions require studies—within species, even within populations—that examine the effects of food when other factors such as taxonomic differences in morphology or physiology are not clouding the picture. Although the rich body of primate field studies that have accumulated during the past several decades have provided tidbits to suggest the profitability of such investigations, in this volume the topic is brought to center stage for the first time, with focus on one major source of food variability: food-enrichment that results from human activities.

The consequences of variability in food supply have often eluded quantitative investigation or clear interpretation. Situations of human-supplied food enrichment offer some unique research opportunities. Because the changes in food supply that result from direct, indirect, or inadvertent provisioning by humans are usually greater than those that result from foraging exclusively on wild foods, effects may be more easily discerned. In addition, changes in food availability that arise from human activities have direct implications for conservation and for coexistence of wildlife and humans because the resultant food enrichment will sometimes entail risks for the long-term survival of these animal populations. As wildlife areas experience more and more encroachment by human habitation, agricultural expansion, and tourism, studies of food-enhanced primates are significant not only for the insight they shed on basic biological questions but also for their potential contribution to urgent issues in conservation and wildlife management.

The immediately striking effects of food-enhancement are the dramatic changes in activity profiles, which are so well documented in this volume for all the species and situations examined thus far. Whereas many of the initial findings have come from interpopulation comparisons, for which several factors may vary, a few within-population comparisons leave no question about the potency of food enrichment alone in producing a doubling of the time wild animals spent resting, halving of the time spent feeding, and smaller increases in the time spent socializing, as had been reported for animals in corrals or larger semi-captivity situations. This robust finding provides only a starting place, however. These effects raise major questions on every level of behavioral and populational biology, and most of these have barely been elucidated, much less answered.

To what extent are changes in activity profiles of biological import in themselves? To what extent does time spent searching for and processing food constrain time available for other activities, and what are the consequences of such limitation? For example, Altmann [1980] suggested that increased foraging demands would result in less time available for socializing, whereas Dunbar [e.g., 1984] argued that maintaining social bonds is so important to group-living primates that socializing time is compromised only after resting time is used up. Recent studies reported in this volume and elsewhere [e.g., Lee et al., 1986; Saunders, 1987; Altmann and Muruthi, 1988] support an intermediate conclusion: socializing or grooming time do decrease with increasing foraging demands, though less so than does resting time, and social bonds are conserved with a reduced number of partners rather than time with each (or only one) being completely protected [Saunders, 1987]. These first results suggest the depth and subtlety of the investigations that will be both necessary and fruitful. It would not be at all surprising, for example, to find that the conservation of social bonds will differ across seasons, or across other conditions that differ in group stability or susceptibility to immigration or emigration, that effort toward maintenance of social relationships will differ by gender or age-class as a function of life-history variables for each species, and that the constraints within groups differ from those between groups.

To what extent do differences in allocation of time translate into differences in nutritional intake and energy expenditure? Most studies indicate that food-enriched groups spend the same or somewhat less time traveling than do totally wild-foraging groups. However, energy expended in travel is a function of the distance traveled, not the time spent traveling; differences in energy expended in travel may be much greater than suggested by time differences if food-enriched animals travel in a more leisurely fashion. That is the case among baboons in Amboseli, where there are no differences between a food-enriched group and wild-foraging ones in time spent walking, but a fourfold difference in distance traveled and a tenfold difference in annual home range size [e.g., Altmann and Muruthi, 1988; Muruthi, unpublished]. So, differences in energy expenditure are sometimes much greater than suggested by differences in activity profiles.

Translation of feeding time to nutritional intake is even more problematical than that for traveling time, and the relationship, even more than for traveling, probably is very different between groups, or even populations, than for comparisons between individuals within a single group. Do food-enriched animals take in less, the same, or more nutrients in their reduced time spent feeding? Judging by the very few data available thus far, the answer seems to depend on the nature of the enrichment conditions: crop-raiding, intentional feeding by managers or researchers, and scrounging of food at garbage dumps differ greatly in risks for the animals, as well as in quality and, to some extent, dispersal of food. From the few studies thus far reported, it is clear and should hardly come as a surprise, that the temporal and spatial distribution of food may be at least as important as the total quantity available.

Most situations of food enhancement are ones in which predictability of food availability is high and the spatial distribution of the food is very restricted. Many of the effects that we are coming to associate with high food availability may result from this clumping and predictability of food rather than its increased quantity. Social and demographic processes may be particularly susceptible to such spatial and temporal components of food enhancement.

Rates of aggression are greater at provisioning sites than elsewhere, but both observational and experimental studies suggest that the effects of differing levels of food availability are qualitatively different within high and low parts of the availability range. At relatively low levels of food availability, increasing food availability increases various forms of social behavior, whereas at high levels of well-dispersed provisioning, competition sometimes decreases [e.g., Southwick, 1967; Belzung and Anderson, 1986; Lee, 1984 and this volume]. Food availability and predictability have effects on parent–offspring interaction as well, both indirectly, through demographic changes, and directly, as reviewed by Berman and by Rosenblum (this volume).

The effects of food availability cannot be considered independent of effects on demographic processes. Although we can be confident that food enhancement will reduce age of first reproduction, rates of infant mortality, and length of interbirth intervals, the effects on adult mortality are more ambiguous, and the effects on the various life history stages are probably not quantitatively the same, even for those animals in which the qualitative effects are similar. Moreover, it is here that the importance of considering the within-group variability in effect rather than just the group mean is highlighted. For example, the data on Japanese macaques, as summarized by several authors in this volume and originally reported by Mori [1979], Sugiyama and Ohsawa [1982], and others, suggest that reproductive parameters such as age of first reproduction exhibit greater variability among individuals, greater differentiation by social rank, when there is provisioning than when there is not. The existence, and generality, of this finding, have considerable implications for the genetic and other aspects of population processes and for social processes that are a function of social and genetic structure within groups (e.g., reviewed by Berman this volume). Development of population models such as those outlined by Lyles and Dobson in this volume and those by Cohen [e.g., 1972] and others earlier, will be an integral part of future work. However, application of such models to real populations and to management issues will depend on incorporating features that detect and are sensitive to fluctuations and recent changes in demographic parameters, not just equilibrium conditions. A large stochastic component and changing conditions have probably always been a major feature of primate life at the level of the individual and the group, but this situation is surely exaggerated in the situations involving human–nonhuman primate interactions that we encounter today.

Some of the best documented effects of food enhancement are those of developmental acceleration, resulting in shorter prereproductive stages. However, we know little or nothing about parallel effects on other parameters during development. Are all aspects of physical development promoted equally? The data for Japanese macaques suggest that skeletal development is less accelerated than is rate of growth in body mass, for example. And what of development in social and other behaviors? Do these accelerate at similar rates? If not, what are the consequences of changed co-occurrences among developmental stages? For example, in Amboseli baboons, wild-foraging females usually attain their place in the dominance hierarchy about a year before reproductive maturation. If under conditions of enrichment reproductive maturation is accelerated by at least a year but social maturation is not comparably advanced, females may be faced with the turmoil of social rank changes at the same time they meet the challenge of caring for their first infant. This provides just one example of the potential costs of apparently advantageous effects of food enrichment when there may be different sensitivities to food enrichment or a negative correlation among relevant reproductive or life-history parameters. These, in addition to ecological factors such as disease transmission discussed by Lyles and Dobson (this volume) among others, may put considerable constraints on the extent to which food is a limiting factor in primate population processes and may indicate that population changes would, in such cases, be made only with concomitant major changes in social structure and behavior.

The topics addressed in the present volume are ones of major import for the study of behavior and population processes in both human and nonhuman primates. Food enrichment is a complex, non-unitary phenomenon, at one end of a continuum of resource variability. Where food enrichment cannot or should not be eliminated, its thorough investigation may ultimately be one of the most illuminating in primate behavioral biology.

REFERENCES

Altmann J (1980): "Baboon Mothers and Infants". Mass: Harvard University Press.

Altmann J and Muruthi P (1988): Differences in daily life between semi-provisioned and wild-feeding baboons. Am J Primatol (in press).

Belzung D and Anderson JR (1986): Social rank and responses to feeding competition in rhesus monkeys. Behav Processes 12:307–316.

Cohen JE (1972): Markov population processes as models of primate social and population dynamics. Theoret Pop Biol 3:119–134.

Dunbar RIM (1984): "Reproductive Decisions." Princeton: Princeton University Press.

Lee PC (1984): Ecological constraints on the development of vervet monkeys. Behaviour 91:254–262.

Lee PC, Brennen EJ, Else JG, Altmann J (1986): Ecology and behaviour of vervet monkeys in a tourist lodge habitat; In Else JG and Lee PC (eds): "Primate Ecology and Conservation." Cambridge, England: Cambridge University Press, pp. 229–235.

Mori A (1979): Analysis of population changes by measurement of body weight in the Koshima troop of Japanese monkeys. Primates 20:371–397.

Saunders CD (1987): Ecological, Social and Evolutionary Aspects of Baboon (*Papio cynocephalus*) Grooming Behavior. Ph.D. Dissertation, Cornell University, Ithaca, N.Y.

Southwick CH (1967): An experimental study of intragroup agonistic behaviour in rhesus monkeys. Behaviour 28:182–209.

Sugiyama Y, Ohsawa H (1982): Population dynamics of Japanese monkeys with special reference to the effect of artificial feeding. Folia Primatol 39:238–263.

Jeanne Altmann
University of Chicago
and Chicago Zoological Society

Introduction

Ecological discussion has centered on whether it is social behavior or the food supply of animal populations which place a strong limit to population growth [see e.g. Watson, 1970]. Food can be considered constraining if it prevents a population from increasing; however, it can seldom account wholly for a given population size because neither mortality nor recruitment is dependent on food alone. Among others, Newton [1980] claims that, although indisputable proof that food limits a population is unattainable, circumstantial evidence supports the idea that there is a link between food and population size. For primates there is evidence that shows that a drop in population numbers coincides with food shortage [see Struhsaker, 1976]. Yet field situations that can demonstrate the fine balance between availability of food and demography are few and far between. The converse situation, in which food supply is artificially increased, is becoming easier to find.

Human population expansion and the concomitant increase in the demand for land are resulting in greater proximity between humans, agriculture, and wildlife. Because of their capacity to learn quickly, primates in particular, are drawn into human areas to new sources of food. Such food is usually of a much higher caloric value and palatability, and is available in greater quantities than natural diet items.

Deliberate feeding of primates by humans is not uncommon. Religious reasons, for example in India and South-East Asia, have traditionally been responsible for human–primate contact. The use of primates for study purposes, as occur when free-ranging colonies of macaques are established on islands, often involves long-term provisioning. Other reasons for provisioning are related to the display of animals to the public, in the Baron de Turckheim Barbary macaque parks for example; or to the keeping of monkeys away from town areas, as in Gibraltar; or to provide better conservation protection as in the case of many groups of *Japanese macaques*.

Field primatologists sometimes refer to monkeys and apes in provisioned conditions as being "artificial" and "unnatural" but "provisioning," "provisionization," and "artificial feeding" are merely terms for the offer of food beyond the animals' environmental supply. Although there is an element of difference in animal behavior brought about by the abundance of food and its static distribution in space, as well as by the animals' proximity to humans, provisioning in itself does not induce unnatural behaviors. The animals do adopt behavioral solutions in response to the more predictable and more abundant food supply. These solutions are relatively temporary consequences of the stability and size of such a resource. No doubt long-term consequences of higher quality and greater quantities of foods must be sought in the physiological and demographic responses of the individual animals and of their populations.

To move away from any possible stigma attached to the word "provisioned" the term "food-enhanced" is used in this book. The term was chosen deliberately to include situations where primates are commensal to human populations. Food-enhanced primate groups are those that receive all, or a substantial part of their food, directly or indirectly from humans.

Most chapters in this book report the results of studies that relate the food supply (quantitatively or qualitatively) of the animals with their behavior and ecology. Wherever possible, food-enhanced groups here are compared with non-provisioned groups. The present volume concentrates on examples of food-enhancement in primates following a thematic sequence in three sections: 1) Food and Energetics; 2) Demography and Life History Patterns; and 3) Behavior and Social Organization. The first section contains chapters on feeding behavior of food-enhanced primates in which the proportion, quality, and impact of provisioned foods are assessed. Section 2 deals with the effects of provisioning on the reproductive potential of food-enhanced animals and its consequences on population demography. Section 3 presents data on some aspects of behavior that become modified by provisioning. The influence of food-enhancement on the social ecology and social organization of primate groups is discussed.

The study of food-enhanced primate groups offers a real opportunity to tease apart proximate ecological and socioecological issues. Moreover, observations on the effects of the immediate environment by focusing on a major element—in this case food—may be carried out more successfully, and often in more detail, than in unprovisioned conditions. The obvious advantage of provisioned groups is that they are normally more accessible for intensive study. To dismiss or not to recognize the potential would be to play down the important contribution such controlled ecological conditions can make to primate studies.

REFERENCES

Newton I (1980): The role of food in limiting bird numbers. Ardea 68:11–30.
Struhsaker TT (1976): A further decline in numbers of Amboseli vervet monkeys. Biotropica 8:211–214.
Watson A (1970): "Animal Populations in Relation to Their Food Resources." Edinburgh: Blackwell Scientific Publications.

John E. Fa
Charles H. Southwick

Acknowledgments

This book is the result of a considerable exchange of information among contributors and between editors and contributors for more than a year. These interactions have often been frustrating because of losses in the mail. We would like to acknowledge the patience of all involved. Despite everything, we are confident that the end product is worth the increased adrenalin levels.

We are grateful to Elvia Esparza for her help in the completion of figures. Our special thanks go also to Linda Kleppinger and Linda Archibald for keeping abreast of developments, and making this book possible in spite of the many vicissitudes encountered. Monique Williamson has been an excellent critic; her editorial advice has been invaluable to the editors of the book.

SECTION I: FOOD AND ENERGETICS

Ecology and Behavior of Food-Enhanced Primate Groups, pages 3–24

1

Feeding Behavior of a Food-Enhanced Troop of Hanuman Langurs *(Presbytis entellus)* in Jodhpur, India

Paul Winkler

Institut für Anthropologie, Universität Göttingen, 3400 Göttingen, Federal Republic of Germany

INTRODUCTION

Hanuman langurs *(Presbytis entellus)* are widely distributed throughout India and adjacent countries and inhabit various climatic and vegetational regions [Oppenheimer, 1977; Roonwal and Mohnot, 1977; Vogel, 1977]. Within most parts of their range they enjoy a special status because of protective religious attitudes. Bishop et al. [1981] made a first attempt to define qualitatively human influence, pointing at its consequences on the demography and behavior of these primates. Their ordinal ratings of human disturbance levels at ten different study sites ranges from no disturbance at Orcha [Jay, 1965] to intense disturbance at Singur [Oppenheimer, 1977, 1978] and Mount Abu [Hrdy, 1977].

For some langur populations the proximity to humans implies diet amelioration because provisioning may become an important addition, either temporary in times of food shortage or permanent. Provisioning includes direct feeding by humans as well as access to cultivated fields and gardens [Oppenheimer, 1977].

Hanuman langurs like other colobine monkeys are predominantly vegetarian. A sacculated stomach, functionally resembling that of ruminants, and a rich bacterial flora permit the digestion of even mature leaves with a high content of cellulose [Amerasinghe et al., 1971; Ayer, 1948; Bauchop and Martucci, 1968; Kuhn, 1964; Oxnard, 1969]. As is typical of vegetarians, the langurs' time budgets are filled to a large degree with feeding and resting; the latter is mainly needed for digestion.

The composition of the langurs' natural diet has been investigated by several authors at various study sites [Hladik and Hladik, 1972; Oppenheimer, 1978; Ripley, 1970; Vogel, 1976; Yoshiba, 1967]. However little

Plate I. Members of the Bijolai troop feeding on groundnuts after provisioning.

Plate II. Female langur feeding on leaves of *Zizyphus nummularia*.

attention has been paid to seasonal and diurnal variation of feeding behavior or to the relationship between natural and provisioned food.

This paper describes the feeding behavior of a food-enhanced Hanuman langur troop and analyzes relevant factors such as feeding and home range use, as well as the quality and quantity of natural and provisioned food.

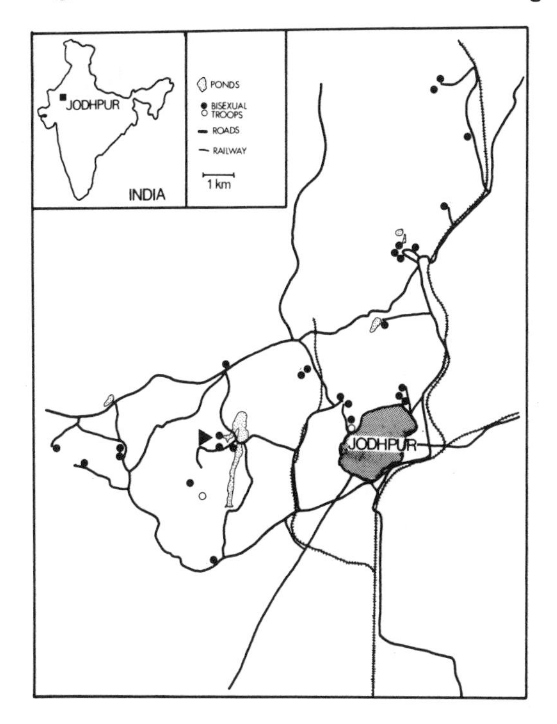

Fig. 1. Location of the study area and of bisexual troops around Jodhpur (counted troops, full circles; uncounted troops, open circles). The Bijolai troop is marked by an arrow [modified from Winkler et al., 1984].

Possible consequences of provisioning on the reproductive behavior of the troop are discussed.

STUDY AREA

Jodhpur, situated west of the Aravalli Range at the eastern fringe of the Thar desert, is one of the extreme habitats in which Hanuman langurs live (Fig. 1). Hot summers with temperatures exceeding 50°C and dry winters with moderate temperatures are typical. The monsoon season, with only limited rainfall (ϕ 360 mm of precipitation), is restricted to the months of July to September.

The vegetation of the area can best be described as a semidesert thorn-scrub community [Winkler, 1981]. Apart from a few cultivated parks and fields maintained by irrigation the bulk of the flora is xerophytic. Only a few plant species are evergreen, since the growth period of most plants is reduced to the time of monsoon. The original vegetation has been drastically altered by introduced plants. Above all, *Prosopis juliflora* (Mimosaceae) has become the most dominant species in this region.

STUDY ANIMALS

The langur population of Jodhpur is organized into two different troop types: bisexual troops, composed of one adult male and 4–35 females with their offspring, and all-male bands, consisting of males of all age classes except infants. According to a census carried out in 1980, there were 29 bisexual troops totaling about 900 animals. The number of all-male bands was about 13 with approximately 120 animals. According to a recent census by Mohnot (personal communication), the total population size in 1984 was around 1,200 animals, spread over an area of approximately 150 km^2. The population is genetically isolated. No other langurs live within a radius of about 100 km (for a detailed description of relevant data of the Jodhpur population see Winkler et al. [1984]).

Since 1977 the Jodhpur langur population has been surveyed by German and Indian primatologists. Ethological studies have concentrated on two main troops, Bijolai and Kailana-I, which have been under nearly constant observation. This report concentrates on data of a pilot study, undertaken in 1977/78. Its aim was to study the comparative ecology and behavior of two langur troops, one of which was the Bijolai troop.

In 1977, the Bijolai troop lived near Kailana Lake, an artificial reservoir about 8 km west of the city, which provides drinking water for Jodhpur. The demographic development of this troop has been fairly well documented since 1967 [Winkler et al., 1984]. In addition several cases of infanticide and adult male changes have been described in detail by Mohnot [1971, 1974], Makwana and Advani [1981], Vogel and Loch [1984], Sommer and Mohnot [1985], and Agoramoorthy [1986]. Some ecological descriptions have been made by Winkler [1981, 1984]. Since August 1977, all members of the troop have been known individually. Kinship relations and reproductive parameters are published elsewhere [Winkler et al., 1984].

MATERIALS AND METHODS

The present study is based on 509 h of observation (focal animal sampling and scan-sampling [Altmann, 1974]) between September 1977 and July 1978 with an interruption of 2 months in May and June 1978. In the course of this study troop size varied between 11 and 17 animals.

The behavior and home range use of the Bijolai troop was documented by scan-sampling [Altmann, 1974] at 15-min intervals. For each visible animal, current behavior (classified as grooming, resting, feeding, monitoring, or locomotion [see Winkler, 1981]) as well as the animals' location was noted. The results of all scans were summarized monthly for every hour of the day. To allow for a comparison of data over the whole observation period (i.e., with respect to every month and every hour of day), the summarized behavior and location frequencies were divided by the total number of scans per hour obtained for every month. Thus, the reported results represent the relative amount of time (in percent) spent by the troop in a particular activity or at a specific place.

RESULTS

Diurnal Variation in Feeding Behavior

Diurnal variation in feeding activity is shown in Figure 2. For the entire observation period, feeding was considerably low in the morning, but increased in the afternoon. A Friedman two-way analysis of variance revealed that the observed diurnal variations are statistically significant ($\chi^2 = 46.51$; $P < 0.05$).

If data for all months are pooled (Fig. 3) two peaks of feeding activity can be discerned: one between 11 and 12 h and another between 17 and 18 h. Such a bimodal shape in feeding activity has also been reported from other langur study sites. In general, there is a morning peak between 0600 and 0900 and an evening peak between 1500 and 1800 [Jay, 1965; Kurup, 1970; Rahaman, 1973; Ripley, 1965, 1970; Starin, 1978; Sugiyama, 1964; Vogel, 1976; Yoshiba, 1967]. In some studies, however, an increased feeding rate around noon has also been reported [Ripley, 1965, 1970; Yoshiba, 1967]. In Sariska [Vogel, 1976] and Dharwar [Yoshiba, 1967] feeding activity is higher in the afternoon, but the morning peak at Gir Forest [Kurup, 1970] and Polonnaruwa [Ripley, 1965] exceeds that of the afternoon.

When compared with these other study sites, the feeding peaks of the Bijolai troop are somewhat later in the day. Such differences may be attributable to the effect of temperature on the monkeys' behavior. This influence can be demonstrated by a comparison of temperatures at those times of day when feeding reaches its maximum. The hourly average temperature of the activity period (6–20 h) for each month was ranked by giving the hour with the lowest temperature rank No. 1 and the hour with the highest temperature rank No. 14. The mean rank of those hours in which the morning and evening peak fell was calculated. A mean temperature rank of 7.5 was obtained for the feeding activity of the Bijolai troop over the whole study. Thus the troop fed especially at those times of the day that had temperatures that were medium as compared with the monthly average.

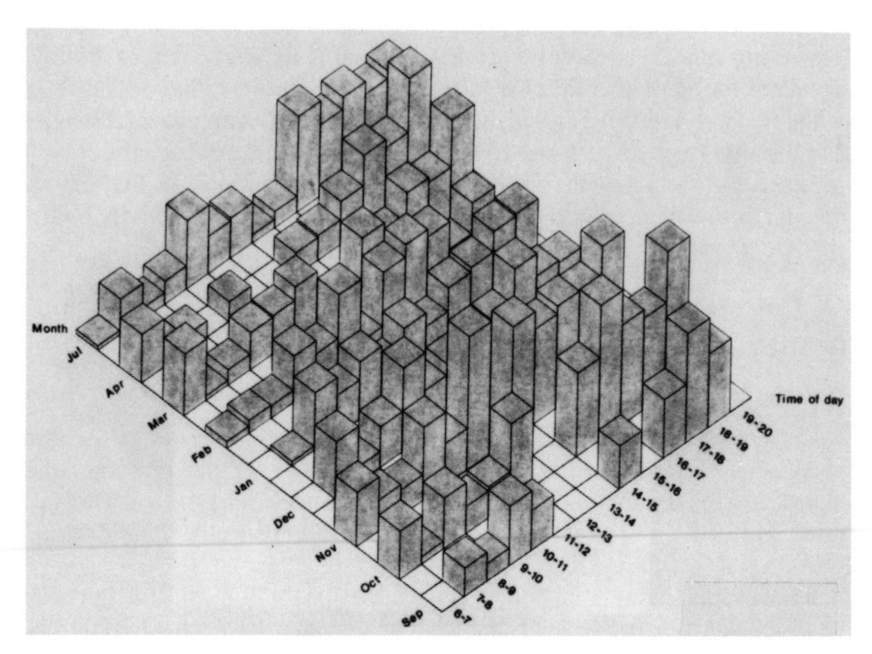

Fig. 2. Diurnal distribution of feeding activity according to the months of the study period (September 1977–July 1978). Each block represents the proportion of time spent feeding by the Bijolai troop.

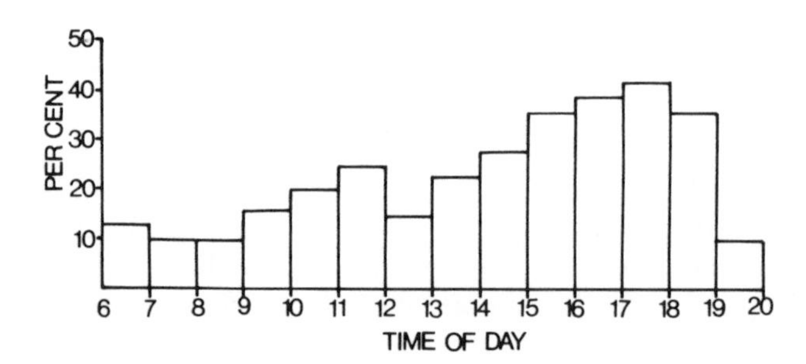

Fig. 3. Diurnal distribution of feeding activity (as a percentage of total activity) for the whole study period. Each column represents the proportion of time spent feeding by the Bijolai troop per time of day.

A similar correlation between temperature and feeding activity can be seen from the study by Vogel [1976]. At his Sariska study site, where temperature changes between day and night are as high as in Jodhpur, the feeding peak

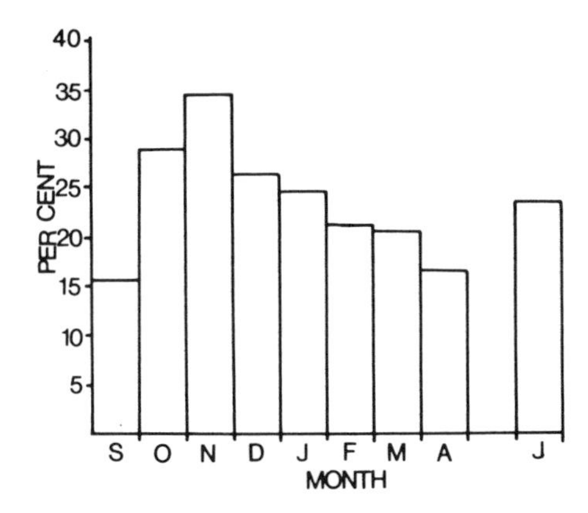

Fig. 4. Monthly distribution of feeding activity (as a percentage of total activity). Each column represents the proportion of time spent feeding per month by the Bijolai troop.

lay between 0900 and 1100 h. These are the hours of the day with almost medium temperatures. However, at his Bhimtal study site where there are very low overall deviations in temperature and where mean temperatures are reached very early in the day, a feeding peak was observed between 0700 and 0800. Temperature could also account for the resting activity of the Bijolai troop, which prevails at times with higher temperatures.

Monthly Variation in Feeding Behavior

Variation in monthly feeding rate is plotted in Figure 4. The mean amount of time spent feeding by the Bijolai troop during the whole period of observation is 23.5%, with a minimum of 15.7% in September and a maximum of 34.3% in November. The graph reveals two different trends. There is an increase in feeding from September to November and a subsequent decrease up to April. The significance of this distribution was tested by a Friedman two-way analysis of variance ($\chi^2 = 21.38$; $P < 0.05$). There was no correlation between feeding activity and any climatic parameter, such as temperature, humidity, etc.

Feeding Localities

Since distribution and abundance of food influence the overall activity of a troop—above all, feeding behavior—the size of a troop's home range as well as the behavior shown in different parts of the home range must be documented. The home range of the Bijolai troop was divided into nine districts (Fig. 5). Borders of the various areas used by the langurs, to be known as districts, were defined by the coherence of the troop, irrespective

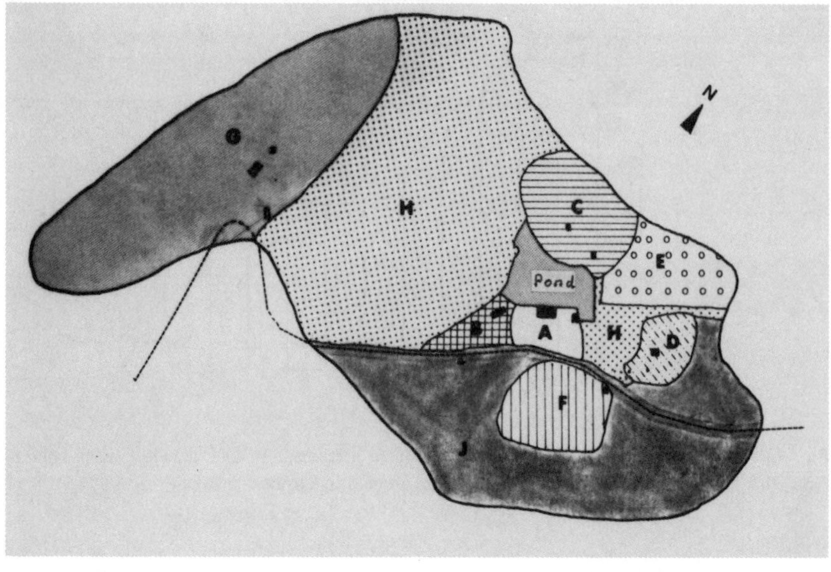

100 200 m

Fig. 5. Home range of the Bijolai troop. The different districts (A–J) are marked by different hatchings. ■, house; ■, temple; ——, street.

of the current behavior. In general, it was apparent that the whole troop occupied a certain area which they left together and rapidly to occupy another.

With the exception of one district (district H) all districts were characterized by one or more prominent landmarks—such as hills, trees, or temples—from which large parts of the home range could be surveyed by the monkeys. Total home range size for the entire observation period was 56.6 ha. Diurnal and monthly variations in the use of each district were tested by means of a Friedman two-way analysis of variance. The differences in monthly variation of home range use are statistically significant for all nine districts, but variations in day time use were significant for only six out of the nine.

The Bijolai troop spent 63.3% of time in three districts (A,C,D; Fig. 5), an area which accounts for only 8.4% of the total range. These data show that there is a preference for certain districts, the reasons for which will be analyzed below.

Preference for a specific district may indirectly be ascertained from the behavior of the troop typical for that district. With reference to feeding behavior alone two main relationships were tested monthwise: 1) the relation between the time the troop spent feeding in a specific district and total time

TABLE I. Results of the Spearman Rank Correlation for Testing the Hypothesis of a District-Specific Feeding Rate ($P \leqq 0.05$)

District	Feeding activity	
	I	II
A	+	+
B	+	−
C	+	−
D	−	−
E	+	−
F	+	−
G	+	−
H	+	+
J	+	−

+, positive correlation; −, no correlation; I, relation between feeding time and duration of stay; II, relation between feeding time in the district and total feeding time.

the troop remained in it, and 2) the relation between the time the troop spent feeding in a specific district and total time the troop spent feeding.

The results of these tests, which were done by a Spearman rank correlation (Table I), should clarify whether the time spent feeding in a district depends on the time the troop stayed in that district or on the overall time spent on this activity. Thus a high feeding value in district A in October may be due to a general high feeding level for that month or due to a higher use of district A. Of course, both influences may be acting simultaneously so that only a combination of the two relationships can provide evidence for a district-specific behavior pattern.

If both relationships reveal a positive correlation (e.g., district H)—i.e., if both the total amount of time the troop spent in a district and the total amount of time the troop spent feeding correlates with the time the troop spent feeding in a district—it can not be identified definitely which influence was responsible for the feeding rate in a district. Both influences are superimposed. If, however, no correlation appears in either case (e.g., district D), the observed feeding rate in a district is district-unspecific as well as activity-unspecific. But if the test reveals a positive correlation between the feeding in a district and the total amount of time the troop stayed in that district, but no correlation with the total amount of time the troop spent feeding (e.g., district B), the feeding rate in this district is district-specific.

Table I shows that in six out of nine districts a district-specific feeding rate can be determined. This means that among other things these districts do have certain peculiarities that make them attractive for feeding. It should be

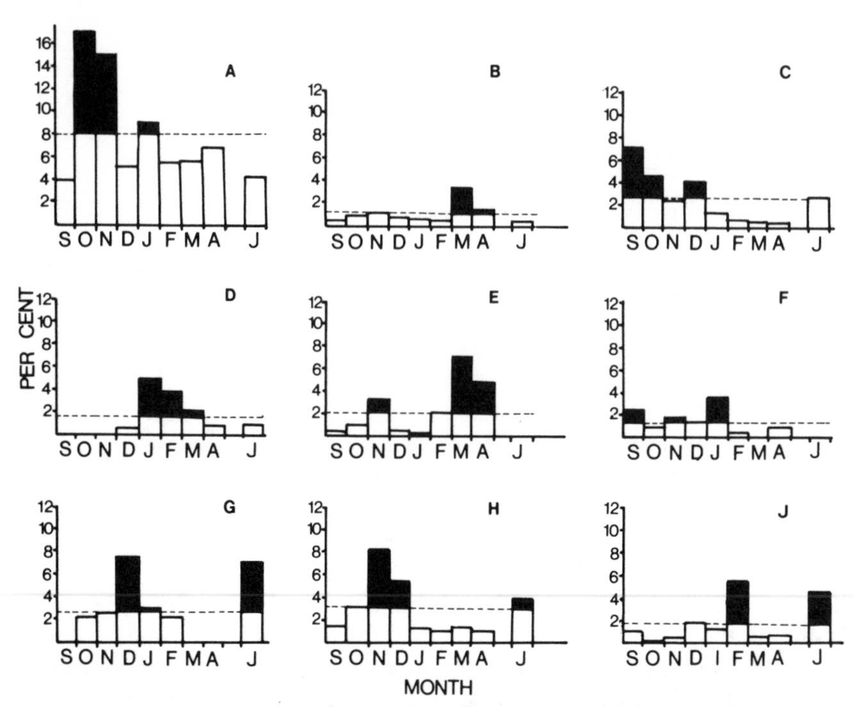

Fig. 6. Monthly distribution of feeding activity within the nine districts of the Bijolai troop's home range: graphical representation of the tendency (= deviation from the mean value of the total observation period). Each column represents the proportion of time spent feeding (in percent) in each month and district. ■, positive deviation from the mean value; □ negative deviation from the mean value. The mean value is indicated by the boundary line between positive and negative deviation.

mentioned, however, that in addition to feeding activity several districts are characterized by a combination of other activities as well [Winkler, 1981].

Figure 6 gives a graphical representation of feeding times in different districts. For each month the feeding times are plotted and for each district a mean feeding value was calculated for the whole observation period. Feeding in districts B (60.3%), F (60.3%), and J (47.0%) accounts for the highest amount of time spent on a specific activity in these three districts. However, only 16.7% of the total feeding time and 7.3% of the total observation time was spent in these, perhaps pointing to some qualitative rather than quantitative aspects of the diet.

Diet of the Bijolai Troop

To analyze the composition of the natural diet all plant species eaten by the Bijolai troop were identified. A total of 35 different species formed part of the Bijolai troop's diet (Table II). Plant species which were eaten on more than 10% (= 17 days) of the total observation days (Table III) were considered as staple food. Table III shows that two of them, *Prosopis juliflora* and *Acacia senegal,* were fed upon more than 50% of the observation days. Both species are also widely distributed within the home range of the Bijolai troop (Table IV). *Prosopis juliflora* is the most prominent plant species in the Jodhpur region. This plant, originally from Mexico, had been introduced to Jodhpur only in 1914 [Bhandari, 1978].

An attempt was made to find a comparative measure of food intake. A plant-specific feeding rate was calculated by using the number of feeding hours on each species relative to the total number of feeding hours observed. Thus, the ratio Q can vary between 0 (species not eaten) and 1 (species eaten within every feeding hour). The comparison of different plant specific feeding rates was done hourly and monthly.

Monthly feeding rates of staple foods are presented in Figure 7. The marked variation of all species over the study period depends only to some degree on the restricted growth period *(Grewia tenax, Trapa natans, Euphorbia caducifolia).* In the other species there is no such correlation. Maximum feeding rates of the species most frequently eaten *(Prosopis juliflora, Acacia senegal, Trapa natans, Anogeissus pendula)* lie in the months of November and December.

Figure 8 demonstrates diurnal feeding rates of staple foods of the Bijolai troop. Here, too, a marked variation throughout the day is evident. In addition, a significant negative correlation exists between the feeding rates of *P. juliflora* and *A. senegal* (Spearman rank correlation, $r_s = -0.579$, $P < 0.05$). While *P. juliflora* is preferred during the early morning and evening hours, *A. senegal* has its maximum rate at noon.

Another correlation can be calculated for the feeding rates of *A. senegal* and *A. pendula* ($r_s = 0.893$, $P < 0.002$). Figure 8 not only demonstrates the importance of the main feeding plants, but also the different composition of the diet across the day. During early-morning and evening hours the diet consists of plants or plant parts with a high water content (fruits of *P. juliflora* and leaves of *T. natans).* At noon the preferred diet consists of plant parts with a low water content (seeds of *A. senegal,* dried leaves of *A. pendula).* The feeding on the natural diet must, however, be seen in connection with provisioning, which very much influences the general feeding behavior of the Bijolai troop.

Provisioned Food Feeding Rates

The Bijolai troop belongs to those troops of the Jodhpur population that are fed almost daily by humans. Local people attract the langurs' attention by loud calling. Sometimes, the animals run for more than 100 m to come to the

TABLE II. Plant Species Eaten by Members of the Bijolai Troop

Family	Species	Local name	Life form	Fl[a]	Fr[b]
Acanthaceae	*Barleria acanthoides* Vahl	Bajardanti, chapari	Shrub	IX	XI
Amaranthaceae	*Achyranthes aspera* Linn. var. *argentea* Hook. f.	Andhi-jalo, undo-kanto, katio-bhuratio, unta ghada	Herb	VIII	XII
Amaranthaceae	*Pupalia lappaceae* (Linn.) Juss.	Undio bhurat	Herb	VIII	I
Apocynaceae	*Wrightia tinctoria* R. Br.	Bhakar-aak, kerno	Tree	III to IV	VIII–XII
Asclepiadaceae	*Calotropis procera* (Ait.) R. Br.	Akaro, aak	Shrub	I–XII	I–XII
Asclepiadaceae	*Pentatropis spiralis* (Forsk.) Decne.	Aakari bel	Shrub	X–I	XII–II
Asteraceae	*Vernonia cinerascens* Schultz.-Bip.	?	Shrub	X	I
Burseraceae	*Commiphora wightii* (Arnott) Bhandari	Guggul	Shrub	V to VI leaves	
Caesalpiniaceae	*Cassia pumila* Lamk.	?	Herb	VIII	XII
Capparaceae	*Capparis decidua* (Forsk.) Edgew.	Ker, kerro	Shrub	III–IV IX–X	V to VI, XI
Celastraceae	*Maytenus emarginata* (Willd.) Ding Hou	Kankero	Tree	X	II
Combretaceae	*Anogeissus pendula* Edgew.	Dhauro, dhau, endruk	Tree	IX	X to XI
Elatinaceae	*Bergia ammannioides* Roxb.	Jal bhangro	Herb	IX	XII
Ehretiaceae	*Cordia gharaf* (Forsk.) Ehrenb. & Aschers.	Goondi	Tree	III to IV	V to VI
Euphorbiaceae	*Euphorbia caducifolia* Haines	Thor, danda thor	Shrub	I to II	II–IV
Fabaceae	*Indigofera cordifolia* Heyne ex Roth	Bekario, bekar	Herb	VIII	XI

Fabaceae	*Indigofera hochstetteri* Baker	Adio-bekario	Herb	VIII	X
Fabaceae	*Indigofera oblongifolia* Forsk.	Goilia, jhil	Shrub	IX	III
Fabaceae	*Rhynchosia minima* DC.	Chiri-motio, kalta	Herb	VIII	X
Fabaceae	*Tephrosia purpurea* (Linn.) Pers.	Biyani	Herb	VII	XII
Fabaceae	*Tephrosia strigosa* (Dalz.) Sant. & Mahesh.	Jhino-biyono	Herb	VIII	X
Liliaceae	*Asparagus racemosus* Willd.	Norkanto, satawar	Shrub	XI	XII
Lythraceae	*Lawsonia inermis* Linn.	Mehndi	Shrub	?	?
Menispermaceae	*Cocculus hirsutus* (Linn.) Diels.	Bajar-bel	Shrub	IX	XII
Mimosaceae	*Acacia nilotica* (Linn.) Del. ssp. *indica* (Benth.) Brenan	Banwal, babul	Tree	V–X	XII–IV
Mimosaceae	*Acacia senegal* (Linn.) Willd.	Kumbat, kumatiyo	Tree	VII	I
Mimosaceae	*Prosopis juliflora* (Swartz) DC.	Angreji bavanlio	Tree	I–XII	I–XII
Moraceae	*Ficus religiosa* Linn.	Pipal	Tree	IV	VI
Poaceae	*Dactyloctenium aegyptium* (Linn.) P. Beauv.	Makaro, manchi	Grass	IX	I
Rhamnaceae	*Zizyphus nummularia* (Burm. f.) Wt. & Arn.	Borti, bordi	Shrub	VIII–X	X–XII
Salvadoraceae	*Salvadora oleoides* Decne.	Kharo jhal	Tree	III to IV	V to VI
Solanaceae	*Solanum albicaule* Kotschy ex Dunal	Nar-kanta	Shrub	VIII	XII
Tiliaceae	*Grewia tenax* (Forsk.) Fiori	Gangerun, gangan kankeran	Shrub	VIII–X	IX–XII
Trapaceae	*Trapa natans* var. *bispinosa* (Roxb.) Makino	Singhara	Herb	IX	XI
Verbenaceae	*Phyla nodiflora* (Linn.) Greene	?	Herb	I–XII	I–XII

[a]Fl, period of flowering (months).
[b]Fr, period of fruiting (months).

TABLE III. Staple Food of the Bijolai Troop

Species	n[a]	%[b]
Prosopis juliflora	97	58.1
Acacia senegal	87	52.1
Trapa natans	49	29.3
Anogeissus pendula	44	26.3
Grewia tenax	33	19.8
Zizyphus nummularia	33	19.8
Euphorbia caducifolia	23	13.8

[a]n, No. of feeding days.
[b]%, feeding days in percentage of observation days.

TABLE IV. Distribution and Frequency of Staple Food Within the Home Range of the Bijolai Troop

Species	*Prosopis juliflora*	*Acacia senegal*	*Trapa natans*	*Anogeissus pendula*	*Grewia tenax*	*Zizyphus nummularia*	*Euphorbia caducifolia*
District							
A	+ + +	+	+ + +	−	+	+	−
B	+	+	+	−	+	+	+
C	+ +	+ + +	+ +	+	+	+	+ +
D	+ +	+ +	−	+	+ +	+ +	+ +
E	+ + +	+ +	−	+	+ +	+	+
F	+ +	+ +	−	+	+	+	+
G	+ +	+ + +	−	+ +	+	−	+ +
H	+ + +	+ + +	+ + +	+ +	+ +	+ +	+ +
J	+ + +	+ + +	−	+ +	+ +	+ +	+ +

−, not existing; +, between 1 and 10 specimens; + +, more than 10 specimens, distributed irregularly throughout the district; + + +, very frequent throughout the district.

feeding site. However, on other occasions, they would not move at all, even if they were very close by.

The overall composition of the provisioned food diet is listed in Table V, showing the high quality of the component items. Details of the provisioning events are listed in Table VI. There is great variation in the total number of provisioning events as well as in feedings per feeding day. January shows the highest and November and April the lowest scores.

Such feedings are concentrated in only a few districts of the langurs' home range (Table VII). A total of 80.6% of all feedings took place in four districts (A, B, C, and D) where the troop spent 65% of its time. Most feedings took place in district A, where the sleeping site of the troop was located.

The artificial feeding rate varies between months but generally takes up a very high proportion of the total feeding time (Fig. 9). On comparing the feeding rates for the natural diet (Fig. 7) with that of provisioning, a close

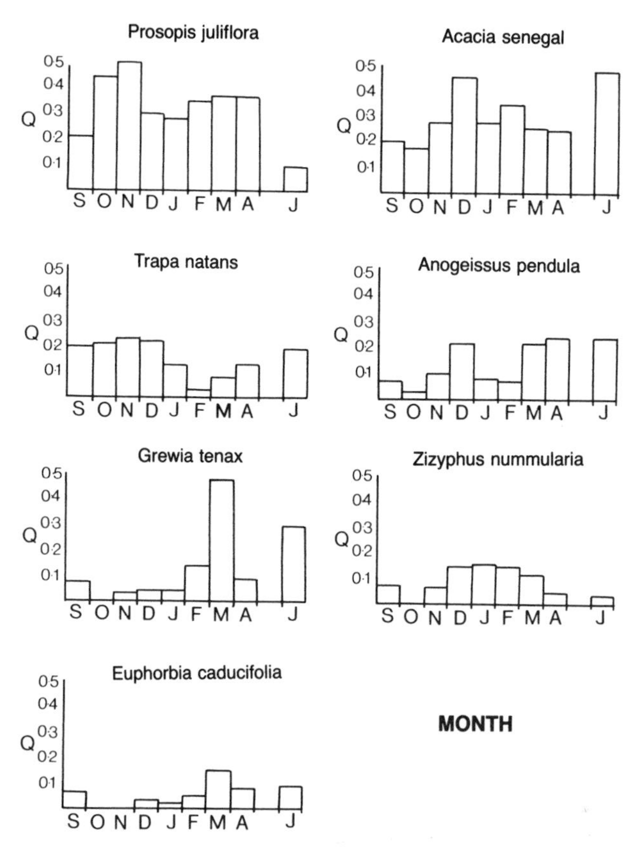

Fig. 7. Monthly distribution of plant-specific feeding rates of the Bijolai troop. Q = number of feeding hours on a specific plant species in relation to the total number of feeding hours.

correspondence between feeding on *P. juliflora* and artificial feeding appears. The maxima of *P. juliflora* (November and April) correspond to the minima in feeding on provisioned food. Likewise the maximum provisioned food feeding for January corresponds with the minimum feeding rate for *P. juliflora*.

DISCUSSION AND CONCLUSIONS

The different aspects of feeding behavior analyzed in this study can be summarized under three main topics.

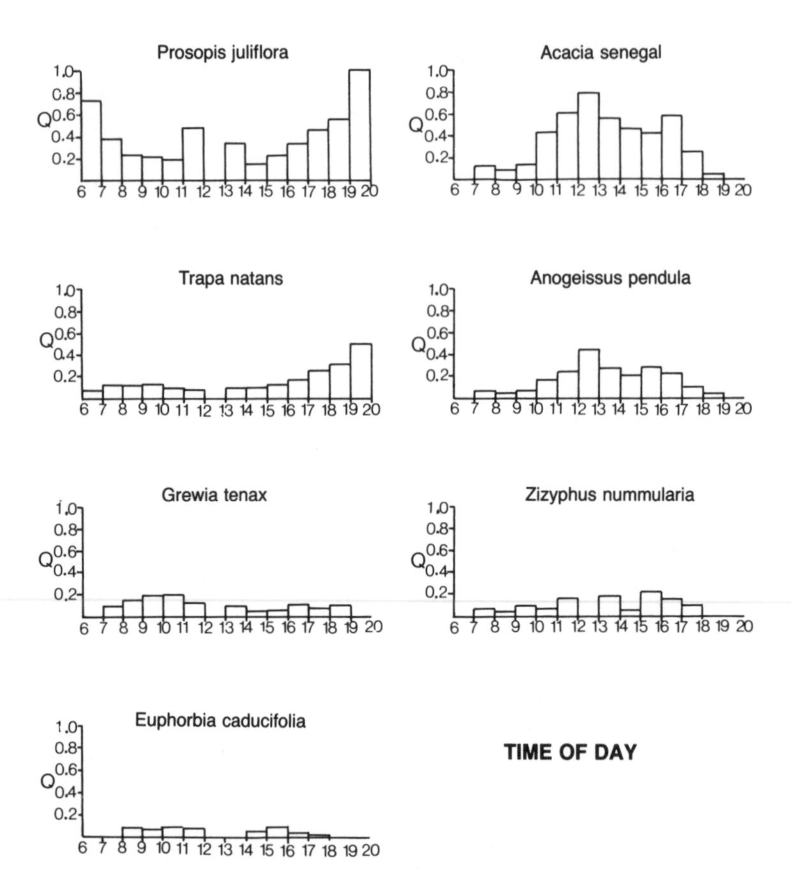

Fig. 8. Diurnal distribution of plant-specific feeding rates of the Bijolai troop. Q = number of feeding hours on a specific plant species in relation to the total number of feeding hours.

The Daytime Course of Feeding

Over the whole period of observation the Bijolai troop showed significant diurnal variations in feeding. This holds true for the total feeding rate as well as for the plant-specific ones. There is good evidence that temperature may be one of the factors responsible for daytime fluctuations in behavior. Feeding behavior is concentrated during hours with medium temperatures—an observation also reported for Sariska and Bhimtal [Vogel, 1976]. Thus, one can assume that langurs even in a comparably rich habitat (such as Jodhpur) try to optimize their energetic budget.

The Bijolai troop exhibited marked selectivity in feeding during the day. Feeding rates of the most preferred plants were closely related. While the rate

TABLE V. List of Provisioned Foods of the Bijolai Troop

Plant species		
Family	Species	Common name
Anacardiaceae	*Mangifera indica* L.	Mango
Apiaceae	*Daucus carota* L.	Carrot
Arecaceae	*Cocos nucifera* L.	Coconut
Brassicaceae	*Brassica oleracea* var. *botrytis* L.	Cauliflower
Brassicaceae	*Brassica oleracea* var. *capitata* L.	White cabbage
Brassicaceae	*Raphanus sativus* L.	Radish
Chenopodiaceae	*Beta vulgaris* var. *crassa* L.	Sweet turnip
Convolvulaceae	*Ipomoea batatas* L.	Sweet potato
Cucurbitaceae	*Cucumis sativus* L.	Cucumber
Fabaceae	*Arachis hypogaea* L.	Groundnut
Liliaceae	*Allium cepa* L.	Onion
Musaceae	*Musa sapientium* L.	Banana
Passifloraceae	*Carica papaya* L.	Papaya
Punicaceae	*Punica granatum* L.	Pomegranate
Solanaceae	*Solanum melongena* L.	Aubergine
Solanaceae	*Solanum tuberosum* L.	Potato
Prepared foods		
Chapatis, kofta, puri, millet cakes, toasted bread, sweets, salt cakes.		

TABLE VI. Details of Artificial Feedings in the Bijolai Troop

Year	Month	OD	FD	FD/OD	AF	AF/FD	FT	ϕ FT/AF
1977	Sep	19	14	0.74	22	1.57	264	12.00
	Oct	20	14	0.70	25	1.79	385	15.24
	Nov	19	6	0.32	8	1.33	114	14.15
	Dec	19	13	0.68	18	1.38	318	17.40
1978	Jan	23	17	0.74	47	2.76	937	19.56
	Feb	23	17	0.74	28	1.65	465	16.37
	Mar	20	13	0.65	24	1.85	486	20.15
	Apr	9	6	0.67	8	1.33	126	15.45
	May	—	—	—	—	—	—	—
	Jun	11	8	0.73	14	1.75	190	13.34
	Jul	15	8	0.53	12	1.50	190	15.50
	Σ	178	116		206		3475	
	ϕ			0.65		1.78		16.52

OD, days of observation; FD, days with artificial feedings; AF, number of artificial feedings; FT, feeding time in minutes.

for *P. juliflora* is U-shaped, that for *A. senegal* shows an opposite trend. From the other feeding rates and the plant parts eaten there seems to be a tendency for the Bijolai troop to feed on a diet with a high water content in the morning and evening and with low water content at noon.

TABLE VII. Proportion (in Percentage) of Artificial Feedings (n = 206) Within the Districts of the Home Range of the Bijolai Troop

District	%
A	35.9
B	14.6
C	11.2
D	18.9
E	—
F	4.4
G	2.4
H	6.8
J	5.8

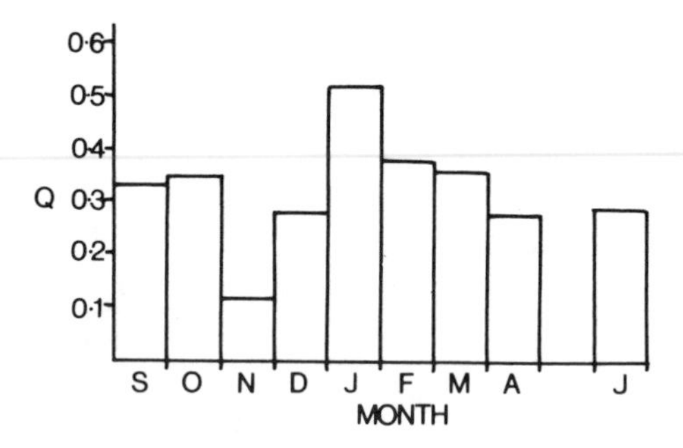

Fig. 9. Monthly distribution of feeding rate based on artificial feeding of the Bijolai troop. Q = number of feeding hours based on artificial feeding in relation to the total number of feeding hours.

The Seasonal Course of Feeding

In contrast to reports from other study sites [Curtin, 1975, for Junbesi, Nepal; and Oppenheimer, 1978, for Singur, India] monthly variations in feeding behavior of the Bijolai troop were not related to temperature or other climatic variables. For the present study, it was not possible to observe any relation between time spent feeding and quality of diet in terms of plant parts eaten, since no data on the diet's energetic value were available.

A negative correlation was found between feeding on *P. juliflora* and on provisioned food. Data from another troop, Daijar, of the Jodhpur population show that in the same period of time, only 76 feedings were performed when compared to the 206 feedings of the Bijolai troop. Although *P. juliflora* was the main diet item, as was observed at Bijolai, no correlation between the rates of feeding on this species and feeding on provisioned foods was found. It is suggested that *Prosopis juliflora,* an evergreen and widespread species, is the main source of nutrition for the Jodhpur langurs. However, this can be substituted for by provisioning.

Although little is known of the influences of climatic variables on langur behavior, the few published results from other study sites show a similar trend. During the cooler months of the year the amount of time spent feeding is greater than during the warmer months of the year. This has been reported for the extreme cold habitat at Junbesi, Nepal [Curtin, 1975; cited in Bishop, 1979] as well as for the savanna-type habitat at Singur near Calcutta. For the latter, Oppenheimer [1978] reported an average amount of daily feeding activity of 29%. This amount varied from 22% to 26% with temperatures above 30°C and 33% to 39% with temperatures below 21°C. The absence of a correlation between climate and feeding at Jodhpur can be explained by the composition of the diet. Throughout the year there seems to be sufficient nutrition available, both in qualitative and quantitative terms.

District-Specific Feeding Behavior

The different use made of the Bijolai troop's home range can be traced to the specific attributes of the component districts. Analyses of feeding data show that six out of nine districts have a district-specific feeding rate, indicating that the diet (whether natural or provisioned) in these districts is the main reason for staying there.

When defining the Bijolai troop as a provisioned troop, it is especially important to assess whether or in which way provisioning influences the behavior of the animals, above all their reproductive behavior. Several consequences of food enhancement are discussed in this book for different species. Such data on langurs, however, are not available yet. There are only two places for which continuous data on individually known females are published—one is the colony in Berkeley, California [Harley, 1985], the other is Jodhpur, India [Winkler et al., 1984; Winkler, 1986]. Although reproductive data from other study sites are also available [e.g., Hrdy, 1977; Jay, 1963; Sugiyama, 1965, 1966], they are mainly based on estimates or calculations on a cross-sectional basis, but such calculations tend to be overestimates. This can be clearly demonstrated for the Jodhpur langurs. Our long-term studies on two troops over a period of 9 years revealed an average birth interval of 14.7 months for the Bijolai troop and 16.9 months for the Kailana-I troop [Winkler, 1986]. When calculating birth intervals on a cross-sectional basis it comes to 20.4 months for the Bijolai troop and 23.5 months for the Kailana-I troop [Winkler et al., 1984]. These intervals correspond with those reported from other field studies on non-seasonal-breeding langurs (mostly 19–24 months; see Vogel and Loch [1984] for a summary).

The long-term study at Berkeley revealed a median birth interval of 15.4 months for a period of 10 years [Harley, 1985] which is nearly identical with that of Jodhpur. Other data, such as age at menarche, first conception, and first live birth, seem to be very similar. Thus the problem is that, on the one hand, the only available long-term data stem from provisioned populations that might differ from nonprovisioned populations; on the other hand there is evidence that data from nonprovisioned and non-seasonal-breeding populations are probably overestimates.

Therefore the question of the influence of provisioning on the reproductive behavior of langurs cannot be answered until long-term data from nonprovisioned and non-seasonal-breeding populations become available.

SUMMARY

In the course of an ecological study on Hanuman langurs *(Presbytis entellus),* living around Jodhpur, India, the feeding behavior of the food-enhanced Bijolai troop was studied. Daytime fluctuations in the total feeding rate seem to be related to temperature. The main feeding bouts were concentrated during hours with medium temperatures. The feeding rates of the most preferred plants were closely related, with a tendency to feed on a diet with a high water content in the early morning and evening and with low water content at noon. Although the feeding time varied from month to month, there was no correlation with climatic variables. A total of 35 different plant species formed the major part of the troop's diet, with *Prosopis juliflora* being the most important one. A negative correlation was found between feeding on this plant and on provisioned food, suggesting that *Prosopis juliflora* is the main source of nutrition, which, however, can be substituted by provisioning. Six out of nine districts of the Bijolai troop's home range have a district-specific feeding rate, indicating that the diet in these districts is the main reason for staying there. The influence of provisioning on the reproductive behavior of the langurs cannot be determined until long-term data from nonprovisioned and non-seasonal-breeding populations become available.

ACKNOWLEDGMENTS

This study was supported by a grant from the Deutscher Akademischer Austauschdienst (DAAD) and the Ministry of Education and Social Welfare, Government of India, under the Indo-German Cultural Exchange Programme. The long-term study was supported by a grant from the Deutsche Forschungsgemeinschaft (DFG), grant No. Vo 124/11-13.

REFERENCES

Agoramoorthy G (1986): Recent observations on 12 cases of infanticide in Hanuman langur, *Presbytis entellus,* around Jodhpur, India. Primate Rep 14:209–210 (abstract).

Altmann J (1974): Observational study of behavior: sampling methods. Behaviour 49:227–267.

Amerasinghe FP, Van Cuylenberg BWB, Hladik, CM (1971): Comparative histology of the alimentary tract of Ceylon primates in correlation with the diet. Ceylon J Sci Biol Sci 9:75–87.

Ayer AA (1948): "The Anatomy of *Semnopithecus entellus*". Madras: The Indian Publishing House.

Bauchop T, Martucci RW (1968): Ruminant-like digestion of the langur monkey. Science 161:698–700.

Bhandari MM (1978): "Flora of the Indian Desert." Jodhpur: Scientific Publishers.

Bishop NH (1979): Himalayan langurs: temperate colobines. J Hum Evol 8:251–281.

Bishop N, Hrdy SB, Teas J, Moore J (1981): Measures of human influence in habitats of south Asian monkeys. Int J Primatol 2:153–167.

Curtin R (1975): "The Socioecology of the Common Langur *(Presbytis entellus)* in the Nepal Himalaya." Ph.D. thesis. Berkeley: University of California.

Harley D (1985): Birth spacing in langur monkeys *(Presbytis entellus).* Int J Primatol 6:227–242.

Hladik CM, Hladik A (1972): Disposibilités alimentaires et domaines vitaux des primates à Ceylan. Terre et Vie 26:149–215.

Hrdy SB (1977): "The Langurs of Abu." Cambridge: Harvard University Press.

Jay PC (1963): "The Social Behavior of the Langur Monkey." Ph.D. thesis. Illinois: University of Chicago.

Jay P (1965): The common langur of north India. In DeVore I (ed): "Primate Behavior—Field Studies of Monkeys and Apes." New York: Holt, Rinehart, and Winston, pp 197–249.

Kuhn HJ (1964): Zur Kenntnis von Bau und Funktion des Magens der Schlankaffen (Colobinae). Folia Primatol (Basel) 2:193–221.

Kurup GU (1970): Field observations on habits of Indian langur, *Presbytis entellus* (Dufresne) in Gir Forest, Gujarat. Rec Zool Surv India 62:5–9.

Makwana SC, Advani R (1981): Social changes in the Hanuman langur, *Presbytis entellus,* around Jodhpur. J Bomb Nat Hist Soc 78:152–154.

Mohnot SM (1971): Some aspects of social changes and infant-killing in the Hanuman langur, *Presbytis entellus* (Primates: Cercopithecidae), in western India. Mammalia 35:175–198.

Mohnot SM (1974): "Ecology and Behaviour of the Common Indian Langur, *Presbytis entellus* Dufresne." Ph.D. thesis. Jodhpur: University of Jodhpur.

Oppenheimer JR (1977): *Presbytis entellus:* the Hanuman langur. In His Serene Highness Prince Rainier III of Monaco, Bourne GH (eds): "Primate Conservation." New York: Academic Press, pp 469–512.

Oppenheimer JR (1978): Aspects of the diet of the Hanuman langur. In Chivers DJ, Herbert J (eds): "Recent Advances in Primatology, Vol. 1: Behaviour." London: Academic Press, pp 337–342.

Oxnard CE (1969): A note on the ruminant-like digestion of langurs. Lab Prim Newslett 8:24–25.

Rahaman H (1973): The langurs of the Gir Sanctuary (Gujarat)—a preliminary survey. J Bombay Nat Hist Soc 70:295–314.

Ripley S (1965): "The Ecology and Social Behavior of the Ceylon Gray Langur *(Presbytis entellus thersites).*" Ph.D. thesis. Berkeley: University of California.

Ripley S (1970): Leaves and leaf-monkeys—the social organization of foraging in gray langurs *Presbytis entellus thersites.* In Napier JR, Napier PH (eds): "Old World Monkeys—Evolution, Systematics and Behavior." New York: Academic Press, pp 481–509.

Roonwal ML, Mohnot SM (1977): "Primates of South Asia—Ecology, Sociobiology and Behavior." Cambridge: Harvard University Press.

Sommer V, Mohnot SM (1985): New observations on infanticides among Hanuman langurs *(Presbytis entellus)* near Jodhpur (Rajasthan/India). Behav Ecol Sociobiol 16:245–248.

Starin ED (1978): A preliminary investigation of home range use in the Gir Forest langur. Primates 19:551–568.

Sugiyama Y (1964): Group composition, population density and some sociological observations of Hanuman langurs *(Presbytis entellus).* Primates 5:7–37.

Sugiyama Y (1965): On the social change of Hanuman langurs *(Presbytis entellus)* in their natural condition. Primates 6:381–418.

Sugiyama Y (1966): An artificial social change in a Hanuman langur troop *(Presbytis entellus).* Primates 7:41–72.

Vogel C (1976): "Ökologie, Lebensweise und Sozialverhalten der grauen Languren in verschiedenen Biotopen Indiens." Berlin: Parey.

Vogel C (1977): Ecology and sociology of *Presbytis entellus.* In Prasad MRN, Anand Kumar TC (eds): "Use of Non-human Primates in Biomedical Research." New Delhi: Indian National Science Academy, pp 24–45.

Vogel C, Loch H (1984): Reproductive parameters, adult male replacements, and infanticide among free ranging langurs *(Presbytis entellus)* at Jodhpur (Rajasthan), India. In Hausfater G, Hrdy SB (eds): "Infanticide: Comparative and Evolutionary Perspectives." New York: Aldine Publishing Company, pp 237–255.

Winkler P (1981): "Zur öko-Ethologie freilebender Hanuman-Languren *(Presbytis entellus entellus* Dufresne, 1797) in Jodhpur (Rajasthan), Indien." Ph.D. thesis. Göttingen: University of Göttingen.

Winkler P (1984): The adaptive capacities of the Hanuman langur, and the categorizing of diet. In Chivers DJ, Wood BA, Bilsborough A (eds): "Food Acquisition and Processing in Primates." New York: Plenum Press, pp 161–168.

Winkler P (1986): Comparative troop dynamics in free-ranging Hanuman langurs. Primate Rep 14:206–207 (abstract).

Winkler P, Loch H, Vogel C (1984): Life history of Hanuman langurs *(Presbytis entellus):* Reproductive parameters, infant mortality, and troop development. Folia Primatol (Basel) 43:1–23.

Yoshiba K (1967): An ecological study of Hanuman langurs, *Presbytis entellus.* Primates 8:127–154.

Ecology and Behavior of Food-Enhanced Primate Groups, pages 25–51

2

Dynamics of Exploitation: Differential Energetic Adaptations of Two Troops of Baboons to Recent Human Contact

Debra L. Forthman Quick and Montague W. Demment

Center for Environmental Studies, Arizona State University, Tempe, Arizona 85287 (D.F.Q.), and Department of Agronomy and Range Science, University of California, Davis, California 95616 (M.W.D.)

INTRODUCTION

Access to environmental resources and avoidance of hazards dictate the movements and activities of most organisms [Altmann, 1974b; Crook, 1970], while geographic and seasonal variations in resource availability account for much of the variation in behavior of individuals or groups of animals [Clutton-Brock, 1977]. However, owing to the complexity inherent in undisturbed ecological systems, it is often difficult to assess the degree to which a given behavior is related to specific resource variables. In such circumstances, natural experiments may be of heuristic value, as alterations in one or two variables may produce specific changes in behavior. Food enhancement, whether it is intentional or incidental, offers an opportunity to assess the effects of varied food quality and availability on behavior, without influence from factors like seasonality, which would normally covary with food quality and abundance.

Free-living primates are among those animals that have been food-enhanced deliberately to facilitate data collection or control [Fa, 1984; Itani, 1958; Kawanaka, 1984; van Lawick-Goodall, 1968; Wrangham, 1974] or because populations transplanted to islands could not be sustained by the foods available there [Koford, 1963; Sade, 1965] and incidentally, as a result of human population expansion [Altmann and Altmann, 1970; Crockett and Wilson, 1980; DeVore and Hall, 1965; Kalter, 1977; Kummer, 1968; Matsuzawa et al., 1983; Teas et al., 1980]. As human populations increase, conditions of incidental food enhancement will occur more frequently. In addition to opportunities for research, these create an obligation to address the fate of primates that become pests as a result of habitat loss from human encroachment.

TABLE I. Instantaneous Sample Behavior Definitions

Passive

Any occasion in which the target was either quietly alert or at rest, or was engaged in individual maintenance, ie., autogrooming, scratching, yawning, urination, or defecation

Move

Any nonsocial locomotion that involved a change in location over a distance of at least one body length

Forage/feed

Any search for, or investigation, preparation, or ingestion of, foodstuffs; neither chewing nor consumption of previously ingested cheek pouch contents were scored as forage/feed

Social

Any behavior in which the target interacted with one or more individuals; contact was not required and passive contact was not included; transport of own infant was considered passive contact; all affiliative, agonistic, and sexual behaviors were scored

Construction of time or activity budgets has often been used to give a basic overview of the behavioral priorities of individual animals or groups. Taken alone, these are of limited utility, but they may be used in comparative paradigms to identify ecological variables responsible for differences in activity patterns within and between species [Anderson and Harwood, 1985; Bercovitch, 1983; Bekoff and Wells, 1981; Clutton-Brock and Harvey, 1977; Harrison, 1985; Milton, 1984; Post, 1981].

In a previous paper [Forthman Quick, 1986], the activity budgets of two troops of olive baboons *(Papio anubis)* at the Gilgil Baboon Project [Harding, 1976] were compared during the crop season 3 years after humans began to settle in the baboons' range [Strum, 1984; Strum and Western, 1982]. By that time (1981) both troops had learned to raid crops, and one had begun to exploit refuse that was freely available at a school and army base within the range. The comparison showed significant differences in the percentage of time the two troops were *passive* and engaged in *forage/feed* (see category definitions in Table I). The larger troop (PHG) was *passive* less often and *foraged/fed* more than did the smaller troop (WBY). In an attempt to account for those differences, human food consumption and ranging were examined and were also found to differ between the troops. It was concluded that, although human food consumption was low in both troops, greater human food consumption by WBY enhanced their nutrition enough to produce the observed difference in activity.

In this study, the same two troops are compared over 2 years to determine the effects of nutritional differences in natural and human foods on the baboons' activity budgets, ranging patterns, and food habits.

Fig. 1. The Gilgil Baboon Project study site. Gilgil is indicated by a dot on the inset of the African continent; Kenya is outlined heavily, with Tanzania to the south and Uganda to the west. Details of the area are shown on a topographical map of Gilgil village and the surrounding area. Dotted lines denote 1-km^2 quadrats. The section of Kekopey utilized by the study troops is outlined.

STUDY AREA

The Gilgil Baboon Project was located on Kekopey, a cattle ranch of approximately 18,211 ha (45,000 acres) west of Gilgil, Kenya, and 105 km northwest of Nairobi (approximately 36° 17′ E, 0° 31′ S; Fig. 1). The ranch lies on the floor of the Central Rift Valley at altitudes between 1,800 and 2,285 m [Blankenship and Qvortrup, 1974; Qvortrup and Blankenship, 1975]. The area is semi-arid, with an average annual rainfall of 670 mm since 1955, but the climate is moderated by altitude; between June 1981 and November 1982, the maximum mean monthly temperature was 26°C (range

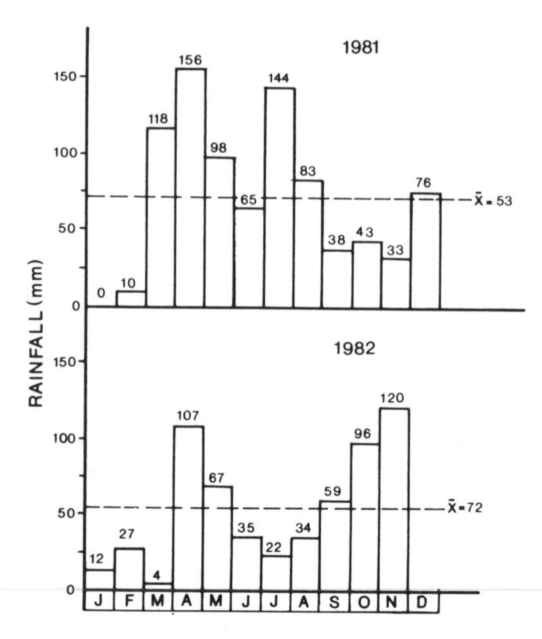

Fig. 2. Gilgil Baboon Project Monthly Rainfall: 1981 and 1982. Monthly rainfall is graphed in millimeters from January 1981 through November 1982. Dotted lines indicate mean rainfall over a 12-month period in 1981 and an 11-month period in 1982 (the 1981 mean is the same when computed over 11 months). 1981 annual rainfall = 864 mm; 1982 11-month total = 583 mm. 1981 data collection ended in October, by which time 745 mm of rain had fallen, compared with 463 mm by October 1982. 1982 data collection ended in July, by which time only 274 mm of rain had fallen, compared with 591 mm in 1981. (Used through courtesy of S.C. Strum.)

18–36°C) and the minimum mean monthly temperature was 10.5°C (range 6–15°C). Measures of rainfall for the period from January 1981 through November 1982 are shown in Figure 2.

The vegetation is best described as Acacia savannah-grasslands. Although a portion of the soda lake Elementeita lies within the ranch, the water is unfit for drinking. Since the 1950s, virtually all of the potable water on Kekopey was found in tanks and troughs supplied by pipeline from a spring 17 km to the northeast. The presence on the ranch of obligate drinkers thus dates from that time. In addition to baboons, there were numerous species of large and small mammals resident during the study [Harding, 1976]. Over the years, most of the large predators have been exterminated. Leopards were still present in small numbers, as were cheetah; lions were seen occasionally. These historical predators rarely threatened Kekopey baboons (this volume, chapter 15). Instead, the principal causes of mortality at the time were humans, domestic dogs, and disease.

TABLE II. Gilgil Baboon Project Age/Sex Classifications

Adult male
A male 10 years of age and older; when age is unknown, animal must have passed behavioral "transition" phase and be an established member of a troop; physically, adult males are filled-out, with well-developed cape and canines

Adult female
Females are classified as adult either when they bear their first infant, or when their first pregnancy is aborted; primipares/multipares may be distinguished from nullipares by elongation of teats

Subadult male
Males are classified as subadult after they reach the "growth spurt" stage: to be subadult, a male must be larger than adult females

Subadult female
Females are subadult from onset of the first menstrual cycle through the end of the first pregnancy

Juvenile male/female
Any animal between 2 years of age and subadult; juveniles are completely weaned and have adult pelage

Infant
Birth to 2 years of age

In most cases, the birthdates for juveniles, subadults, and some adult females were known.
After S.C. Strum [1984].

More detailed treatments of the geology and ecology of the area may be found in Blankenship and Qvortrup [1974], Harding [1976], and Qvortrup and Blankenship [1975]. Since those studies, there have been a few major physical changes on the ranch. The dominant leleshwa bush *(Tarconanthus camphoratus)*, which has not been cleared extensively since Harding [1976] wrote, has recovered. A transcontinental highway now cuts through part of the ranch. Finally, the influx of settlers since the ranch was sold in 1978 has resulted in considerable clearing and cultivation.

SUBJECTS

The troops studied were PHG and WBY, which split from PHG shortly before observations began. From the original troop of 82 animals, 40 adults, subadults, and juveniles were chosen for individual sampling (see Table II for age definitions). All baboons known to raid maize crops were chosen, and others were selected according to a stratified random-sampling scheme by age/sex class. By this method, approximately equal numbers of raiders and nonraiders were selected. No infants were sampled.

During the fission of PHG, subjects split almost evenly between the troops, although raiders tended to join WBY; in October 1981, PHG consisted of 67 animals (28 subjects), WBY of 28 animals (20 subjects). In

September 1982, PHG had 57 members (24 subjects), while WBY had 40 members (20 subjects).[1]

MATERIALS AND METHODS

Subject Sampling

During the 1981 crop season, which extended from June through October, 240 hr of data (6 hr/subject) were collected on activity, ranging, and food consumption. In 1982, 120 hr (3 hr/subject) of the same data were collected during July. At that time, observations were discontinued to enable data collection of a different nature and because there was pressure from local people to contain PHG in unpopulated portions of its range. This was begun in July (see Results: Ranging Patterns), but was enforced so stringently thereafter that assessments of normal feeding and ranging were impossible.[2]

Subjects were randomly assigned a sampling position in each of three time blocks (0600–1000, 1001–1400, 1401–1800) and were observed for 1 hr during each block until all subjects had been observed.[3] Instantaneous

[1]The numbers do not sum to 40 for two reasons. Some baboons that were promoted during data collection occupy two age/sex classes. Also, while membership was reasonably stable within a given year, an animal was designated a member of the troop it was with during the sampling period. In 1981, two subadult males and one juvenile divided their time between PHG and WBY. In 1982, one adult male, subadult male, subadult female, and juvenile did the same. Yearly activity budgets were computed as percentages of instantaneous samples with data from those animals included and excluded. Results are presented as follows:

	Passive	Move	Forage/feed	Social
PHG				
1981	23(22)[a]	20(21)	47(47)	10(10)
1982	40(38)	25(26)	26(28)	9(8)
WBY				
1981	39(38)	25(25)	29(29)	7(7)
1982	47(46)	25(26)	12(13)	16(16)

[a]Data included (data excluded).

[2]When the original PHG troop began to raid farms on a plateau above the sleeping cliffs, 3–5 local people were hired as "chasers." They stood on the clifftops and deterred troop members with noise and rocks whenever they tried to ascend. Initially, "chasers" never descended the cliffs or herded the troop. Beginning in July 1982, the "chasers" were told to round up PHG from the army/school complex each morning and herd them down the cliffs and southwest onto the plains, where they stayed with the troop and prevented them from returning to the army for as long as possible.

[3]If a subject eluded the observer after more than 15 min of observation, the number of intervals required to complete an hour was made up at the end of the data block, in the original sampling order whenever possible. If the subject was lost during the first 15 min, the entire hour was redone.

TABLE III. Food-Category Definitions

Wild (cf. Table IV)

Corms/roots

Any vegetable material consumed that required excavation

Fruits

Any uncultivated fruits, such as berries on shrubs, melons on vines, cactus fruits, etc.

Leaves/seeds

Included all identifiable leaves, seeds, pods, flowers, ranging from grass seed to opuntia pads and acacia flowers

Animal

Any animal material caught or obtained from a wild source: animal food obtained from refuse pits was excluded; insects and their eggs and larvae, plant galls, snails, scorpions, eggs, birds, hares, gazelle fawns are all examples

Other

Any noncultivated edible material not included in the above categories, such as plant exudates, soil, etc.

Human (cf. Table IV)

Maize

Includes husks, cob, and kernels obtained fresh from fields or in refuse, raw or roasted on the cob

Onion

Both onions obtained from refuse and those excavated in fields

Potato

Similarly, both potatoes and potato skins from refuse and those obtained in fields

Other

All other cultigens and unidentifiable human refuse: beans, tomatoes, and so on (See Table IV)

samples of subject activity and location in 1-km^2 quadrats (see Fig. 1) were scored at 5-min intervals. The activity categories used were *passive, move, forage/feed* and *social* (cf. Table I).

Additionally, one-zero sampling (in which one is scored when a behavior occurs one or more times, zero when it does not occur) was used during each 5-min interval to determine the relative proportions of the food types eaten [Altmann, 1974a; Rhine and Ender, 1983; Rhine and Linville, 1980; Suen and Ary, 1984]. Five classes of *wild* foods (*corms/roots, fruits, leaves/seeds, animal,* and *other*) and four classes of *human* foods (*maize, onion, potato,* and *other*) were recorded (see Table III for category definitions).

Wild Foods Identification

The list of wild foods eaten (Table IV) was compiled from observations of feeding baboons made between October 1976 and December 1977 (approximately 2400 hr). PHG numbered 95–102 members at that time. Plant specimens were identified by Demment, with verification by the National Herbarium in Nairobi; animal identifications were verified by the National

Museum. Food samples were collected and dried during this period, then were analyzed for fiber using the procedures of Goering and Van Soest [1970] and for nitrogen by the Kjeldahl method.

Data Analysis

Of primary interest were analyses of the activity budgets of the troops over time, the relative frequency with which *wild* and *human* foods were consumed, range use in relation to the availability of *human* foods, and the relationships among these variables. As the result of mortality or maturation, only 29 of the 40 subjects were observed in both years; as not all cells could be filled for a three-way ANOVA with year as a repeated factor, each year was analyzed separately with 2×5 ANOVAs for each of the four activities, with troop and age/sex class as the independent variables and percentage of intervals as the dependent variable. To examine within-troop changes over time, age/sex class was disregarded and four 2×2 ANOVAs were performed, with troop as a fixed factor and year as a repeated factor. Because data were not normalized by transformation, alpha was set at 0.02. In all analyses, data from juveniles of both sexes were combined because there was no significant effect of age/sex class on the *forage/feed* category in 1981 ($F_{4,36}=1.01$, $P=0.41$) and only a marginally significant effect in 1982 ($F_{4,34}=3.02$, $P=0.03$). For the same reason, age/sex was disregarded in analyses of feeding.[4]

Two-way ANOVAs with troop as a fixed factor and food as a repeated factor were also performed for each year to compare differences between *wild* (2 × 5) and *human* (2 × 4) food consumption, respectively. The dependent variable was the number of one-zero samples for each subject. Percentage or amount of "time" will be used throughout to refer to measures of the dependent variables.

Ranging was summarized as follows. Five-minute samples of map location were tabulated for each troop in each year, then converted to percentages of the total number of samples and graphed. Two other variables were of interest because of their potential correlation with activity and feeding. The number of 5-min samples collected per subject during a day when the troop was known to have spent some *portion* of that day foraging in fields or the army/school complex was tabulated and divided by the total number of samples for each subject; this is referred to as "army" utilization. The number of quadrats utilized by each subject was also tallied.[5]

[4]That there were only slight differences in the amount of time each age/sex class engaged in *forage/feed* does not necessarily indicate that every age/sex class ate each food category in equal proportions. However, that question is not addressed in this paper.

[5]The number of samples from which these measures were calculated is not the same for each

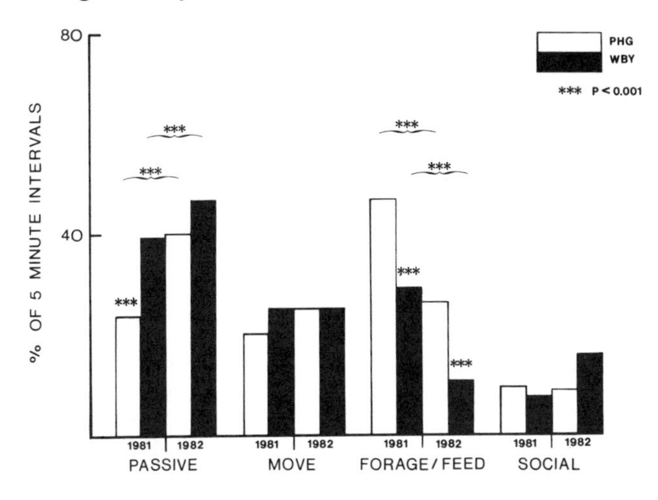

Fig. 3. Activity budgets for PHG and WBY in 1981 and 1982. Each behavior is graphed as a percentage of the total number of 5-min instantaneous samples (N = 4,320; significant results of three ANOVAs performed on each behavior category are indicated).

Correlations were used to identify significant associations between food types and measures of *passive, move,* and quadrat and "army" utilization. In both years, all subjects had scores for each food category and for the four "dependent" variables above; these were subjected to pairwise correlations.

RESULTS

Activity Budgets

Figure 3 shows activity budgets for both troops and years.

1981. In comparisons between troops, PHG spent significantly more time in *forage/feed* ($F_{1,36} = 19.19$, $P < 0.001$), and less time *passive* than WBY ($F_{1,36} = 23.04$, $P < 0.001$). There was also a significant effect of age/sex class on *passive* ($F_{4,36} = 5.69$, $P < 0.001$; see Fig. 4).

1982. The pattern of activity was the same for both troops, as PHG's priority switched from *forage/feed* to *passive.* However, PHG still *foraged/fed* more than WBY ($F_{1,34} = 9.51$, $P < 0.001$). There were marginal effects of age/sex class on *passive* and *forage/feed* ($F_{4,34} = 2.91$ and 3.02, $P = 0.04$ and 0.03, respectively; see Fig. 4).

variable, as definitions of missing data differed. For example, a subject's instantaneous sample might be missed because a clear view of the subject was not obtained "on the instant." However, the subject's location and feeding might very well be observed during that same interval.

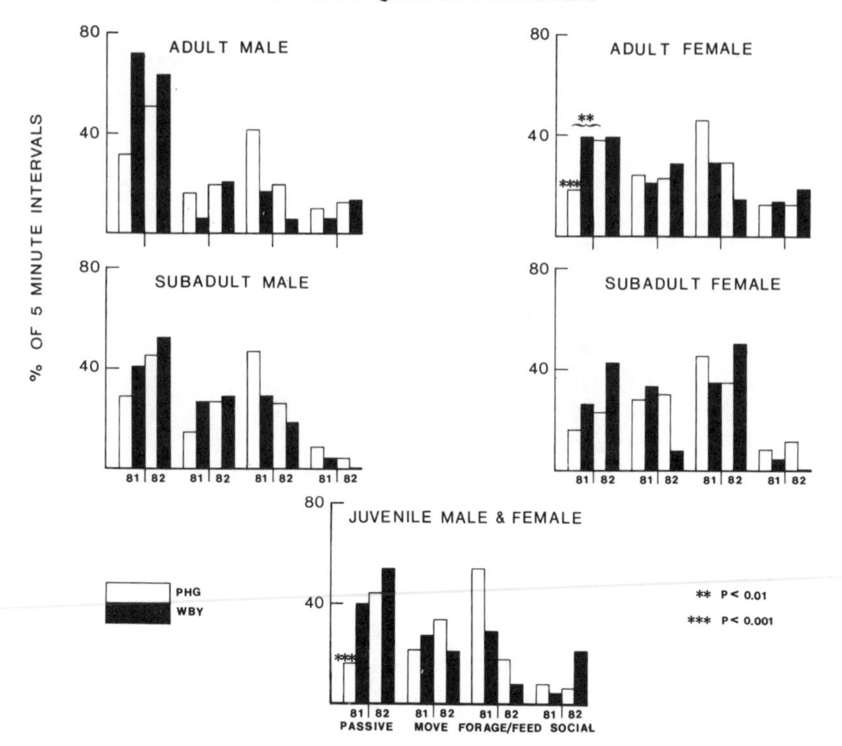

Fig. 4. Activity budgets for each age/sex class in PHG and WBY in 1981 and 1982. (Significant results of post hoc tests on 1981 *passive* behavior are indicated.)

Across years. Analysis of troop changes over time showed that for both, time *passive* increased ($F_{1,27}=14.22$, $P<0.001$), while *forage/feed* decreased ($F_{1,27}=35.22$, $P<0.001$); still, PHG always fed significantly more than WBY. The changes in PHG were greater than those in WBY ($F_{1,27}=16.06$ for *passive,* 20.37 for *forage/feed,* $P<0.001$ for both).

Age-Sex effects. There was a significant age/sex effect for *passive* in 1981, with marginal effects for *passive* and *forage/feed* in 1982. Post hoc comparisons were performed for all three year/activity combinations and showed that in both years, adult and subadult males were *passive* more than adult and subadult females (1981: $t_{17,20}=2.71$, $P<0.02$; 1982: $t_{17,20}=2.89$, $P<0.01$); in 1982, the difference between adults alone was also significant ($t_{7,15}=2.78$, $P<0.02$). In 1982, adult males and juveniles *foraged/fed* less than adult females and subadults of both sexes ($t_{12,30}=2.53$, $P<0.02$).

Feeding was next examined in detail to determine which factors accounted for the striking differences in activity between troops and over time. For

TABLE IV. Gilgil Baboon Project Foods List: 1976–1977/1981–1982

Wild		
Plants[a]		
Amaranthaceae		
Amaranthus graecizans	F [L,S]	(2,1)
Amaryllidaceae		
Crinum kirkii	B [T]	(1,3)
Anacardiaceae		
Rhus natalensis	S [R]	(2,2)
Schinus molle	S [R]	(1,2)
Boraginaceae		
Cordia ovalis	S,T [F,R]	(1,2)
Vaupelia hispida	F [P,L]	(2,1)
Cactaceae		
Opuntia vulgaris	S [R,T]	(1,2)
Commelinaceae		
Commelina reptans	F [Tb]	(1,2)
Compositae		
Lactuca capensis	F [L,F]	(2,2)
Launea cornuta	F [L,F]	(2,2)
Osteospermum vaillantii	F [R]	(1,2)
Sonchus sp.	F [L]	(2,2)
Cruciferae		
Erucrastrum arabicum	F [P]	(1,1)
Cucurbitaceae		
Coccinea grandis	F [R]	(1,3)
Kedrostis foetidissima	F [R]	(1,3)
K. hirtella	F [R]	(1,3)
Peponium vogelii	F [R]	(1,3)
Cyperaceae		
Cyperus stuhlmannii	F [C]	(1,1)
Geraniaceae		
Monsonia augustifolia	F [T]	(2,2)
Gramineae		
Cynodon nlemfuensis	G [L]	(2,1)
C. plectostrachyus	G [L]	(2,1)
Eleusine multiflora	G [I,L]	(2,1)
Eragnostis papposa	G [I,L]	(1,1)
Panicum maximum	G [I]	(2,2)
Tragus berteronianus	G [I]	(1,1)
Labiatae		
Plectranthus cylindracaus	F [L]	(2,2)
Liliaceae		
Aloe graminicola	F,S [L]	(2,2)
Chlorophytum bakeri	F [L,Ro]	(2,3)
Chlorophytum sp.	F [R]	(2,2)
Trachyandra saltii	F [F,L,T]	(2,2)
Malvaceae		
Hibiscus aponeurus	F,S [Fb]	(2,2)
H. flavifolius	F,S [Fb]	(2,2)
Sida ovata	F,S [R]	(2,1)

(continued)

TABLE IV. Gilgil Baboon Project Foods List: 1976–1977/1981–1982 *(continued)*

Wild		
Mimosaceae		
Acacia drepanolobium	T [I]	(2,2)
A. xanathophloea	[R]	(2,2)
Moraceae		
Ficus sonderi	T [R]	(2,1)
Nyctaginaceae		
Commicarpus pedunculosus	F [R]	(2,3)
Papillionaceae		
Medicago laciniata	F [P]	(1,1)
Rhynuchosia densiflora	F [R]	(2,3)
R. elegans	F [L]	(2,3)
Trifolium semipilosum	F [P]	(2,2)
Polygonaceae		
Oxygonum sinuatum	F [P]	(2,2)
Rhamnaceae		
Scutia myrtina	S [R]	(2,2)
Rubiaceae		
Oldenlandia wiedemannii	T [R]	(2,2)
Solanaceae		
Lycium europaeum	F [L,F]	(2,2)
Solanum incanum	F [Ig]	(2,2)
Withania somnifera	F,S [R]	(2,2)
Tilliaceae		
Grewia similis	S [F,R]	(2,2)
G. tembensis	S [F,R]	(2,2)
Zygophyllaceae		
Tribulus terrestris	F [P,R]	(2,1)
Fungi		
Mushroom spp.	F [P]	(2,3)
Animals		
Insecta		
Coleoptera		
Carabidae		
Anthia artemis	Adults	(2,3)
Chilanthia cavernosa	Adults	(2,3)
Isoptera	Winged adults	(2,3)
Lepidoptera		
Nymphalidae		
Precis oenone	Larvae	(1,3)
Psychidae	Larvae	(2,3)
Orthoptera		
Acrididae	Adults	(2,1)
Phasmatidae	Adults	(2,2)
Tanaoceridae	Adults	(2,3)
Gastropoda		
Achatinidae	Adults	(2,1)
Aves		
Motacillidae		
Anthus novaeseelandiae	Eggs, nestlings	(3,2)
Anthus sp.	Eggs, nestlings	(3,2)

(continued)

TABLE IV. Gilgil Baboon Project Foods List: 1976–1977/1981–1982 ***(continued)***

Wild		
Phasianidae		
Coturnix delegorguci	Adults, young	(3,3)
Numida mitnata	Young	(2,3)
Mammalia		
Lagomorpha		
Leporidae		
Lepus capensis	Adults, young	(3,1)
Hyracoidea		
Procaviidae		
Procavia capensis	Young	(3,1)
Artiodactyla		
Antilopinae		
Aepyceros melampus	Young	(3,1)
Gazella thomsoni	Young	(3,1)
Neotraginae		
Raphicerus campestris	Young	(3,2)
Madoquinae		
Rhynchotragus binki	Young	(3,2)
Human		
Banana peel		
Bread		
Cabbage		
Chapati—Indian unleavened bread		
Coconut		
Colgate toothpaste (n = 2)		
Kale		
Maize, Whole—fresh, dried, roasted, canned (cob and husk also)		
Maize, Ground—meal, cooked, or uncooked		
Mango seed		
Onion—rarely		
Orange peel		
Paper (n = 1)		
Pineapple peel		
Pineapple leaves		
Potato—usually peels, sometimes whole		
Rice—cooked or uncooked		
Squash plants		
String beans—raw or cooked		
Sugarcane		
Tomato—generally fresh		
Wheat flour		

The first No. in parentheses indicates the estimated frequency of the item in the diet. Only those plants that were heavily (1) or commonly (2) utilized by the Gilgil baboons are listed here. Plants that were rarely eaten comprised at least an additional 56 species from 27 families. A complete list will be published elsewhere. The second number indicates the estimated spatial and temporal density of the item within the baboons' range (1, common; 2, uncommon; 3, rare). All animal species known to be eaten are listed, regardless of the frequency of consumption.

[a]*Code for Plant growth form:* B = bulb; F = forb; G = grass; S = shrub; T = tree. *Code for Plant Parts Eaten* [ ]: C = corm; Dt = dead twig; E = exudate; F = flower; Fb = flower bud; I = inflorescence; Ig = insect gall; L = leaf; O = stolon; P = entire above-ground plant; R = fruit; Rh = rhizome; Ro = root; S = seed; T = stem.

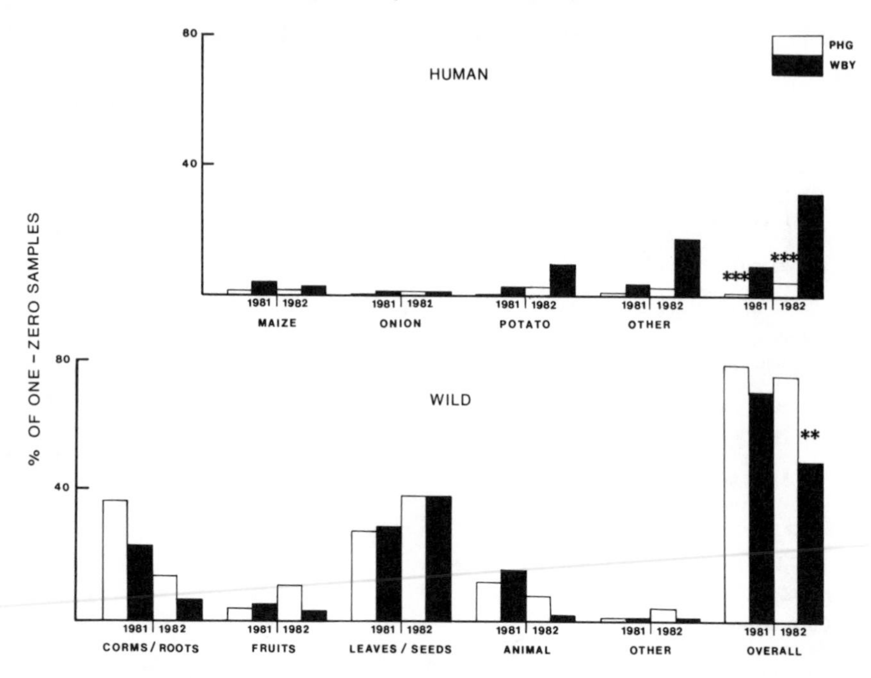

Fig. 5. Human and wild food consumption for PHG and WBY in 1981 and 1982. Each food category is graphed as a percentage of the total number of one-zero ingestion samples during 5-min intervals (N=4,569; for each year, significant troop differences in overall human and wild food consumption are indicated).

descriptive purposes, *wild* and *human* food consumption is graphed by category in Figure 5. (Table IV gives lists of *wild* and *human* foods commonly eaten.)

First, however, the relationship between *forage/feed* and the other dependent variables was assessed, as negative correlations have been found between feeding and both passive [Harrison, 1985] and social behavior [Post, 1981]. In both troops and years, significant negative correlations occurred primarily between *forage/feed* and *passive* (1981: PHG: $r=-.66$, $P<0.001$, $df=27$; WBY: $r=-.79$, $P<0.001$, $df=19$; 1982: PHG: $r=-.76$, $P<0.001$, $df=24$; WBY: $r=-.53$, $P<0.02$, $df=20$). Therefore, the relationships between *social* behavior and food categories were not examined further.

Food Consumption: 1981

Human foods. PHG consumed *human* foods 1% of the time, compared with 12% for WBY ($F_{1,46}=20.87$, $P<0.001$). There was also a significant effect of food type, as intake of *maize* and *human other* predominated ($F_{3,138}=8.83$, $P<0.001$) and a significant troop × food interaction ($F_{3,138}=10.03$, $P<0.001$). PHG ate only *maize* and *human other;* WBY ate foods from all categories. For WBY in particular, the most important food in the *human other* category was pieces of cooked maize porridge. Thus, although *human* foods constituted a relatively small proportion of the diet for both troops in 1981, PHG was only slightly influenced by *human* foods in comparison with WBY.

Wild foods. There was no troop effect on *wild* food consumption ($F_{1,46}=3.02$, n.s.), because both troops primarily consumed *wild* foods; however, there was an effect of food type and a troop × food interaction ($F_{4,184}=102.47$ and 9.66, $P<0.001$ for both). *Corms/roots* and *leaves/seeds* were the most important items in the diet; PHG ate *corms/roots* more than any other category in 1981, followed by *leaves/seeds*. WBY members ate *leaves/seeds* about as often as PHG, but ate far fewer *corms/roots*.

For PHG, the negative correlation between *forage/feed* and *passive* reported above was primarily attributable to *wild* food ingestion, specifically *corms/roots,* which required extensive manipulation and *leaves/seeds* (see Table V for 1981 correlations of food type with activity and ranging). WBY's *passive* behavior increased only with decreases in *animal* ingestion.

Food Consumption: 1982

Human foods. As in 1981, there were significant troop and food effects on *human* food consumption ($F_{1,42}=20.99$; $F_{3,126}=22.98$, $P<0.001$ for both) and a troop × food interaction ($F_{3,126}=14.41$, $P<0.001$). PHG's *human* food consumption increased to 6%, while WBY's jumped to 39%. Although *wild* foods still dominated the diet of both troops, *human* food consumption was much more important in 1982 (see below). Both troops consumed all types of *human* foods in 1982, but while PHG ate mostly *potato* and *human other,* WBY concentrated on *human other*.

In WBY there was a significant negative correlation between *passive* and *potato* consumption, perhaps because *potato* required more harvest time than maize porridge (see Table VI for 1982 correlations).[6]

Wild foods. In 1982, there were significant effects of troop ($F_{1,42}=14.1$, $P<0.01$) and food on *wild* food consumption ($F_{4,168}=74.74$, $P<0.001$), but no troop × food interaction. Although both troops ate *wild* foods much less frequently in 1982, PHG still ate them more often than WBY. *Corms/roots* consumption was greatly reduced in both troops; instead, the baboons shifted

[6]Consumption of *wild other* and *onion* was omitted from the table because both categories were consumed so rarely (see Fig. 3).

TABLE V. Pearson Correlations Between Activity and Food Variables: 1981

	Passive		Move		No. quadrats[a]		Army utilization[b]	
	PHG	WBY	PHG	WBY	PHG	WBY	PHG	WBY
Corms/ roots	−.49**	−.53	−.46*	.14	.23	.33	−.37	−.51
Fruit	−.05	.05	.05	.13	.58**	.14	−.34	−.35
Leaves/ seeds	−.51**	−.42	.16	.46	.63***	.41	−.24	−.29
Animal	−.33	−.56*	.28	.52	.52**	.35	−.28	−.11
Total wild	−.58**	−.60**	−.11	.49	.63***	.49	−.43	−.47
Maize	.16	−.11	.04	.43	−.04	.05	.46*	.34
Potato	—[c]	.02	—	.13	—	−.05	—	.21
Human other	.16	.03	.04	.05	−.04	−.02	.46*	.28
Total human	.16	−.04	.04	.26	−.04	.00	.46*	.33
df	27	19	27	19	27	19	27	19

[a]Quadrats = No. Map quadrats utilized/subject.
[b]Army utilization = % of total intervals recorded on days of visits to army/subject.
[c]— = never observed.
*P<0.02.
**P<0.01.
***P<0.001.

to *leaves/seeds*. Table VII shows that *human* foods and corms have similar protein content and that all *wild* foods have considerably higher fiber content.

In both troops, correlates of *passive* behavior scarcely changed from 1981.

Ranging Patterns: PHG

Figure 6 shows PHG's range use during both years. The lower grid is a simplified version of the range shown in Figure 1, with the 1-km^2 quadrats superimposed. Sleeping sites and sources of water and human foods are indicated. The upper grid shows the percentage of time PHG spent in each quadrat during the two years.

1981. PHG was observed in 17 quadrats, but spent 73% of its time in 5 quadrats. Quadrat utilization was positively correlated with every category of *wild* food except *corms/roots* (see Table VI). Similarly, *move* was inversely related to *corms/roots,* as sedge corms are a spatially dense and localized resource that are best exploited by the "sit and scoot" method. PHG was never recorded in any of the three quadrats that had the greatest concentration of humans and refuse (4640, 4650, and 4740), and "army" utilization (in this case farms only) by the troop or individual subjects was only 10%.

TABLE VI. Pearson Correlations Between Activity and Food Variables: 1982

	Passive		Move		No. quadrats[a]		Army utilization[b]	
	PHG	WBY	PHG	WBY	PHG	WBY	PHG	WBY
Corms/ roots	−.55**	−.52*	.15	.16	.44	.09	−.10	///[c]
Fruit	−.23	−.06	−.15	−.07	.35	−.05	−.53**	///
Leaves/ seeds	−.64***	−.59**	.00	.63**	.63**	.09	−.25	///
Animal	−.24	−.49	.06	.36	.19	.17	−.55**	///
Total wild	−.64***	−.66**	.00	.57**	.62**	.09	−.46	///
Maize	.02	.12	.09	−.02	.23	.30	.22	///
Potato	−.15	−.58**	.38	.56*	−.09	.11	.19	///
Human other	.10	−.14	.43	.21	−.08	.06	.41	///
Total human	.00	−.19	.45	.24	.00	.18	.38	///
df	27	19	27	19	27	19	27	19

[a]No. Quadrats = No. map quadrats utilized/subject.
[b]Army utilization = % of total intervals recorded on days of visits to army/subject.
[c]/// = always observed.
*P<0.02.
**P<0.01.
***P<0.001.

Nonetheless, "army" utilization was positively correlated with *human* food consumption.

1982. PHG's movement patterns altered considerably during the study. Members were still observed in 17 quadrats and spent 70% of their time in 6. However, 40% of the time they were in 4640, 4650, and 4740 and "army" utilization was 63%.[7] Quadrat use was positively correlated with *leaves/seeds* (see Table VI). There was a marginal positive correlation between "army" utilization and *human/other* consumption.

Ranging Patterns: WBY

Figure 7 shows the same range utilization data for WBY during the 2 years.

1981. WBY ranged in 14 quadrats and spent 72% of the time in 4. Troop members spent 28% of their time in quadrats 4640, 4650, and 4740, during which they consumed 50% of all the *human* foods they ate, including 63%

[7]That figure would have been higher, but during the remaining 37% of observations, PHG was actively herded (see footnote 2).

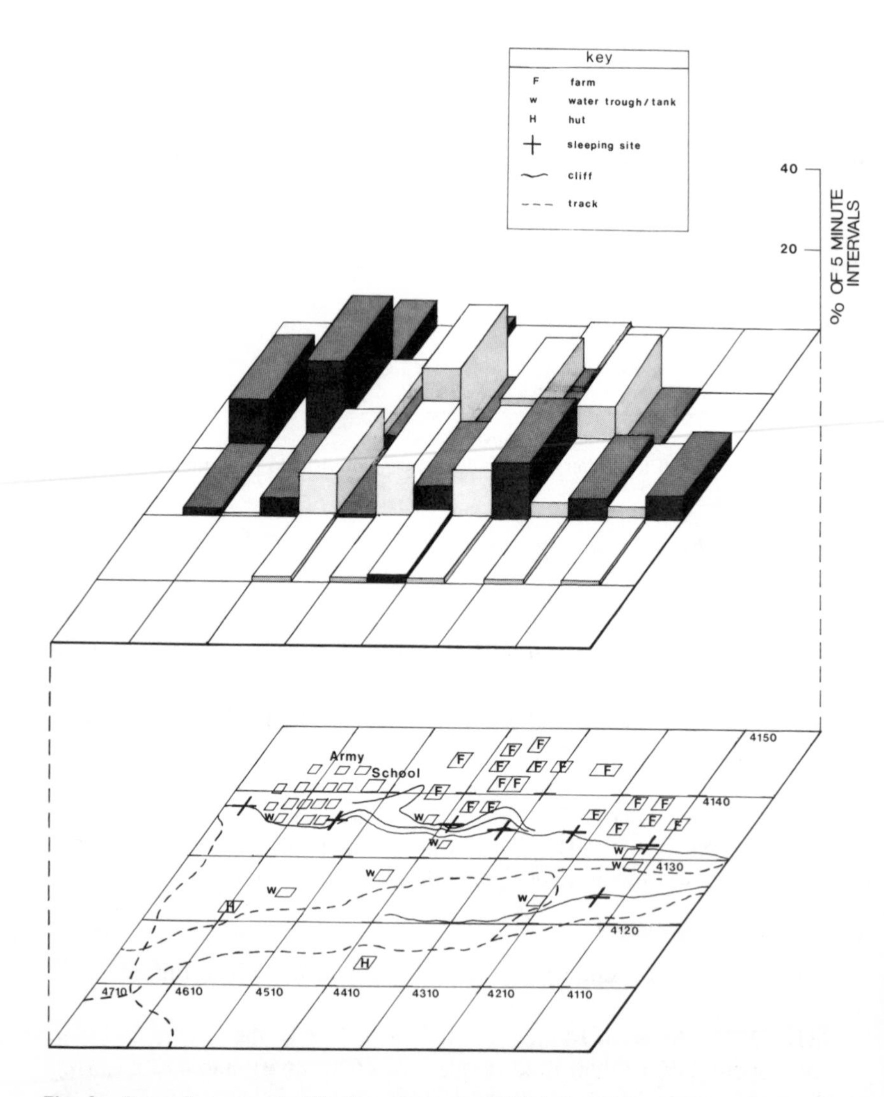

Fig. 6. Range features and utilization patterns for PHG during 1981 and 1982. The lower grid represents the range map of the two troops; 1-km² quadrats and major resources are denoted. The upper grid is a superimposition that illustrates ranging as percentages of the total number of 5-min intervals in which each of the 1-km² quadrats below was occupied. 1981: light hexahedrons; 1982: dark hexahedrons.

TABLE VII. Nutritional Measures of Wild and Human Foods Consumed by Gilgil Baboons

	n[a]	Crude protein	Cell wall	Cellulose	Hemicellulose
Wild foods[b]					
Corms	2	5.1	35.4	7.9	25.2
Leaves (grass)	4	26.7	43.6	13.9	25.1
Seeds (grass)	4	11.8	65.5	23.2	30.5
Flowers	3	13.7	25.4	16.0	5.3
Fruits	4	14.8	30.5	10.1	15.8
Human foods[c]					
Maize		8.9	7.9	2.0	4.4
Potato		8.7	4.7	1.8	2.7

[a]No. of species averaged for category mean.
[b]Collected at Gilgil, Kenya.
[c]Data from Van Soest and Robertson [1977].

of human other. ''Army'' utilization in 1981 was 73% for WBY or individual subjects. Taken together, these pieces of evidence suggest that, even in 1981, WBY had begun to switch from a staple of *corms/roots, leaves/seeds,* and *animal* to a diet of maize meal porridge, other human foods and *leaves/seeds*.

1982. WBY's range contracted sharply in 1982, and troop members were observed exclusively in quadrats 4640, 4650, and 4740. The troop occupied quadrat 4740 72% of the time; thus, variance in quadrat utilization was reduced almost to zero. Similarly, *move* was positively associated only with ingestion of *leaves/seeds* and *potato* (see Table VI), the only forage that required significant linear travel. ''Army'' utilization could not be correlated with other variables in 1982 because every WBY subject visited the army/school complex every day.

DISCUSSION

The amount of human food consumed appeared to have a strong influence on the activity patterns of these baboons. The percentage of time spent feeding was lowest in WBY, the troop that ate the most human foods, and it decreased in both troops over time, as human foods were eaten more frequently. PHG spent approximately 50% of its time feeding in 1981 when troop members ate little human food. That figure is consistent with levels measured in troops that eat only wild foods [Altmann, 1980; Bercovitch, 1983; Post et al., 1980]. In contrast, WBY fed only 12% of the time in 1982 when human foods were a significant part of the diet. Diet composition also differed; PHG ate more corms than WBY in both years, but both troops

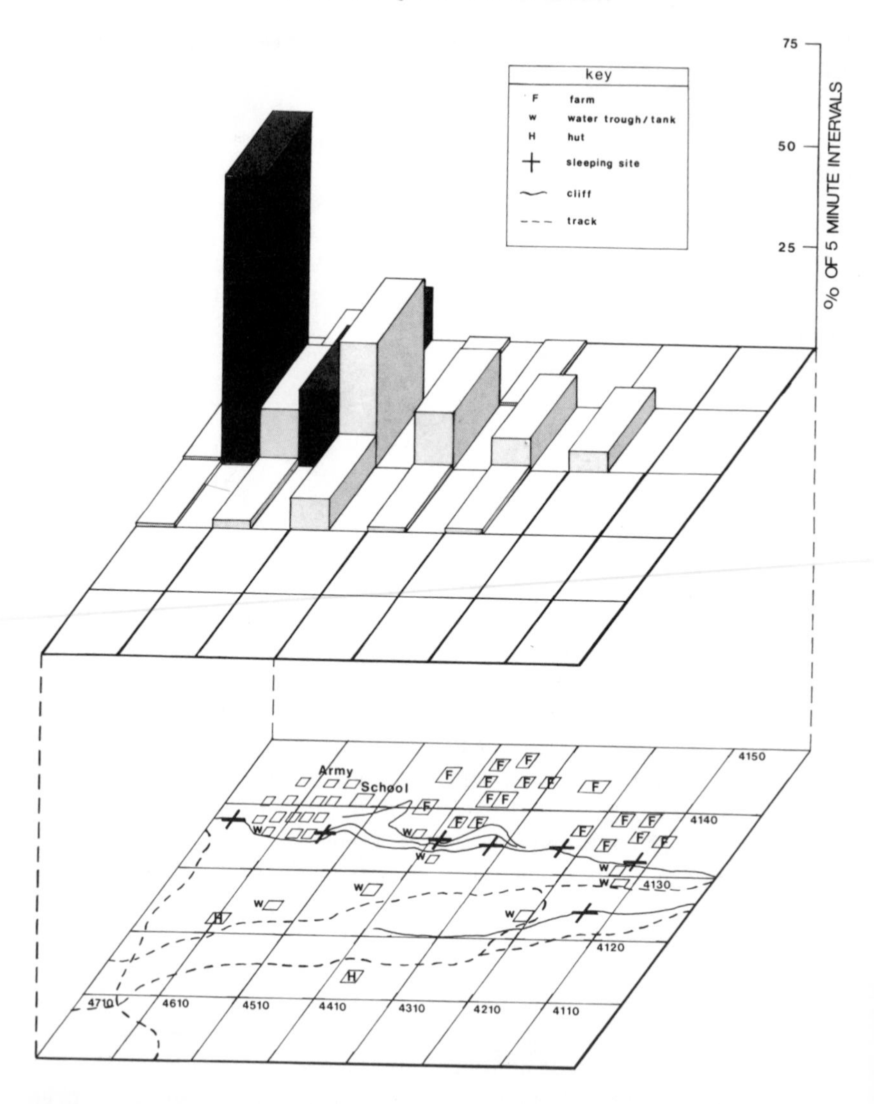

Fig. 7. Range features and utilization patterns for WBY during 1981 and 1982. Explanation is the same as for Figure 6.

decreased corm consumption and increased intake of leaves/seeds in 1982. The decrease in time spent feeding was reflected almost entirely in increased passive behavior.

Caloric density and the amount of work required to obtain food are among the variables that influence animals' learned strategies of diet selection

[Lang, 1970]. These factors characterize the two principal advantages of human over wild foods. Baboons are relatively small omnivores that primarily consume plant material. Because their metabolism and gut size are scaled to body weight, they are constrained in their ability to eat plant material that contains high concentrations of cell wall [Demment, 1983]. Starch is the main component of the human foods consumed in this study. The digestibility of maize starch is improved by refinement and heat treatment [Knapka and Morin, 1979; Mason et al., 1976], which also reduce or eliminate toxic elements [Harris, 1970]. Because baboons digest soluble plant material enzymatically in the foregut and ferment cell wall and indigestible starches microbially in the hindgut, rapidly digested high-energy foods, like starches, have significant nutritional value. It is no coincidence that WBY members, which ate more human foods, weighed more (Strum, personal communication; Forthman Quick, unpublished data) and had heavier pelages.

In an "ancestral" environment, high-quality foods (those that are rapidly digested and low in cell-wall content) are rare [Demment and Van Soest, 1985], and the primary price baboons must pay to acquire such foods is time. The observed age/sex effects, in which adult males were passive significantly more than females and females fed more than males, but not significantly so, is consistent with the different nutritional ecology of the sexes and findings of others [Bercovitch, 1983; Dunbar, 1977; Rose, 1977]. Females, with their constraints on gut capacity and high costs of reproduction, are more active than males, in part because they forage for high-quality foods that are difficult to locate or harvest, while males eat abundant low-quality forage that requires considerable time to digest [Demment, 1983]. The finding that adult males and juveniles fed less than adult females and subadults of both sexes is not as clear, but preserves the dichotomy between adult males and females.

Human foods violate the inverse relationship between quality and abundance that governs wild foods; thus, their second advantage for both sexes is one of foraging efficiency. Although the baboons ate wild foods 60% of the time, because the human foods they chose often occurred in large, dense packages that could be harvested rapidly to produce high intake rates relative to wild foods, neither frequency nor duration of consumption is closely related to the volume of food ingested [Hladik, 1978]. For example, an adult male Gilgil baboon consumes 3–5,000 0.1 g wild food items daily. An average corm weighs 0.05 g and can be eaten at a rate of up to 1.25 g/min. In contrast, a handful of maize porridge or a potato may provide 20–50 g/min (Demment, unpublished data), which allows an animal to ingest more before feeding is interrupted by competitors or predators. Furthermore, because the principal human food eaten (maize porridge) is both refined and of small particle size, digestion and assimilation rates for human foods are much higher per unit of intake than for natural foods [Heller et al., 1980].

Thus, human foods provided more energy for less effort than the same amount of wild forage. However, they have lower concentrations of protein than many natural foods [Inglett, 1975]. Increased carbohydrate intake requires high protein consumption either to maintain constant protein/carbohydrate ratios or to compensate for potentially higher fecal nitrogen losses that can occur in animals on high starch diets [Mason et al., 1976].

The baboons' external and internal environments interacted to alter the composition of the wild diet in 1982. First, the relatively sparse "long rains" from March through June (see rainfall measures in Fig. 2) reduced the availability of normal sources of protein (flowers, leaves, seeds). Because the army/school complex was watered continually by effluent and open sink drains, the baboons (particularly PHG) switched from dry grasses to the succulents (opuntia, aloe, and mint) abundant there. This facilitated habituation to humans and initial exploitation of their foods. Secondarily, PHG found that the area provided a rich source of foods high in carbohydrates and low in cell wall content (see Table VII for nutritional comparisons of selected wild and human foods). Positive nutritional feedback from the carbohydrate-rich foods [Green and Garcia, 1970; Rozin and Kalat, 1971] established increased preference for them; as a result, corm (carbohydrate) consumption decreased, and leaf/seed (protein and water) intake increased. This also resulted in the alteration (PHG) or restriction (WBY) of their ranges.

Locomotion in both troops was associated with consumption of leaves/seeds. This suggests that animals are sensitive to nutrient concentrations when they make foraging decisions [Belovsky, 1981; Owen-Smith and Novellie, 1982; Westoby, 1974]. The patterns of nutritional response in this study parallel those observed by Milton [1979] in howler monkeys *(Alouatta)*. Howler feeding was judged to be a balance between consumption of energy-rich fruits and protein-rich leaves. Both fiber and protein content of foods were major factors responsible for food selection and, in turn, the monkeys' food selection determined activity and ranging patterns [Milton, 1980].

Interestingly, the primary effect of leaf/seed consumption in the smaller troop (WBY) was on movement, while in PHG the effect was on quadrat utilization. Small troops may meet their requirements in a few patches within their core area. WBY members moved short distances to eat leaves/seeds at frequent intervals during the day and were few enough that resources in the core area were never depleted. Because the core area was more heavily exploited by PHG, rapid long-distance movements brought them to resources farther away. Therefore, leaf/seed consumption in PHG was more closely associated with the number of quadrats used than the amount of time spent moving. The fact that the smaller troop also utilized human foods most and wild foods least would accentuate this pattern.

Because the present observations indicate that maximal energy intake is not the only factor affecting fitness in these baboons [Caraco and Chasin, 1984; Harrison, 1984, 1985; Hodges, 1981; Pyke, 1981; Vickery, 1984], it is possible to make a few speculations about foraging and its constraints.

While WBY's behavior almost seems consistent with the prediction that animals maximize their net energy gain, PHG largely maintained its historical range and diet in 1981. If a long-term optimal strategy is to minimize time spent foraging or handling food, as long as starvation is avoided [Krebs and McCleery, 1984], one wonders why PHG was so reluctant to exploit an abundant and predictable source of energy-rich foods.

However, predation risk may account for some of WBY's departures from simplicity, and it may have been a factor in PHG's initial conservatism. When in the army base, both troops typically dispersed into small parties or ranged alone through yards and refuse pits. Throughout these forays, the baboons were highly vigilant and prone to startle and avoidance responses, probably because the risks of encounters with dogs, vehicles, and armed humans were greater there. Recent work on the trade-off between foraging and predation risk parallels the behavior of WBY members foraging at the periphery of the army base: rather than eat small items in the open, they often carried bulky pieces of maize porridge to cover; similarly, when far from cover, they moved quickly and either ate small items or stored them in their pouches [Lima et al., 1985; Lima and Valone, 1986].

Surprisingly, overall mortality among PHG subjects that foraged sporadically in the army/school complex was almost as high as for WBY habitues; 6% of PHG and 7% of WBY subjects were either confirmed to have died on power lines or from one of the above causes, or were presumed to on the basis of circumstantial evidence. It is possible that the relatively high mortality in PHG is also dependent on troop size. People were far less tolerant when conspicuously large numbers of baboons foraged through their refuse heaps, fields, gardens, and kitchens, particularly as they more often frightened women and children and harassed livestock.

The results of this study suggest that human food consumption strongly affects the activity, natural diet, and range use of baboons in ways that are consistent with the concept that animals respond to the nutritional quality of available food. While it is clear as well that troops differ in their willingness to exploit human foods, the demographic and social correlates of this higher nutritional plane coupled with increased predation risk remain to be studied.

SUMMARY

Studies of food-enhanced primates are useful in determining the effects of food supply and quality on behavior. In this study, the activities and ranging patterns of two troops of olive baboons (PHG and WBY) on Kekopey ranch near Gilgil, Kenya, were observed during 1981–1982, three years after the ranch was sold and baboons began to raid the fields, gardens, and homes of people who had settled within the baboons' range. The behavior of a total of 40 adult, subadult, and juvenile subjects was recorded with both instantaneous and one-zero sampling during the 1981 and 1982 crop seasons. There were significant differences in activity budgets both between and within troops over the two years. PHG fed much more than WBY, although feeding decreased and time passive increased in both troops over time. These differences were correlated with different food habits and ranging patterns. Compared with PHG, WBY had a smaller home range centered on an area of high human population density. In that area WBY exploited more human foods and consumed wild foods in different proportions when compared with PHG. However, in 1982, PHG's behavior approximated WBY's, perhaps in part owing to environmental pressures that year. These results suggest that the baboons learned, through positive nutritional feedback, to exploit varied sources of high-quality foods. Incorporation of these foods into the diet required compensatory changes in the selection of wild foods and alterations in ranging. Factors that may contribute to the differential responses of the two troops to human foods were discussed.

ACKNOWLEDGMENTS

We would like to thank the Office of the President and the National Council for Science and Technology of Kenya for permission to conduct our research. We also thank J. Garcia, S. Strum, J. Else, the Institute of Primate Research, the National Museum, and the National Herbarium for their aid and sponsorship; S. Azango, F. Bercovitch, Beth Singoi, Festus Adero, J. Musau, H. Ooga, Rebecca, and Reuben for their assistance in the field; and R. Harding, H. Hendy, C. Gustavson, D. Manzolillo, D. Rasmussen, R. Rhine, and E. Sherwood for helpful criticisms of earlier drafts of the manuscript. D.F.Q. was supported by Fulbright-Hays award 5048211, Sigma Xi, a Hortense Fishbaugh scholarship, UCLA patent/travel grants, and by J. Garcia's NIH # NS11618. M.W.D. was supported by the Wenner-Gren Foundation and an NIMH Research Service award.

REFERENCES

Altmann J (1974a): Observational study of behavior: Sampling methods. Behaviour 49:227–267.

Altmann J (1980): "Baboon Mothers and Infants." Cambridge: Harvard University Press.

Altmann SA (1974b): Baboons, space, time, and energy. Am Zool 14:221–248.

Altmann SA, Altmann J (1970): "Baboon Ecology." Chicago: University of Chicago Press.

Anderson SS, Harwood J (1985): Time budgets and topography: How energy reserves and terrain determine the breeding behaviour of grey seals. Anim Behav 33:1343–1348.

Bekoff M, Wells MC (1981): Behavioural budgeting by wild coyotes: The influence of food resources and social organization. Anim Behav 29:794–801.

Belovsky GE (1981): Food plant selection by a generalist herbivore: The moose. Ecology 62:1020–1030.

Bercovitch FB (1983): Time budgets and consortships in olive baboons *(Papio anubis)*. Folia Primatol (Basel) 41:180–190.

Blankenship LH, Qvortrup SA (1974): Resource management on a Kenya ranch. J S Afr Wildlife Manag Assoc 4:185–190.

Caraco T, Chasin M (1984): Foraging preferences: Response to reward skew. Anim Behav 32:76–85.

Clutton-Brock TH (1977): Some aspects of intraspecific variation in feeding and ranging behaviour in primates. In Clutton-Brock TH (ed): "Primate Ecology: Studies of Feeding and Ranging Behaviour in Lemurs, Monkeys and Apes." New York: Academic Press, pp 539–556.

Clutton-Brock TH, Harvey PH (1977): Species differences in feeding and ranging behaviour in primates. In Clutton-Brock TH (ed): "Primate Ecology: Studies of Feeding and Ranging Behaviour in Lemurs, Monkeys and Apes." New York: Academic Press, pp 557–579.

Crockett CM, Wilson WL (1980): The ecological separation of *Macaca nemestrina* and *M. fascicularis* in Sumatra. In Lindburg DG (ed): "The Macaques: Studies in Ecology, Behavior and Evolution." New York: Van Nostrand Reinhold, pp 148–181.

Crook JH (1970): The socio-ecology of primates. In Crook JH (ed): "Social Behaviour in Birds and Mammals." New York: Academic Press, pp 103–166.

Demment MW (1983): Feeding ecology and the evolution of body size of baboons. Afr J Ecol 21:219–233.

Demment MW, Van Soest PJ (1985): A nutritional explanation for body-size patterns of ruminant and nonruminant herbivores. Am Nat 125:641–672.

DeVore I, Hall KRL (1965): Baboon ecology. In DeVore I (ed): "Primate Behaviour: Field Studies of Monkeys and Apes." New York: Holt, Rinehart and Winston, pp 20–52.

Dunbar RIM (1977): Feeding ecology of gelada baboons. A preliminary report. In Clutton-Brock TH (ed): "Primate Ecology: Studies of Feeding and Ranging Behaviour in Lemurs, Monkeys and Apes." London: Academic Press, pp 251–273.

Fa JE (1984): Structure and dynamics of the Barbary macaque in Gibraltar. In Fa JE (ed): "The Barbary Macaque: A Case Study in Conservation." New York: Plenum Press, pp 263–306.

Forthman Quick DL (1986): Activity budgets and the consumption of human food in two troops of baboons, *Papio anubis,* at Gilgil, Kenya. In Else JG, Lee PC (eds): "Primate Ecology and Conservation." Cambridge: Cambridge University Press, pp 221–228.

Goering HK, Van Soest PJ (1970): Forage fiber analysis. Handbook No. 379. ARS, USDA. Washington, D.C.

Green KF, Garcia J (1970): Recuperation from illness: Flavor enhancement for rats. Science 173:749–751.
Harding RSO (1976): Ranging patterns of a troop of baboons *(Papio anubis)* in Kenya. Folia Primatol (Basel) 25:143–185.
Harris RS (1970): Natural vs. purified diets in research with nonhuman primates. In Harris RS (ed): "Feeding and Nutrition of Nonhuman Primates." New York: Academic Press, pp 251–262.
Harrison MJS (1984): Optimal foraging strategies in the diet of the green monkey, *Cercopithecus sabaeus,* at Mt. Assirik, Senegal. Int J Primatol 5:435–472.
Harrison MJS (1985): Time budget of the green monkey, *Cercopithecus sabaeus:* Some optimal strategies. Int J Primatol 6:351–376.
Heller SN, Hackler LR, Rivers JM, Van Soest PJ, Roe DA, Lewis BA, Robertson JB (1980): Dietary fiber: The effect of particle size of wheat bran on colonic function in young adult men. Am J Clin Nutr 33:1734–1744.
Hladik CM (1978): Adaptive strategies of primates in relation to leaf-eating. In Montgomery GG (ed): "The Ecology of Arboreal Folivores." Washington: Smithsonian Institution Press, pp 373–395.
Hodges CM (1981): Optimal foraging in bumblebees: Hunting by expectation. Anim Behav 29:1166–1171.
Inglett GE (1975): Effects of refining operations on cereals. In Harris RS, Karmas E (eds): "Nutritional Evaluation of Food Processing." Westport: The Avi Publishing Co, pp 139–158.
Itani J (1958): On the acquisition and propagation of a new food habit in the troop of Japanese monkeys at Takasakiyama. Primates 1:84–98.
Kalter SS (1977): The baboon. In HSH Prince Ranier III of Monaco, Bourne GH (eds): "Primate Conservation." New York: Academic Press, pp 385–419.
Kawanaka K (1984): Association, ranging, and the social unit in chimpanzees of the Mahale mountains, Tanzania. Int J Primatol 5:411–434.
Knapka JJ, Morin ML (1979): Open formula natural ingredient diets for nonhuman primates. In Hayes KC (ed): "Primates in Nutritional Research." New York: Academic Press, pp 121–138.
Koford CB (1963): Group relations in an island colony of rhesus monkeys. In Southwick CH (ed): "Primate Social Behavior." New York: Van Nostrand, pp 136–152.
Krebs JR, McCleery RH (1984): Optimization in behavioral ecology. In Krebs JR, Davies NB (eds): "Behavioral Ecology: An Evolutionary Approach." Sunderland, MA: Sinauer Associates, pp 91–121.
Kummer H (1968): "Social Organization of Hamadrayas Baboons: A Field Study." Chicago: University of Chicago Press.
Lang CM (1970): Organoleptic and other characteristics of diet which influence acceptance by nonhuman primates. In Harris RS (ed): "Feeding and Nutrition of Nonhuman Primates." New York: Academic Press, pp 263–275.
Lima SL, Valone TJ (1986): Influence of predation risk on diet selection: A simple example in the grey squirrel. Anim Behav 34:536–544.
Lima SL, Valone TJ, Caraco T (1985): Foraging-efficiency—predation risk trade-off in the grey squirrel. Anim Behav 33:155–165.
Mason VC, Just A, Bech-Andersen S (1976): Bacterial activity in the hind-gut of pigs 2. Its influence on the apparent digestibility of nitrogen and amino acids. Z Tierphysiol Tier Futtermittelkde 36:310–324.
Matsuzawa T, Hasegawa Y, Gotoh S, Wada K (1983): One-trial long-lasting food-aversion learning in wild Japanese monkeys *(Macaca fuscata)*. Behav Neural Biol 39:155–159.

Milton K (1979): Factors influencing leaf choice by howler monkeys: A test of some hypotheses of food selection by generalist herbivores. Am Nat 114:362–378.

Milton K (1980): "The Foraging Strategy of Howler Monkeys: A Study in Primate Economics." New York: Columbia University Press.

Milton K (1984): Habitat, diet, and activity patterns of free-ranging woolly spider monkeys (*Brachyteles arachnoides* E. Geoffroy 1806). Int J Primatol 5:491–514.

Owen-Smith N, Novellie P (1982): What should a clever ungulate eat? Am Nat 119:151–178.

Post DG (1981): Activity patterns of yellow baboons *(Papio cynocephalus)* in the Amboseli National Park, Kenya. Anim Behav 29:357–374.

Post DG, Hausfater G, McCuskey SA (1980): Feeding behavior of yellow baboons *(Papio cynocephalus)*: Relationship to age, gender and dominance rank. Folia Primatol (Basel) 34:170–195.

Pyke GH (1981): Honeyeater foraging: A test of optimal foraging theory. Anim Behav 29:878–888.

Qvortrup SA, Blankenship LH (1975): Vegetation of Kekopey, a Kenya cattle ranch. East Afr Agr Forestry J 40:439–452.

Rhine RJ, Ender PB (1983): Comparability of methods used in the sampling of primate behavior. Am J Primatol 5:1–15.

Rhine RJ, Linville AK (1980): Properties of one-zero scores in observational studies of primate social behavior: The effect of assumptions on empirical analyses. Primates 21:111–122.

Rose M (1977): Positional behaviour of olive baboons *(Papio anubis)* and its relationship to maintenance and social activities. Primates 18:59–116.

Rozin P, Kalat JW (1971): Specific hungers and poison avoidance as adaptive specializations of learning. Psychol Rev 78:459–486.

Sade DS (1965): Some aspects of parent-offspring and sibling relations in a group of rhesus monkeys, with a discussion of grooming. Am J Phys Anthropol 23:1–17.

Strum SC (1984): The Pumphouse Gang and the great crop raids. Animal Kingdom 87:36–43.

Strum SC, Western JD (1982): Variations in fecundity with age and environment in olive baboons *(Papio anubis)*. Am J Primatol 3:61–76.

Suen HK, Ary D (1984): Variables influencing one-zero and instantaneous time sampling outcomes. Primates 25:89–94.

Teas J, Richie T, Taylor H, Southwick C (1980): Population patterns and behavioral ecology of rhesus monkeys *(Macaca mulatta)* in Nepal. In Lindburg DG (ed): "The Macaques: Studies in Ecology, Behavior and Evolution." New York: Van Nostrand Reinhold, pp 247–262.

van Lawick-Goodall J (1968): Expressive movements and communication in chimpanzees. In Jay PC (ed): "Primates: Studies in Adaptation and Variability." New York: Holt, Rinehart & Winston, Inc., pp 313–374.

Van Soest PJ, Robertson JB (1977): Analytical problems of fiber. In Hood LF, Wardrip EK, Bollenback GN (eds): "Carbohydrates and Health." Westport, CT: AVI, pp 69–83.

Vickery WL (1984): Optimal diet models and rodent food consumption. Anim Behav 32:340–348.

Westoby M (1974): An analysis of diet selection by large generalist herbivores. Am Nat 108: 290–304.

Wrangham RW (1974): Artificial feeding of chimpanzees and baboons in their natural habitats. Anim Behav 22:83–93.

Ecology and Behavior of Food-Enhanced Primate Groups, pages 53–78

3

Supplemental Food as an Extranormal Stimulus in Barbary Macaques (*Macaca sylvanus*) at Gibraltar—Its Impact on Activity Budgets

John E. Fa

Animal Ecology Research Group, Department of Zoology, University of Oxford, Oxford OX1-3PS, United Kingdom; and Departamento de Ecologia, Instituto de Biologia, Universidad Nacional Autonoma de Mexico, Mexico City, Mexico 04510 D.F.

INTRODUCTION

The use of time by primate groups can be related to the dispersion and predictability of food resources as well as to energy gain per unit weight of food. Observations of food-enhanced troops provide valuable insight into these relationships since it is usually possible to measure the abundance, distribution, and quality of food. Differences in the response of animals to varying food supplies may be attributed to how energy availability changes relative to the number of monkeys competing for the resource. Provisioned foods, which are normally higher in energy than wild ones and more palatable, predictable, and highly abundant, can influence the time an animal needs to devote to feeding. Thus, food-enhanced monkey troops may spend proportionately less time feeding and moving, but more time resting and socializing than monkeys in the wild [see Baulu and Redmond, 1980; Fa, 1986a; Post, 1978; Post and Baulu, 1978; Rasmussen and Rasmussen, 1979; Siddiqi and Southwick, this volume, chapter 6]. The underlying paradigm is that proportions of time spent on the various activities form a mutually dependent set, in that time spent on one activity state—feeding, for example—is not necessarily available for the performance of others [Altmann SA, 1974]. However, the relationship between food and activity may not always be a clear ecological one; i.e., the quantity and/or distribution of food given may be less important than the attractiveness of the items offered. In certain situations, a frequent offer of highly attractive food may cause alterations in the time spent on nonfeeding activities. This could perhaps be compared with a "supernormal stimulus" effect, although the term cannot be used here in the classical ethological sense, as a stimulus which is even more effective in releasing fixed action patterns than the original ones [Burghardt, 1973], since the behaviors involved are more complex than fixed action patterns. Nonetheless, the attractiveness of such foods resembles supernormal stimuli in that the releasers are more effective than the natural ones. Their magnitude is larger than normal because the offer of particular foods is identified by the animals as a complex of sounds and images from people approaching them. In the case of sweets, the reinforcement received by the animals from the pleasantness of sweetness can lead to an addiction to these items, where the search image becomes people. The constant presence of people close to animals that know the benefits of such foods can have a significant impact on their ecology.

Temporal fluctuations in birth rates in the two Barbary macaque (*Macaca sylvanus*) troops at Middle Hill and Queen's Gate in Gibraltar are described in Fa [1984a]. Mean annual birth rate for Queen's Gate is 48.51% and 63.24% for Middle Hill. The explanation is also forwarded that it is the level and manner of feeding, and the food types given by passersby, that cause disruptions in the animals' activity patterns and have consequences in dampening breeding. Across nearly five decades of demographic data for the Gibraltar troops, a significant fall in birth rates appears in association with an increase in tourist numbers during a boom tourist period from 1960 to 1968. Fa [1984a] arrives at this tentative causal explanation for the fall in birth rates from observations on how humans interact with the monkeys. Hypotheses such as a common climatic effect, changes in group structure, density, or the high relatedness of the troops do not explain satisfactorily such a drop in fertility. For example, if the high consanguinity values for both Middle Hill and Queen's Gate [see Burton and Sawchuk, 1984; Fa, 1986a] were to be responsible there should have been a progressive decline in birth rates across years [Goodman, 1980]. Yet the pattern observed is one of high annual birth rates for decades preceding 1960–1970 but a drastic fall during the 1960–1970 period in both troops. Moreover, it is unlikely that the deleterious effects of inbreeding, such as a fall in reproductive fitness, could have been expressed in a population of animals that had been established for less than 30 years when the low birth rates were observed.[1] That a greater degree of relatedness confers lower fertility is totally contradicted by the fact that the most fertile troop, Middle Hill, has always been the more inbred, but also the less affected by visiting humans. This group has annual birth rates comparable to those of provisioned rhesus (*Macaca mulatta*) and Japanese macaque (*Macaca fuscata*) troops (this volume, chapter 9).

[1]The present Barbary macaque population in Gibraltar is descended from a last introduction of monkeys from Morocco that took place from 1936 to 1940. By 1939 there was only one old female from the original Gibraltar monkeys that did not breed. The Barbary macaques on the Rock at the time of the study were descendants of a very varied founder population since animals were brought to Gibraltar from various parts of Morocco.

Fa [1984a] has suggested that the mechanism of disruption is through people frequently feeding attractive foods, usually peanuts and sweets. The offer of such foods distracts animals from any other activity so that an increase in the number of interruptions during the start of the breeding period leads to a corresponding decrease in the monkeys' breeding performance. The claim is made here that the statistically significant negative correlation between human interference (taken as an index using the number of tourists visiting Gibraltar annually) and yearly birth rates as reported in Fa [1984a] is not an artifact but a true consequence of the type and level of human disruption. Given the difference in the level of human interference that exists between the two troops of macaques in Gibraltar, this important variable should be reflected in changes in their activity patterns, although aspects of home range use could also be affected. This paper will analyze the activity budgets and touch briefly on the use of space by the two troops of Barbary macaques in Gibraltar. A more detailed comparison of use of space and energy by the two troops is intended elsewhere. For the purpose of this paper temporal distribution of activities (diurnal and seasonal) and the variability among individual animals will be assessed in relation to the different levels of provisioning and human disturbance.

STUDY AREA

Gibraltar lies at latitude 36° 09′ N and longitude 05° 21′ W at the eastern end of the north shore of the Strait of Gibraltar (Fig. 1). The Rock, of primarily Jurassic limestone, is a peninsula approximately 4 km long and 1.5 km wide, joined to the Spanish mainland by a low, sandy isthmus 1 km wide by 1 km long. An eastern scarp slope crowned by cliffs forms a north–south running ridge with a gentler dip slope on the west. This reaches a maximum height of about 424 m.

The general climate regime in Gibraltar is Mediterranean: mild wet winters and hot dry summers.

SUBJECTS AND METHODS

Troops' History

The subjects of this study are members of the Middle Hill and Queen's Gate troops of Barbary macaques in Gibraltar. The history and demography of the troops are documented in Fa [1981, 1984a, 1986a]. The British Army and the Gibraltar Regiment have officially managed both groups since 1915 [Fa, 1981], being responsible for feeding the monkeys and culling numbers since that date [see Fa, 1984a].

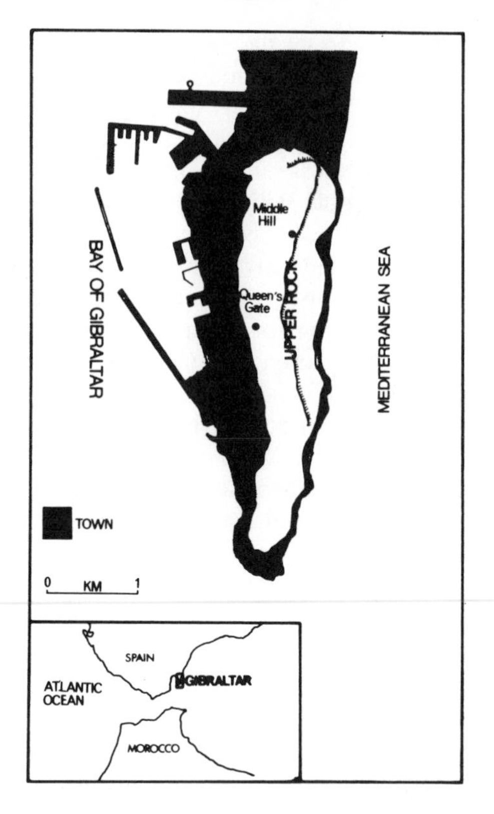

Fig. 1. Location of Gibraltar and study areas.

Troops' Membership

At the start of the study, the Middle Hill troop comprised 14 animals—2 fully mature subadult males, 4 adult females and 8 younger animals—and Queen's Gate had 21—7 adult females, 2 adult males, and 12 juvenile and immature animals. Numbers varied through the study period as a result of three births, one death, and one removal at Middle Hill and six births, one death, and one removal at Queen's Gate.

Subjects' Age and Agonistic Rank

Records of births and deaths of the Gibraltar monkeys have been kept consistently by the army since 1936. Each monkey's age was accurately known for the present study. Age-sex class definitions follow Fa [1984b].

Data on the direction of supplanting interactions between individuals were used to construct agonistic hierarchies for both troops [see Fa, 1986a for more details]. Agonistic rank is significantly correlated with age in both troops (Pearson's correlation QG. $r = 0.655$, $P<0.001$; M.H. $r = 0.586$, $P<0.001$).

Data Collection

An intensive study on the behavioral ecology of the troops lasted from January 1979 to March 1980. Before the start of the study, two 5-day observation periods per month per troop were selected at random. Whenever possible, observations were made during the preselected study days. The length of time for which the study groups were observed each month was affected by the degree of disturbance caused by extraneous factors. Although moderate disruption by the army's ape-keeper during feeding times was ignored, observations had to be stopped on days when animals were being trapped or when construction work was in progress in the area. Tourists feeding and interacting with the monkeys at Queen's Gate was considered a normal event, typical for that troop since its start [Fa, 1984a]. Counts were made of the number of people visiting the monkeys, interacting with them and of monkeys responding to people. The Middle Hill troop was never in contact with people except for the ape-keeper and his assistant.

A total of 832 hr, ranging from 45 to 94 hr per month and covering a mean of 71.27% (SD = 12.34) of the possible daylight hours was collected for Middle Hill. At Queen's Gate, a total of 503 hr, with a range of 20 to 56 hr per month and covering a mean of 73.50% (SD = 10.52) of the daylight hours was obtained. Only data from March 1979 to February 1980 (inclusive) are used in the present analyses. The months of January and February 1979 were spent trying out recording methods.

Recording Methods

On each sample day, the behavior of all animals (except dependent young) was recorded in both groups. Scan samples [sensu Altmann J., 1974a] were performed every 15 min, and a record was taken of the animals' behavior state (feeding, moving, inactive, playing and allo- and auto-grooming) and its location (in 5-m quadrats). Totals of 42,448 and 38,228 records were made for Middle Hill and Queen's Gate, respectively.

To be sure that the level of accuracy and consistency in observation could be maintained for both troops, 2 months were devoted to testing recording techniques prior to the start of the study. Full-day censuses during February 1979 showed that accuracy in recognition and recording of individuals and activities increased from 60% to about 100% by the end of the month [Fa, 1986a]. Because the composition of the groups was exactly known, each scan record made could be matched against the true composition of the troops.

TABLE I. Definitions of Activity States

Feeding	Activities associated with gathering, handling and masticating food, including digging, cleaning, and manipulating potential food items whether ingested later or not; chewing presented certain problems in that it did not necessarily preclude the simultaneous performance of other behaviors that were being distinguished from feeding, namely, inactive and moving; chewing while inactive was not considered feeding since in terms of overall energy use the dominant activity was resting; practical considerations largely dictated the placing of chewing while moving within the category of moving
Moving	All locomotor activities, including walking, running, climbing, and leaping between arboreal supports, were considered moving; the only exceptions were short movements during feeding (less than one full stride) and movement during social behavior
Inactive	Solitary behaviors when an animal was sitting, looking about, or sleeping
Allogrooming	Grooming or being groomed by a conspecific
Autogrooming	Scratching, examination, or removal of skin or other items from own fur
Play	Chasing, wrestling, and frisking; solitary play not included

The time scheduling technique employed for the present study followed Struhsaker's [1975] duration sampling. This technique records the first activity to last uninterrupted for more than 5 sec for each individual animal observed per scan. According to Struhsaker [1975], this has the advantage of distinguishing between short bouts of behavior and essentially important ones in duration. From a bioenergetic point of view, such a sampling technique is more realistic since it tends to approximate the actual time spent on the different activities.

Definitions of Activity States

Exclusive (nonoverlapping activity categories) and exhaustive (all possible behaviors fall within one activity state) categories of activities were chosen. Six activity states were defined (Table I).

Calculation of Time Spent on Different Activities

Each monthly estimate was computed by a procedure similar to that used by Post [1978] to allow for potential bias arising from the unequal distribution of observation time at different hours. The correction is analogous to the method used by Altmann and Altmann [1970: pp. 120, 121]. The following statistics were calculated from the scan data:

1. the time spent by each monkey on each activity state during each hour for the days sampled per month;

2. the mean of each hourly estimate as the estimate of the proportion of time spent by an individual monkey on each activity during an hour that month;
3. the mean proportion of time spent by each monkey per month on each activity state using data derived from No. 2;
4. the mean and SD of time spent on each activity by employing results obtained in No. 3 for all monkeys in the troop.

RESULTS

The Provisioning Background

Throughout the year fresh fruit and vegetables were given daily to both troops of monkeys. This food was obtained from the local market and distributed by the ape-keeper and/or assistant, both soldiers of the Gibraltar Regiment [Fa, 1981], at one or more sites within the home ranges of the Barbary macaques. It was spread around a radius of no more than 2 or 3 m at the feeding point.

Provisioning time for both troops was usually around 0900, irrespective of seasonal changes in sunrise times, when the monkeys first became active. At Queen's Gate, mean feeding time was 0952 but earlier at Middle Hill (mean = 0906) (Fig. 2).

A daily mean of 18.96 ± 12.98 kilos of food was offered to the Queen's Gate troop, but Middle Hill received significantly more (25.10 ± 9.50 kilos) (Mann-Whitney U-test, U = 62.50, $P<0.002$). The amount of food given to the animals varied significantly between months (χ^2 one-sample test, Middle Hill, $\chi^2 = 39.59$, $P<0.001$; Queen's Gate, $\chi^2 = 31.30$, $P<0.001$), but not between seasons.[2] This represents a monthly average of 494 kilos for Queen's Gate and 764 kilos for Middle Hill with annual totals of 5,929 and 9,167 kilos, respectively (Fig. 2). In terms of caloric offer of food per monkey, each animal is given daily 12,201 kcal at Queen's Gate (24 times more than the calculated energy requirement for a wild Barbary macaque [see Drucker, 1984]) and 144,672 kcals at Middle Hill (289 times more).[3]

[2]Seasons are defined as spring (March–May), summer (June–August), autumn (September–November), and winter (December–February).

[3]Conversion of daily food weight potentially available for each animal to caloric availability was made using the energy equivalents of foods given in Fa [1986a]. A mean of all provisioned foods offered was calculated for the conversion.

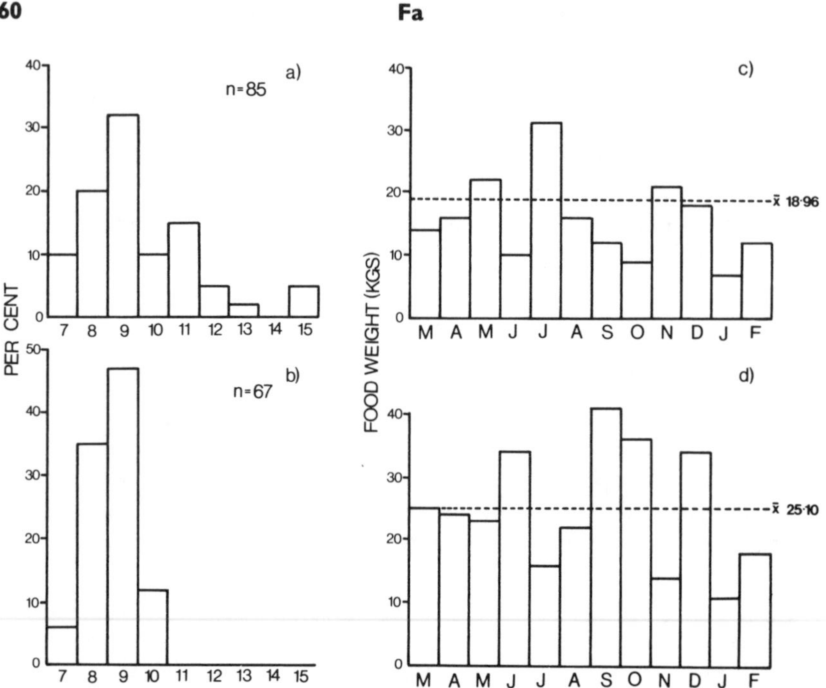

Fig. 2. Distribution of feeding time for **a)** Queen's Gate and **b)** Middle Hill troops in Gibraltar, and monthly changes in mean food weight offered to the two groups: **c)** Queen's Gate; **d)** Middle Hill.

Visitor Patterns

A major difference between the two monkey troops in Gibraltar has been the people visiting them. Because since 1972 Middle Hill troop has been in restricted access Ministry of Defence land, visitors are not allowed in the area. However, the monkeys at Queen's Gate have always featured as one of the most important tourist attractions in Gibraltar [see Fa, 1984a].

During the study period, an estimated total of 42,139 people visited Queen's Gate in a closed frontier situation.[4] Although there were people visiting during all months of the year, there was a considerable peak of tourists from August through October. Monthly changes in numbers of people visiting per hour and per day each month is shown in Figure 3. These range from a low of 30 people per day, or 2 persons per hour, in November, to 240 people per day, or 13 persons per hour, in September. Patterns in

[4]At the time of the study, the land frontier between Gibraltar and Spain had been closed to tourism for 11 years. During these years the maximum annual number of people visiting the Rock was around 140,000 in contrast to 800,000 before the rupture of communications.

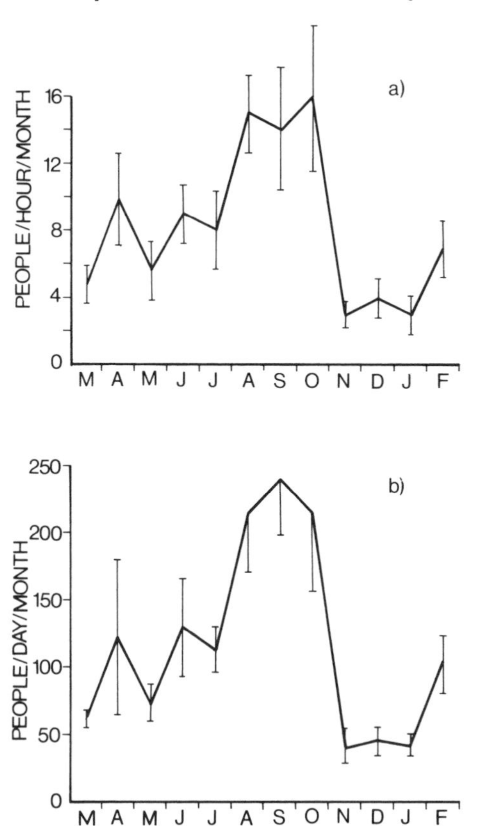

Fig. 3. Monthly distribution of human visitors (± SE) to the Queen's Gate troop expressed as **a)** number of people visiting per hour and **b)** people per day per month.

temporal changes in visitor pressure throughout the year are shown in Figure 4. A concentration of visitors exists from March to November primarily as a biphasic pattern of morning and evening visits. By seasons, means of 0.64 to 0.81 people interacted hourly with the monkeys and from 0.16 to 1.0 monkeys per hour were fed by tourists (Fig. 5). For the entire study period, a total of 221 incidents of humans feeding the monkeys was recorded (Fig. 6). Thirty-eight percent of these consisted of people feeding peanuts to the animals; 30% sweets, primarily sugar-coated chocolate sweets or chocolate-covered nuts; 23% fruit (banana, apple, mandarin orange, and grapes); 9% biscuits; and 0.6% sugar cubes. At each feeding incident from one to ten monkeys were involved, and in 85.97% (n = 189) of the recorded

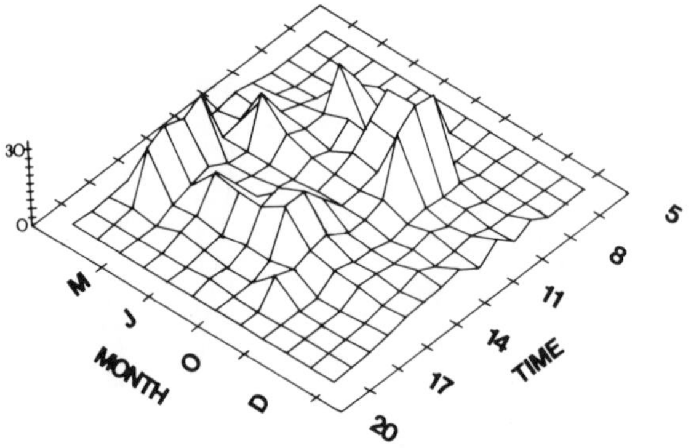

Fig. 4. Diurnal distribution of human visitors to the Queen's Gate troop during the study period.

interactions aggression between individuals to obtain the offered food was typical. Aggression in relation to food type offered was more frequent for sweets (24% of feeding incidents involved competition), less for peanuts (3%), and was nonexistent for biscuits and fruits.

All monkeys except infants obtained food from people. Feeding, however, was not random but concentrated on more dominant individuals, such as the adult males (Ji, Sa) and two adult females (Ra, Ve), which would remain in wait of passersby, even when the other troop members had departed from the site (Fig. 7). A detailed analysis of the different strategies for receiving food from humans and their consequences on individual animals' behaviors will be published elsewhere (Fa, in prep.).

Overall Ranging Patterns

Queen's Gate troop was observed in a home range of 4 ha, but spent more than 80% of its time in less than 1 ha. In contrast, the Middle Hill troop moved around 21 ha. Home ranges of both troops encompassed natural vegetation zones as well as man-made constructions (World War II gun emplacements and buildings). Middle Hill used the crest of the Rock to move long distances north or south, often covering up to 2 km in a day. Queen's Gate would spend much time on Queen's Road, where tourists pass, only making occasional sallies from this point to shaded areas no more than 45 m away.

Groups' Annual Budgets

Estimates of the proportion of time spent by both monkey troops on the different activity states are displayed by month in Figure 8. Over the 12-month study period, monkeys at Middle Hill spent 64% of the daylight

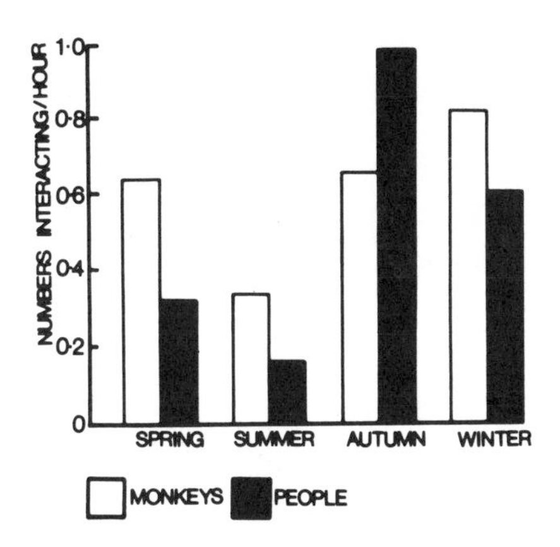

Fig. 5. Seasonal interaction rates between monkeys and people at Queen's Gate troop.

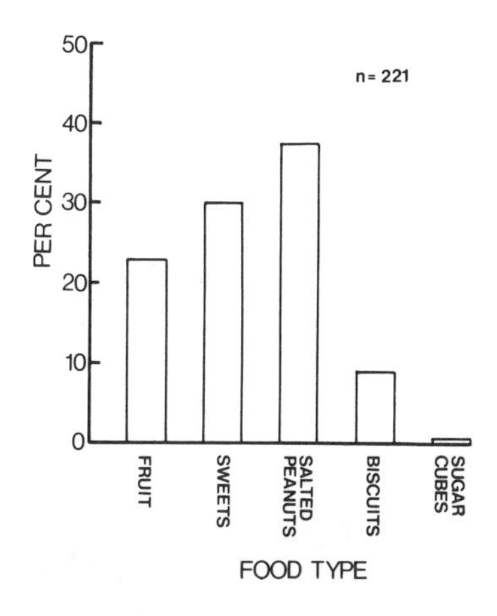

Fig. 6. Distribution of feeds given by human visitors to monkeys expressed as a percentage of the total number of occurrences per food item.

hours active (feeding, moving, allogrooming, autogrooming, and playing), whereas Queen's Gate troop was significantly less active (56.50%) ($U = 82.50$, $P<0.002$).

Feeding on both provisioned and natural foods took up 7% and 4.8% of time in Middle Hill and Queen's Gate, respectively ($U = 22$, $P<0.002$). Time spent moving by Queen's Gate troop was 35.09% ± 5.16% and 37.94% ± 4.04% ($U = 50$, $P<0.002$), but the former (35.43% ± 9.37%) spent more time inactive (33.47% ± 9.47%) ($U = 64$, $P<0.002$). Allogrooming proportions, however, were 15.84% ± 7.31% for Queen's

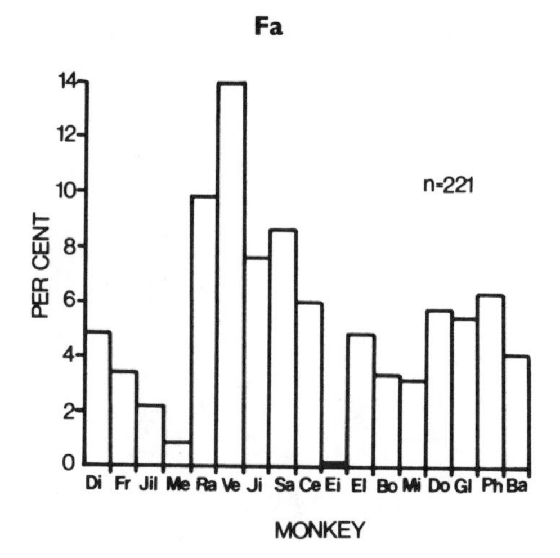

Fig. 7. Percentage number of feeds from tourists received by each member of the Queen's Gate troop.

Gate but lower for Middle Hill (11.07% ± 4.96%) (U = 34, $P<0.002$). Play took up 8.03% ± 2.88% at Queen's Gate and 4.31% ± 2.25% at Middle Hill (U = 18, not significant [ns]). Monthly changes for each activity were only positively correlated between troops for inactivity (Spearman rank correlation, $r_s = 0.613$, $P<0.001$). For both troops there was an appreciable decline in inactive proportions towards the end of the year, inversely mirrored by allogrooming and autogrooming at Middle Hill and by allogrooming, autogrooming, and playing at Queen's Gate. Playing increased towards the end of the year at Queen's Gate, but the inverse appears for Middle Hill. No obvious pattern was observed for moving. Changes across months were significant (χ^2 one-sample tests) for all activities except time spent feeding and moving for both troops and for playing at Middle Hill (allogroom—χ^2 QG = 49.13, $P<0.001$; MH = 24.47, *P* 0.02–0.01; inactive—χ^2 QG = 26.09, *P* 0.01–0.001; MH = 22.53, *P* 0.05–0.02; autogroom—χ^2 QG = 32.96, $P<0.001$; MH = 86.40, $P<0.001$; play—χ^2 QG = 29.18, $P<0.001$). Total active time spent per month was positively correlated between troops ($r_s = 0.902$, $P<0.001$).

Individual and Age-Sex Class Differences

Interindividual differences within each troop were significant for all activities (results of two-way nested ANOVAs). Comparison of activity proportions for each age-sex class between troops were tested with a three-way analysis of variance with replication. The proportion of time spent in each activity per month (using the arcsine transformation for proportions)

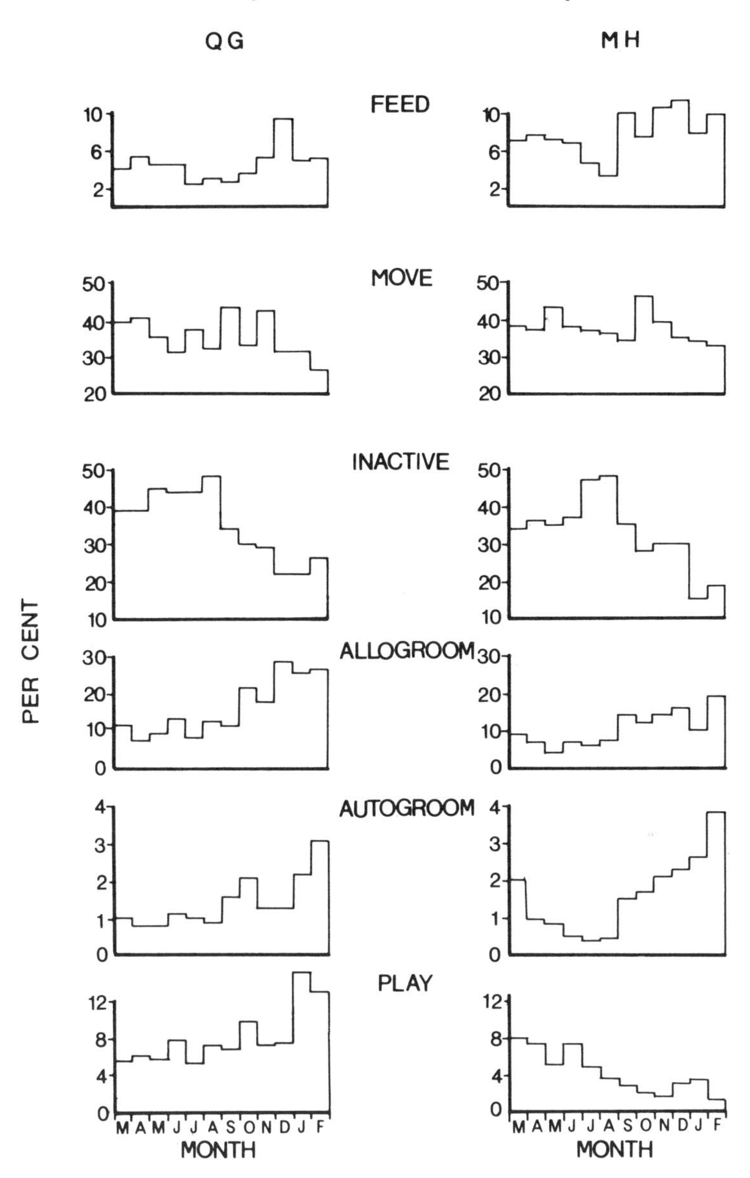

Fig. 8. Monthly distribution of troop activity proportions for Queen's Gate and Middle Hill.

was the dependent variable with troop and age-sex classes as the treatments. Intertroop differences in age-sex class proportions for all activities were significant except for autogroom (feed—F = 10.61, d.f. 1,33, $P<0.001$;

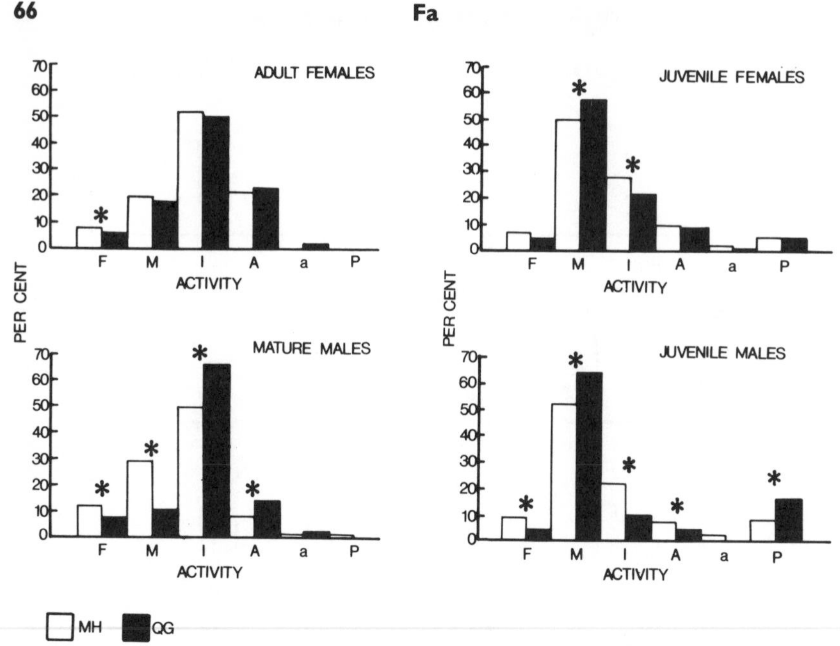

Fig. 9. Comparison of activity budgets of Queen's Gate and Middle Hill troops across age-sex classes. Significant results of analyses of variance are indicated.

move—F = 10.17, d.f. 1,33, $P<0.001$; inactive—F = 10.07, d.f. 1,33, $P<0.001$; allogroom—F = 10.60, d.f. 1,33, ns; autogroom—F = 10.19, d.f. 1,33, $P<0.001$; play—F = 10.21, d.f. 1,33, $P<0.001$). Significant differences between age-sex classes within troop emerged (feed—F = 11.86, d.f. 3,33, $P<0.001$; move—F = 11.35, d.f. 3,33, $P<0.001$; inactive—F = 11.09, d.f. 3,33, $P<0.001$; allogroom—F = 11.86, d.f. 3,33, $P<0.001$; autogroom—F = 10.35, d.f. 3,33, $P<0.001$; play—F = 40.81, d.f. 3,33, $P<0.001$).

Mean troop proportions of time spent in each activity per age-sex class are presented in Figure 9. Post hoc comparisons between troops for each activity state per age-sex class were made using two-way ANOVAs with unequal but proportional subclass numbers [Sokal and Rohlf, 1981: p. 361–363]. The results of these are indicated in Figure 9. Significant differences appear between troops in feeding for adult females, moving and inactive for juvenile females and in all activities for the mature males. For both troops, juveniles of both sexes spent more time moving and playing than adults. There were contrasting differences between the mature males at both troops. Subadult males at Middle Hill were more active (moved more, spent less time inactive)

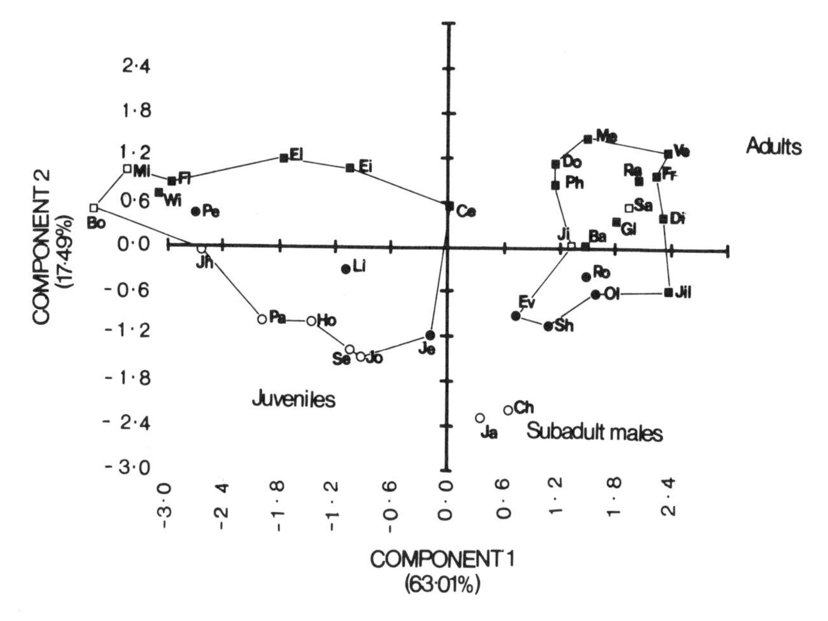

Fig. 10. Principal component analysis of annual activity proportions for all monkeys of both troops.

males at Middle Hill were more active (moved more, spent less time inactive) but fed more and groomed less than the adult males at Queen's Gate. Juvenile males at Queen's Gate fed significantly less, spent less time inactive, allogroomed less, and played more than those at Middle Hill. Queen's Gate juvenile females were more active (moved more, spent less time inactive) than the same age-sex class at Middle Hill.

Data for the individual monkeys composing each troop were submitted to a principal component analysis (PCA) to discriminate troop differences. The PCA generated separates all individuals into three main groups, juveniles, adult females and males and subadult males, but no separation on a troop basis appears (Fig. 10). A log plot of age against total active time for all individuals reveals a significant negative correlation (QG $r = -0.894$, $P<0.001$; MH $r = -0.844$, $P<0.001$; both Queen's Gate and Middle Hill monkeys fall on the same regression line (Fig. 11). For Queen's Gate troop members the relationship between activity and age, social rank, and feeding by humans was tested in a multiple regression analysis (Table II). Overall time spent active per month as well as feeding, moving, inactive, and allogrooming was more positively correlated with age. Rank was important only in feeding proportions.

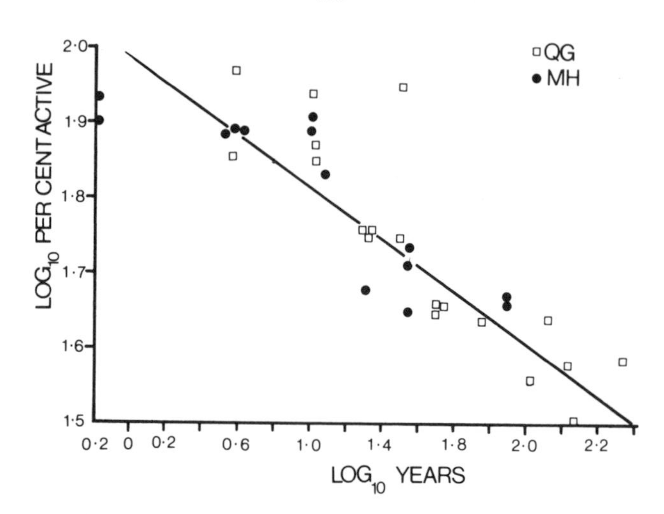

Fig. 11. Log plot regression between age of monkeys and overall time spent active. Monkey identity is not included.

TABLE II. Results of Multiple Regression Analysis Among Age, Agonistic Rank, Feeding by Humans, and Main Activity Proportions for the Queen's Gate Troop

	Activity				
Variable	Feed	Move	Inactive	Allogroom	Total active
Age	23.97	24.95	26.38	4.07	36.82
P	<0.001	<0.001	<0.001	<0.001	<0.001
Rank	18.51	2.56	1.53	1.38	3.83
P	<0.001	NS	NS	NS	NS
Feeding	0.33	0.43	0.19	0.16	0.52
by humans	NS	NS	NS	NS	NS

Values are F-values.

Diurnal Distribution of Activities

The diurnal distribution of activities for both troops throughout the study period is shown in Figures 12 and 13.

At Middle Hill, for all seasons, feeding behavior approximates a diphasic pattern, with morning and middle to late afternoon peaks. The timing of these peaks, in terms of actual "local time," varies from month to month but is consistently similar between seasons. The morning peak in feeding is usually at 0900, which corresponds to provisioning times. Moving also shows a clear diphasic pattern with an early morning high activity period when the animals emerge from their sleeping sites and just before sunset, corresponding to the return to these. Daily routine remains the same through when the animals emerge from their sleeping sites and just before sunset, corresponding to the return to these. Daily routine remains the same through the seasons with only a slight variation in home range area covered by the group. However, during spring and summer the troop tended to move longer distances away from the sleeping site before coming down to feed at the provisioning site. These were usually early morning (sunrise to 2 hr after) excursions to other areas largely spent feeding on natural foods. During spring feeding on natural foods rises to a morning peak around 0700 just before provisioning occurred. There is another peak of natural food feeding in the evening, but these peaks become reduced to one main midday peak during autumn and winter. This shift is due to the shorter time interval between sunrise and provisioning time. There is still a brief early-morning peak in natural feeding which occurs immediately after awakening. A midday peak in inactivity is typical for all seasons. Allogrooming also rises during the middle of the day but is more pronounced in autumn and winter.

Diurnal activity patterns of the Queen's Gate troop differ markedly from Middle Hill. In contrast to Middle Hill there is no clear diphasic pattern in feeding and moving, and only significant increases in feeding in the morning and evening are typical in autumn and winter. No dramatic increases in inactivity proportions during the midday hours, as seen at Middle Hill, appear during the summer months. Time spent moving and inactive is similar throughout the day in all seasons.

Queen's Gate troop's daily routine was relatively constant during all months of the year. Unlike Middle Hill, after rising from the main sleeping sites on cliffs some 50 m away, the animals would move to the Queen's Gate area where they would spend most of the morning. Occasionally the main group of monkeys would move back to vegetation near the sleeping sites if no passersby offered food, except for some individuals. However, most of the daylight hours would be spent in the Queen's Gate area by all monkeys. Feeding on natural foods was infrequent and did not follow any clear diurnal pattern.

The relation between diurnal weather changes and the diurnal distribution of activity levels (a percentage composed of all activity proportions except inactive) was tested by Spearman rank correlation tests. The correlations are significant for the two most fluctuating variables, temperature and humidity. The results for Middle Hill have already been published in Fa [1986a]. For Queen's Gate, however, no significant correlations appear. At Middle Hill the value of r_s increases from the start of the study period, peaks in July and falls thereafter in response to changes in temperature. The converse is true for humidity values.

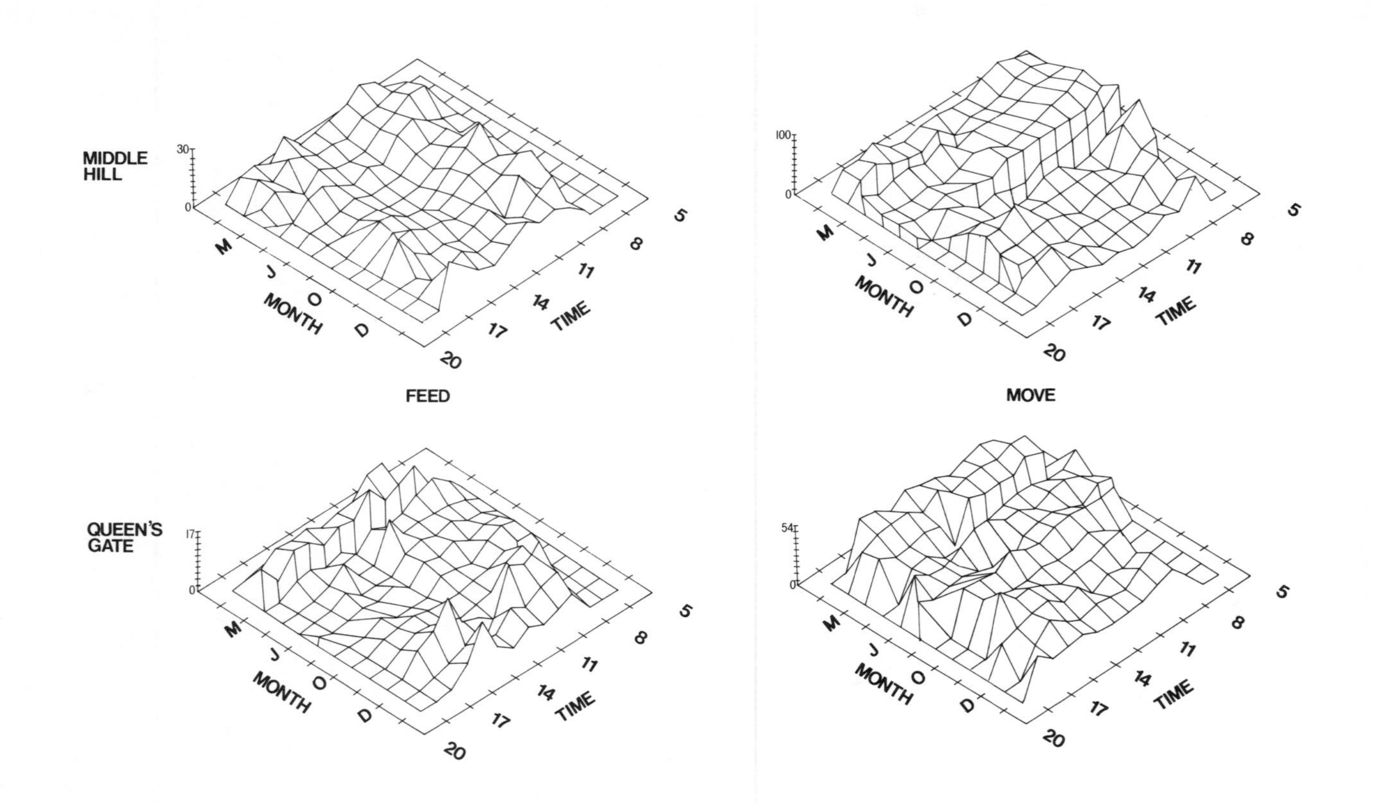
MIDDLE HILL
QUEEN'S GATE
FEED
MOVE
MONTH
TIME
M
J
O
D
5
8
11
14
17
20
30
100
17
54
0

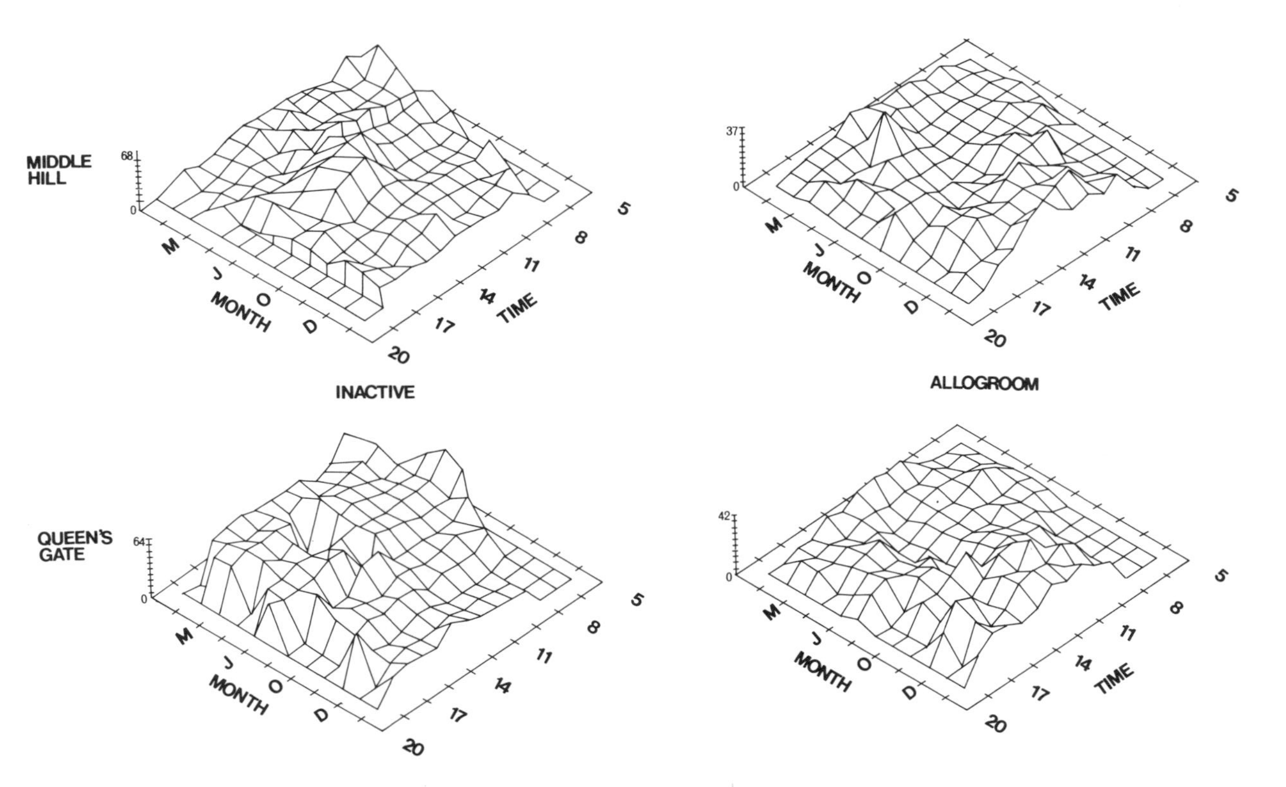

Fig. 12. Three-dimensional plots of activity proportions per hour per month of the study period. Only the main activities are included.

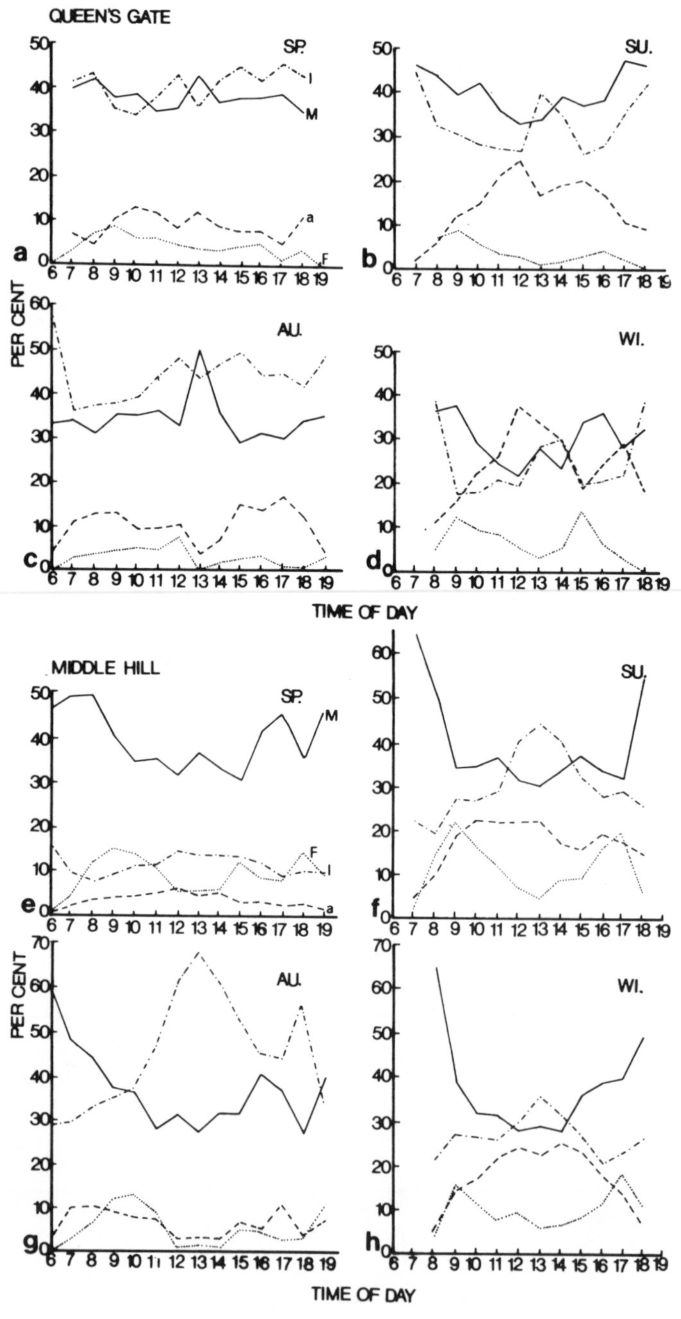

Fig. 13. Diurnal changes in main activity proportions per season for (**a–d**) Queen's Gate and (**e–h**) Middle Hill. F = feed; M = move; I = inactive; a = allogroom.

DISCUSSION

A comparison of provisioned monkeys with wild ones reveals that feeding animals with a clumped and high-calorie food source can have important consequences on activity patterns. While wild Barbary macaques spend almost half their daytime feeding (Deag, 1974, personal communication) [Fa, 1986a], those in Gibraltar spent 7% and 4.8% feeding at Middle Hill and Queen's Gate, respectively. Studies of food-enhanced rhesus macaques at Puerto Rico [Baulu and Redmond, 1980; Post, 1978; Post and Baulu, 1978] (this volume, chapter 7) and of human-fed rhesus in India [Seth and Seth, 1986] (this volume, chapters 5 and 6) have shown that a decrease in feeding and moving time is associated with provisioning. An explanation for this may lie in that the overall caloric gain per unit time and weight is far greater for provisioned foods than for most natural ones. In Gibraltar, provisioned foods had a mean of 134.08 kcals/g in contrast to 0.04 kcal/g for natural forage [Fa, 1986a]. Mean weight of provisioned food items was 101.92 g as opposed to 0.44 g for natural foods. Moreover, because of the static nature in time and space of provisioned food, the expense in time comes only from handling food items and not from searching for them.

The present study investigated the hypothesis that it is the action of tourists feeding and disturbing the monkeys at Queen's Gate that has had a profound effect on the ecology and behavior of this group and at a previous stage on the Middle Hill group [see Fa, 1984a, 1986a]. Human visitors at the Queen's Gate troop are the principal, though not the only, element of difference between the troops on the Rock. Table III summarizes the similarities and differences between Queen's Gate and Middle Hill troops.

Queen's Gate troop was generally less active than Middle Hill. This might be related to the presence of tourists feeding animals, which affects the need to forage. However, in order to generate causal explanations for this phenomenon it is important to test between alternative hypotheses. This will hopefully allow for the determination of the most likely causal explanation rather than rely on the ad hoc speculation that overall activity and use of space differences are controlled by human presence at Queen's Gate.

A first alternative hypothesis is that differences in troop activity are related to provisioning levels and not to human presence. In terms of monthly food weight given to each monkey by the army, Middle Hill received more than double (54.57 kilos) the amount received by Queen's Gate (23.52 kilos). Because Queen's Gate has consistently obtained less food from the army than Middle Hill, the expectation is that Middle Hill would be more dependent on provisioned foods, feed less on natural foods, and thus have a smaller home range than Queen's Gate. The latter, on the other hand, would be expected to have larger home ranges in view of the lower dependence on the more

TABLE III. Summary of Similarities and Differences Between the Barbary Macaque Troops in Gibraltar

Variable	Queen's Gate	Middle Hill
Provisioning by the army	Lower	Higher
Human visitors	Present, feed and harass monkeys	Not present
Group size	21	14
Age structure	Older (mean age 6.75; range 1.53–18.53)	Younger (mean age 4.15; range 1.48–8.74)
Mean annual birth rate[a]	48.51%	63.24%
Home range	Very small	Large
Troop activity	Lower	Slightly more active
Differences in activity with age	Older animals less active than young ones	
Sex differences in activity	Males usually more active than females	
Diurnal activity patterns	Not clear	Clear
Environmental effects on troop activity	No correlation	Correlation with temperature and humidity

[a]Data from Fa [1984a] for 1936–1981 for Queen's Gate and for 1946–1981 for Middle Hill.

energetic foods provided by the army. The results of the year's study conducted simultaneously on both troops show that the exact opposite is true. Queen's Gate troop has a smaller home range than Middle Hill, and Middle Hill feeds more on natural foods (although this is barely 1% of the total ingested food weight [see Fa, 1986a for details]). There is no correlation then between provisioning level and use of space in either troop.

A second alternative hypothesis is that the lower activity profile of the Queen's Gate troop could reflect a difference in group composition. The present study has shown that there is a significant negative correlation between age and activity, so that the lower Queen's Gate activity budgets could be related to the fact that there are older animals in this troop. Indeed, in the multiple regression performed for the Queen's Gate data (Table II), age accounts for most of the variance in the animals' activity. Comparison of activities made between age-sex classes suggests that it could be the older mature males in Queen's Gate that lower the group's overall budget, since there is no difference in adult females between troops and juveniles at Queen's Gate are significantly more active.

A salient feature of Queen's Gate ecology is that the troop is concentrated in an extremely small area. A reduction of home range is explicable in terms of the concentrated food resources and this may be the reason for the reduced home ranges at Gibraltar when compared to wild Barbary macaque troops [see Fa, 1986b, for data on wild troops]. However, the much smaller home range at Queen's Gate does not correspond to provisioning level, as seen above. Furthermore, no environmental differences such as climate and vegetation can explain the dissimilar activity proportions and use of space since the distance between the two group's home ranges is less than half a kilometer.

A major difference between the Gibraltar troops is in the synchrony of activities among troop members. There are more consistently clear diurnal patterns exhibited by the Middle Hill troop than Queen's Gate. More individuals at Middle Hill are engaged in the same activity at given times of the day than at Queen's Gate. Can these differences be explained by the presence of humans at Queen's Gate? The critical characteristic of Queen's Gate visitors is that they feed the monkeys with attractive foods. Of these, sweets are most fought for (see above). Observations on rhesus monkeys have revealed that they drink sugar solutions in preference to plain water [Kemnitz et al., 1981] corresponding to a similar affinity for the sweetness and pleasantness of sugar solutions seen in human subjects [Moskowitz, 1971]. Such a gustatory stimulus appears to be biologically determined as the "instigator of consummatory responses" [Moskowitz, 1971] and is widespread in animal behavior. Because sweets represent a gustatory reward that can condition the monkeys at Queen's Gate to stay around a specific site in wait of passersby, the phenomenon is appreciably linked to an addiction response. Since the animals do not appear to be seeking the food for energetic reasons (the animals are not hungry and are in fact obese, unpublished data), their reaction to passersby food may certainly be mediated by presence of a sweetness reward. Activities such as mating can be postponed when animals are offered such foods (this volume, chapter 5). Since the offer of food is regular (people visit the monkeys daily and throughout the day) the sit-and-wait behavior is continually reinforced. Activity proportions of the adult females and males at Queen's Gate may be explained by the animals' spending most of their time waiting. In contrast, Middle Hill females spend similar activity proportions largely feeding and being inactive around the provisioning site, perhaps related to the larger food supply. The higher resting behavior values at Middle Hill can also be attributed to the greater availability of provisioned foods. The more active Queen's Gate juvenile animals are responding to humans offering food as well as to the larger peer group in playing and moving. Because the monkeys that are waiting for people on the road are likely to be offered food first, and these are normally the adult males and females, juveniles adopt strategies of snatching food from people, unattended bags or even from cars to compensate for the monopolizing by older (and more dominant) individuals. The result is that Queen's Gate juveniles spend much more time moving and less time inactive than those at Middle Hill. The disruptions caused by humans across the day clearly accounts for the absence of a bimodal pattern of feeding at Queen's Gate. In both troops the higher midday temperatures are similar but only Middle Hill shows a correlation between climatic variables and activity. In general, Middle Hill animals engaged in the same behavior as a unit. Access to one main source of food at the provisioning site results in the Middle Hill troop's feeding, moving, resting, and allogrooming more as a single entity, than the Queen's Gate troop, whose members are enticed by other foods.

From the present study there is evidence to support the observation that the Queen's Gate troop has a pattern of activity critically different from the other troop on the Rock. Such differences in use of time and space can be related to the constant presence of humans feeding monkeys with attractive foods at Queen's Gate forcing animals to an area where this reward is expected. These disruptions can have a detrimental effect on the mating activities of the monkeys, as shown in Fa [1984a]. Although this point still needs further study in a situation where numbers of tourists are higher than in 1979, the observed competition for supplemental food between animals and the persistent occurrence of such feeds by tourists can explain the variation in birth rates noted for the 1960–1970 boom tourist period in Gibraltar [Fa, 1984a].

SUMMARY

The relationship between food and activity in food-enhanced primate groups is not always a clear ecological one since quantity and/or distribution of food given may be less important than the attractiveness of the items offered. In some situations frequent exposure to highly attractive foods may cause alterations in the time spent not feeding since animals may actively seek these at the expense of other activities. The present study focuses on the two troops of Barbary macaques in Gibraltar whose provisioning regimes differ. The Middle Hill troop is provisioned more but is not visited by tourists. The Queen's Gate troop is given less provisioned food but has been an important tourist attraction in Gibraltar for some decades and receives supplemental food items from tourists. Both groups have been under the custody of the British Army and subsequently the Gibraltar Regiment since 1915 and are fed daily. This paper analyzes the activity budgets of both Gibraltar troops in relation to the different levels of food enhancement and human disturbance. Results show that overall activity of the Queen's Gate troop is lower than Middle Hill's, and the explanation for this appears to lie in the different age structure and not in the presence of tourists. Because Queen's Gate has older animals than Middle Hill and a strong negative correlation between age and activity is shown in the present work, the lower activity profiles at Queen's Gate may reflect an age-dependent effect. A drastic difference in home ranges is typical, with Queen's Gate troop being restricted to 1 ha in contrast to 21 ha for Middle Hill. Synchrony of activities differed between troops. Queen's Gate showed no regular patterns of diurnal activity, while Middle Hill experienced a morning and an evening activity peak with a clear rest period at midday. These schedules responded to weather conditions at Middle Hill, but no correlation was found for Queen's Gate. The irregular activity patterns at Queen's Gate is explained by the constant presence of people feeding the monkeys tasteful, high calorie foods. Although further study is needed during high tourist years, the present paper supports the notion that the lowered birth rates at Queen's Gate during the 1960–1970 tourist boom can be attributed to the disruptions caused by thousands of visitors passing by and interacting with the monkeys.

ACKNOWLEDGMENTS

I would like to thank Major E. Guerrero, Lt. Col. D. Collado, Sgt. A. Holmes, and Pte. E. Asquez of the Gibraltar Regiment for their help.

I am most grateful to Jaime Ramirez and Carolina Valdespino for their help in sorting out raw data. To Daniel Piñero, I extend my thanks for performing the PCA analysis, and to my wife Monique for her support and encouragement during the study and preparation of this article. Valuable comments on the manuscript were made by Charles H. Southwick and Francisca Feekes. The Ferrary Trust in Gibraltar is acknowledged for financial help during this research. Bland Ltd. (Gibraltar) permitted free use of the Cable Car to the Ape's Den during the study.

REFERENCES

Altmann J (1974): Observational study of behaviour: Sampling methods. Behaviour 49: 227–267.

Altmann SA (1974): Baboons, space, time and energy. Am Zool 14:221–248.

Altmann SA, Altmann J (1970): Baboon Ecology: African field research. Bibl Primatol 12:21–220.

Baulu J, Redmond DE (1980): Some sampling considerations in the quantitation of monkey behaviour under field and captive conditions. Primates 19(2):391–399.

Burghardt G (1973): Instinct and innate behavior: Toward an ethological psychology. In Reynolds GS (ed): "The Study of Behavior: Learning, Motivation, Emotion and Instinct." Chicago: Scott, Foresman and Glenview, pp 323–400.

Burton FD, Sawchuck LA (1984): The genetic implications of effective population size for the *Barbary macaque* in Gibralter. In Fa JE (ed): ''The Barbary Macaque: A Case Study in Conservation.'' New York and London: Plenum Press, pp 307–315.

Drucker GR (1984): The feeding ecology of the Barbary macaque and cedar forest conservation in the Moroccan Moyen Atlas. In Fa JE (ed): ''The Barbary Macaque—A Case Study in Conservation.'' New York and London: Plenum Press, pp 135–164.

Fa JE (1981): The apes on the Rock. Oryx 16:73–76.

Fa JE (1984a): Structure and dynamics of the Barbary macaque population in Gibraltar. In Fa JE (ed): ''The Barbary Macaque—A Case Study in Conservation.'' New York and London: Plenum Press, pp 263–306.

Fa JE (1984b): Definition of age-sex classes for the Barbary macaque. In Fa JE (ed): ''The Barbary Macaque—A Case Study in Conservation.'' New York and London: Plenum Press, pp 335–346.

Fa JE (1986a): ''Use of Time and Resources by Provisioned Troops of Monkeys: Social Behaviour, Time and Energy in the Barbary Macaque (*Macaca sylvanus* L.) at Gibraltar.'' Contributions to Primatology 23. Basle: Karger.

Fa JE (1986b): Balancing the wild/captive equation: the case of the Barbary macaque (*Macaca sylvanus* L.). In Bernisckhe K (ed): ''Primates: The Road to Self-Sustaining Populations.'' New York: Springer. pp 197–211.

Goodman D (1980): Demographic intervention for closely managed populations. In Soulé M, Wilcox BA (eds): ''Conservation Biology: An Evolutionary-Ecological Perspective.'' Massachusetts: Sinauer, pp 171–198.

Kemnitz JW, Gibber JR, Lindsay KA, Brot MD (1981): Preference for sweet and regulation of caloric intake. Am J Primatol 1:313–314.

Moskowitz MR (1971): The sweetness and pleasantness of sugars. Am J Psychiatry 84:387–406.

Post D (1978): ''Feeding and Ranging Behavior of Yellow Baboons (*Papio cynocephalus*).'' Yale University: Ph.D. Thesis.

Post W, Baulu J (1978): Time budgets of *Macaca mulatta*. Primates 19:125–139.

Rasmussen DR, Rasmussen KL (1979): Social ecology of adult males in a confined troop of Japanese macaques (*Macaca fuscata*). Anim Behav 27:434–445.

Seth PK, Seth S (1986): Ecology and behaviour of rhesus monkeys in India. In Else JG, Lee PC (eds): ''Primate Ecology and Conservation.'' Cambridge: Cambridge University Press, pp 89–104.

Sokal RR, Rohlf FJ (1981): ''Biometry.'' San Francisco: WH Freeman and Company. 2nd edition.

Struhsaker T (1975): ''The Red Colobus Monkey.'' Chicago: University of Chicago Press.

Ecology and Behavior of Food-Enhanced Primate Groups, pages 79–94

4

Food and Energetics of Provisioned Wild Japanese Macaques (*Macaca fuscata*)

Toshitaka Iwamoto

Department of Biology, Miyazaki University, Miyazaki-shi 880, Japan

INTRODUCTION

Provisioning has promoted population increases in several troops of Japanese macaques (*Macaca fuscata*) [Takasakiyama: Masui et al., 1973; Arashiyama: Koyama et al., 1975; Ryozen: Sugiyama and Ohsawa, 1974, 1982; Koshima: Mori, 1979a,b]. Sugiyama and Ohsawa [1974] reported that, among population parameters, primiparous age, birth rate, and infant mortality exerted important effects on the increase in the size of Ryozen-A troop during the provisioned feeding period. They also found rank-dependent differences in the effect of provisioning on the above parameters. Mori [1979b] examined body weights of monkeys of the Koshima troop during both provisioned and unprovisioned feeding periods, and found that juveniles of dominant kin groups achieved a heavier body weight than those of subordinate ones during the former period. These results coincide with the finding that troop members shared provisioned foods in a biased manner depending on their dominance-subordinate positions [Iwamoto, 1974, 1978]. Usually, provisioned foods were supplied in a confined area, which increased competition for food between the monkeys. Since the Japanese monkey troops mentioned above also move around freely in natural forests, natural foods can compensate for a lack of provisioned foods in subordinate kin groups. However, the disproportionate distribution of provisioned food resources can account for the dramatic difference in the observed reproductive performance between kin groups. It can, therefore, be inferred that provisioned foods have important nutritional advantages over natural ones that improve reproduction of the monkeys. The purpose of this study is to clarify these observations and to discuss how artificial feeding can enhance the nutritional condition of Japanese monkeys.

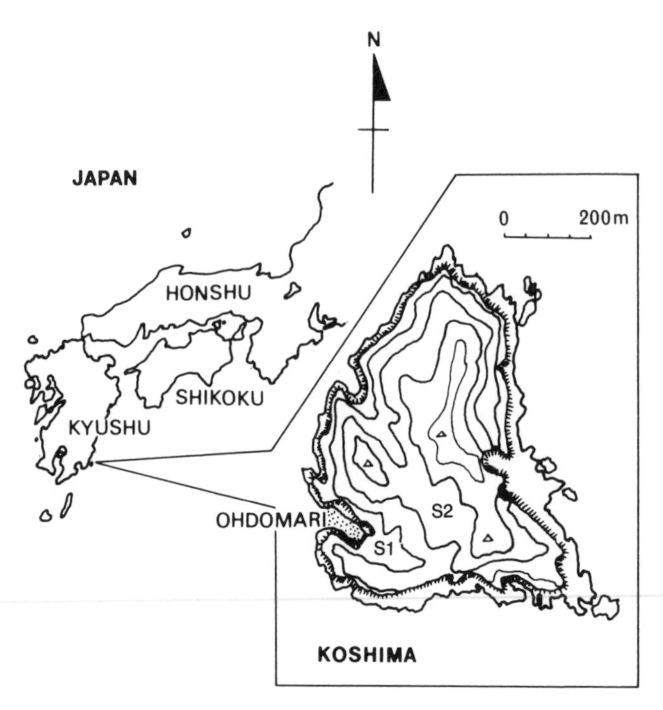

Fig. 1. Koshima Islet. The islet is situated about 300 m off the Coast of Nichinan, Miyazaki Pref. Artificial foods were supplied on **S1** or **S2**. The island is covered by a dense evergreen, broad-leaved forest.

STUDY TROOP AND METHODS

Koshima (Fig. 1) is a small island about 4 km in circumference covered by dense, evergreen, broad-leaved forest. This island (31° 17′ N and 131° 22′ W) is situated about 300 m off the coast of Kyushu. The climate is included in the warm-temperate zone, with 2,500 mm annual rainfall. Monthly mean temperatures range from 6.8°C in January to 27.0°C in August.

The Koshima troop of monkeys has the longest history of provisioning among Japanese macaques (*M. fuscata fuscata*), having started in 1950. Although the intensity of provisioning has not been constant, the population increased beginning in 1950, to reach a maximum of 112 monkeys in 1971. Thereafter, provisioning has been largely restricted, causing a decline in the population. In March 1985, the population was about 94, including a main troop of 74 monkeys, a splinter group of 12, and 8 solitary males. Details of population changes and of the history of provisioning are given by Mori [1979b]. In this study, field observations of the

Koshima troop were carried out during the following three periods: 1) May and September 1972 and February 1973, 2) monthly from October 1977 to March 1978, 3) monthly from April 1980 to September 1981. During the first period, 8 kg of wheat were supplied daily to the group at four separate feeding times. Four individuals (one adult male, two adult females, and one juvenile female) were chosen for focal animal sampling of their feeding behavior. Of these, however, only data for females, Satsuki (top ranked, 10 years old, 9 kg body weight) and Nasi (lower-middle ranked, 15 years old, 9 kg body weight) were used to make comparisons with data from the two later periods in this study. During the last two periods, no food was provided except during September 1981. Focal animal observations were carried out monthly on an adult female, Siba (lower-middle ranked, 12–16 year old, 8 kg body weight) and the data for May and September 1980 and February 1978 were used in this study. Parts of these data have been presented elsewhere [Iwamoto, 1978, 1982]. The feeding behavior of sample individuals in both natural forest and feeding ground was recorded continuously for a complete day in each month. When the observer lost his subject, he tried tracking again from that hour another day. In that case, data obtained in 2–3 days were combined to make a complete sequence of daytime hours. Because of the difficulty of continuous tracking in the forest, only one day's sequential data was obtained for each individual each month. Therefore, statistical tests for comparisons between individuals and months were not applicable. When a subject monkey fed, the observer recorded time spent feeding and type and amount of food eaten. Food samples were collected after each observation, weighed fresh, and subsequently dried in an oven at 60°C for more than 48 hr. They were later analyzed for crude protein, lipids, crude fiber, ash, and soluble carbohydrate [for methods, see Iwamoto, 1982]. Nutritional contents of provisioned foods were obtained from the tables of "Standard Nutritional Composition of Japanese Diets" [Japan Diet Association Corporate, 1975]. Values for monkey chow were obtained from the producer company. The caloric content of each food item was thus calculated by multiplying each value of carbohydrate, protein, and lipid by 4, 4, and 9 kcal, respectively. With this method, the total amounts of nutritional and energy intake for each monkey were calculated based on the focal animal observation data.

RESULTS

Nutritional Contents of Provisioned and Natural Foods

Foods (including provisioned foods) of the Koshima troop were grouped into ten categories (Table I). The mean and 95% confidence limits of each nutritional content are shown in Table I. Each plant part shows the following characteristic nutritional composition.

Leaves, in general, contain much structural cellulose to prevent grazing by herbivores [Van Soest, 1982]. Although no study has measured digestibility of fiber for Japanese macaques, it may be reasonable to assume that they cannot use cellulose as a direct energy source. Evergreen broad-leaved tree leaves have higher crude fiber content than other plants (Table I), but contain less crude protein. Thus, these leaves seem not to be nutritionally profitable as foods for monkeys, yet become very important for the Koshima troop during winter when other plants lose their leaves and invertebrates disappear [Iwamoto, 1982]. A common feature of fruit and seeds is their high proportion of lipid. Their nutrient contents varies widely depending on the amount of pulp relative to seeds. Deciduous tree fruits usually contain tough but fat-rich seeds and less pulp, so that they show a high lipid and crude fiber content. Berries of vines have high amounts of lipid and protein (Plate I). Fruits and seeds of evergreen trees vary widely in their nutritional contents, both owing to the variety of fruit structures and to the differences in parts monkeys utilize. For example, monkeys ate only seeds of *Michelia compressa* and *Cinnamomum camphora* fruits, but abandoned seeds and ate the pulp and skin of *Neolitsea sericea* and *Ilex integra*. Shoots, stalks, and rhizomes were grouped in the same category in Table I, because of their similar nutritional contents—less lipid and protein but abundant crude fiber, from which monkeys might not extract much energy. However, monkeys spent considerable time searching for these foods, especially in winter. As Casimir [1975] described for the foods of gorillas (*Gorilla gorilla*), these items may contain some essential amino acids. Flowers show high protein but less crude fiber. This food category is relatively rich in nutritional terms, but the flowering season is short and the biomass consumed low. Animal foods such as insects, worms, spiders, snails, and limpets have remarkably high crude protein and less indigestible materials. However, monkeys were not able to harvest significant proportions of this food type in a given feeding time or to prey upon them throughout the year. A prominent feature of provisioned foods is their small proportion of crude fiber, whereas soluble carbohydrate (sweet potato, apple, wheat, and rice), crude protein (soybean), or lipid (peanut) are high. The main provisioned items supplied to the Koshima troop were soybean, wheat, and sweet potato, with wheat being the most common. The nutritional pattern of each food is summarized in Figure 2, where the x-axis represents (% soluble carbohydrate) − (% fiber + % ash), and the y-axis 1.732 × (% protein + % lipid). This ternary plot has been used for grouping diverse data samples [see Kay and Covert, 1984; Chivers and Hladik, 1980]. Scores in the x-axis can be regarded as somewhat analogous to the digestibility of foods; and those of the y-axis, pointing at the

TABLE I. Nutritional Contents of Natural and Artificial Foods

Food	Sample size	Water (%)	Protein (%)	Lipid (%)	Fiber (%)	Scarb (%)	Ash (%)	Cal (kcal/g dw)
Leaf								
Deciduous tree	5	75.33±10.35	12.32± 1.96	3.48± 0.84	17.19± 5.80	54.00± 5.88	13.01±7.16	3.65±0.28
Evergreen tree	4	63.28± 9.58	6.93± 2.01	5.09± 3.01	25.82±11.32	53.05± 8.99	9.11±6.81	3.89±0.32
Other plant	5	69.31±13.54	9.85± 3.18	2.57± 0.48	18.90± 6.51	55.38± 5.91	13.29±5.26	3.59±0.16
Fruit and seed								
Deciduous tree	5	36.72±32.34	8.15± 2.30	16.95±10.65	37.85±21.07	33.95±23.51	3.10±1.64	4.72±0.53
Evergreen tree	18	60.70± 7.43	7.45± 2.09	10.11± 5.33	21.05± 6.63	56.97±10.27	4.42±0.91	4.33±0.27
Vine	5	65.73±11.21	9.93± 5.46	17.67±13.89	25.57±15.34	41.02± 8.52	5.82±2.41	4.65±0.68
Shoot, stalk, rhizome and twig	8	73.17± 7.67	6.93± 2.37	2.05± 1.05	30.76± 6.78	51.47± 7.83	8.80±2.07	3.75±0.08
Flower	6	81.59± 6.05	16.28± 6.15	3.76± 1.02	15.83± 4.40	58.27± 8.30	5.87±1.96	3.95±0.06
Animal								
Insect (mixed)	1	67.22	65.23	7.86	13.08	9.67	4.16	4.23
Limpet	1	77.96	69.37	3.72	0.29	11.88	14.74	3.60
Artificial foods								
Sweet potato	1	68.20	3.77	0.63	2.20	90.25	3.14	3.91
Apple	1	85.80	14.00	0.70	3.50	92.30	2.10	4.46
Rice grain	1	13.00	11.38	4.25	8.05	72.99	3.30	4.08
Wheat grain	1	13.50	12.14	3.47	2.43	80.12	1.85	4.10
Monkey chow	1	10.00	22.20	5.60	7.80	53.30	11.10	3.84
Peanut	1	8.50	4.70	10.70	0.40	83.70	0.40	4.52
Soybean	1	7.50	37.70	19.60	5.40	32.20	5.10	4.78
Average	7	29.50±28.18	15.13±10.02	6.42± 5.77	4.25± 2.51	72.12±18.77	3.86±3.01	4.24±0.30

Averages and 95% confidence limits shown. Raw data on animal and artificial foods are shown. ''Other plant'' includes grass, herb, and vine. ''Insect'' data were obtained from a mixed sample of cricket, grasshopper, cicada, fly, and ant.

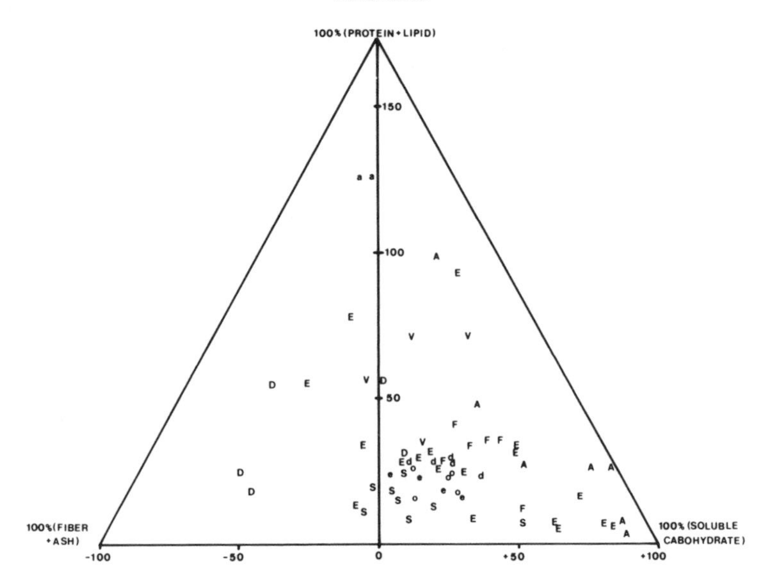

Fig. 2. Ternary plot of nutritional contents for natural and artificial foods of the Japanese monkey. The x-axis represents (%soluble carbon-%fiber-%ash content); y-axis, 1.732 × (%protein + %lipid content). Note that all of the artificial foods distribute near the top-right line. Abbreviations for foods are as follows: d = leaves of deciduous trees; e = leaves of evergreen trees; o = leaves of other plants; D = fruit and seeds of deciduous trees; E = fruit and seeds of evergreen trees; V = fruit and seeds of vines; S = shoots, stalks, and rhizomes; F = flowers; a = animal foods; A = artificial foods.

concentration of main nutrients, protein and lipid. Therefore, foods falling in the upper corner are nutritionally and energetically most suitable for the monkeys, while those at the right-hand corner provide readily utilizable energy. For the Japanese macaque, which has no fermentation chamber in its gut, foods at the left-hand corner may have very little nutritive value.

Leaves fall near the baseline of the graph and close to the right of the central vertical line. Each category of leaves forms a thin layer along the baseline. Leaves of deciduous trees seem to be the most nutritious (protein/lipid rich). Shoots, stalks, and rhizomes appear below the leaves. Flowers were more nutritious and digestible than the above two. Fruits and seeds of both deciduous trees and vines were protein/lipid rich, but the former were less digestible than the latter. Evergreen tree fruits and seeds showed the greater distribution range. All the provisioned foods fall near the top-right line of the graph, which indicates that these are either protein/lipid rich or highly digestible. Soy-bean had extremely high protein, while sweet potato and apple had much more soluble carbohydrate. The main food item, wheat,

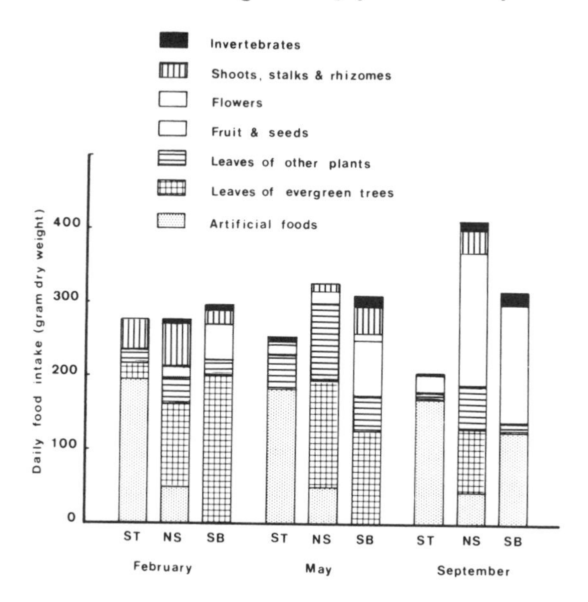

Fig. 3. Amounts of daily food intake of three adult females in February, May, and September. ST, NS, and SB represent names of monkeys, Satsuki, Nasi, and Siba, respectively.

showed a protein/lipid content similar to leaves, but there were marked differences in digestibility.

Daily Amounts of Natural and Provisioned Food Intake

Daily food intake for individual monkeys was around 300 g dry weight (mean = 298.60, n = 9), although there were some variations among individuals and seasons. The proportion of each natural food in the monkeys' diet (Fig. 3) reflects the phenology of food plants. Monkeys mainly fed on leaves of evergreen trees (*Daphiniphyllum teijsmanni* and *Machilus thunbergii*) in winter (February). In May, young leaves of deciduous trees sprout, and monkeys increase their dependence on these. However, they did not completely change over to deciduous leaves but continued to eat evergreen ones. This is probably due to the lower biomass of deciduous leaves in relation to evergreen ones in the forest. Most fruits were produced from September on and the Koshima troop continued to eat these until late November. Fruits and seeds contained more rich protein/lipid or soluble carbohydrate than the other non-reproductive parts (see Fig. 2). These nutritiously rich foods are important in allowing the monkeys to gain body weight for the winter season [Mori, 1979a; Wada, 1964], so monkeys' staple foods change in relation to what becomes available in each season. The proportion of provisioned food in the individual animal's diet was similar for all seasons. The exception was Siba in September, when a larger quantity of sweet potato was supplied. Provisioned foods accounted for 70%, 15%, and 0% (except for September) in the average dry weight intake for Satsuki, Nasi, and Siba. As Iwamoto [1974] has already shown, females belonging to the higher-ranking kin groups obtain more wheat than the lower ones; Satsuki fed 4.7 times more wheat than Nasi.

Provisioning affected the pattern of natural food intake. Firstly, Satsuki, which ate more wheat, tended to feed less (mean = 246.9, n = 3) on natural foods than the other monkeys (mean = 324.5, n = 6). This may be due to the lower crude fiber content in wheat compared to natural foods (Fig. 2). Wheat may allow monkeys to obtain a higher energy and nutritional extract per unit of dry weight. Secondly, the food item that changed most sensitively relative to the amount of wheat eaten was evergreen tree leaves. Especially in February, the amount of evergreen tree leaves in the diet was inversely correlated with that of wheat. Nasi, which fed the least on provisioned food among the females in September, fed the most on evergreen tree leaves. The same relation was also observed between the amount of natural fruit consumed and provisioned food (wheat and sweet potato) intake in September, even though fruits and seeds were rich in nutrients and more digestible during this season (see Fig. 2). Thus, when given provisioned foods, monkeys would reduce their intake of the lowest-ranking food categories. Moreover, if food was supplied in some excess, the animals would even reduce their intake of the higher-ranking food items in their natural diet. These results suggest that the leaves and fruits/seeds categories were less preferred than wheat and probably than any of the provisioned items listed in Table 1.

Digestibility

Since population increase and improvement in reproduction of higher-ranking monkeys seems to result from provisioning, it may be assumed that provisioned foods supply more energy and nutrients than the same amount of natural foods. A comparison of the two food types in terms of their relative energetic contribution will be examined here. Although specific nutrients may also be important, they will not be dealt with here because there have not been any experimental studies on the individual daily minimum nutrient requirements nor on their digestibility for the Japanese macaque. Only a few studies have attempted to estimate the digestive efficiency of natural and provisioned foods for the Japanese macaque. Iwamoto [1978] measured apparent digestive efficiencies for wheat, leaves of a vine (*Pueraria thunbergiana*), and mixed leaves of three tree species (mainly *Ficus erecta* with *Daphiniphyllum teijsmanni* and *Callicarpa japonica*) to be 87%, 60%, and 50%, respectively. The last digestive efficiency was measured from only a small amount of leaves the subject monkey took (31–115 grams dry weight (g. dw)/day, mean = 71). These leaves were probably not preferred. Mori [1979b] also measured apparent digestive efficiencies of wheat and leaves of a deciduous tree, *Ficus erecta,* and found these to be 86% and 67%, respectively. In his experiments, monkeys ate much leaf (196.2–590.2 g dw/day, mean = 325) in a day. Thus, the real digestive efficiencies of the above leaves probably lies between 60% and 70%, while that of wheat may be around 86% and 87%. Goodall [1977] estimated the digestive efficiencies of a number of plant foods for the gorilla, employing a relationship based on cellulose content and digestive efficiency given by Mitchell [1964] for the domestic pig (*Sus scrofa*). His values ranged from 38% to 78% with an average of 65%. These are very similar to the above estimates of natural leaves. Using Mitchell's [1964] relationship, the digestive efficiencies of wheat and leaves of *Ficus erecta* were calculated as 91% and 73%, respectively. These are somewhat higher than those measured directly by Iwamoto [1978] and Mori [1979b]. Therefore, a new relationship was employed assuming that the results of the above two experiments reflected a relationship between crude fiber content and digestive efficiency of food for Japanese monkeys, though data are still scarce. Digestive efficiencies for wheat and *Ficus erecta* leaves were 87% [Iwamoto, 1978] and 67% [Mori, 1979b] with corresponding crude fiber contents of 2% and 16%, respectively. This gives the regression equation $Y = 1.44X + 90.50$, where Y represents the apparent digestive efficiency (%) of a food item which contains X% crude fiber. Using this equation, the assimilated energy was calculated for each monkey (Fig. 4). Only in September, when much fruit was available in the forest, could lower-ranking females Nasi and Siba exceed the caloric intake of the top-ranking female, Satsuki. In February, however, females that could not obtain any wheat, or only small amounts, showed remarkably low values of total assimilated energy. This may be due to the low digestibility of evergreen leaves, the staple food during winter.

Using digestive efficiencies of 53% for evergreen leaves (calculated from the above equation where $X = 25.82$ from Table I), and 87% for wheat, the difference in the energy extracted from the same amount of wheat and leaves can be calculated. The caloric contents of wheat and evergreen leaves were on average 4.10 and 3.89 kcal/g dw, respectively. Thus, monkeys are able to extract 1.72 times more energy per gram from wheat than from natural leaves. Wheat highly exceeded natural leaves as an energy source, in fact, leading to a disproportionate energy intake among differently food-enhanced females in the most severe season, winter [see also Iwamoto, 1982, for energy intakes under the unprovisioned condition].

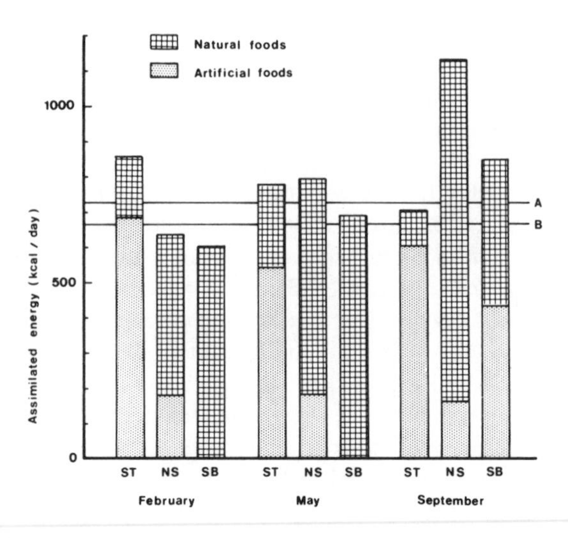

Fig. 4. Daily assimilated energy of three adult females in February, May, and September. Abbreviations for monkey names are the same as Figure 3. Line A shows daily energy expenditure (DEE) for monkeys with 9-kg body weight (applicable to ST and NS); line B, for 8-kg body weight (applicable to SB). Note that NS and SB, which could not eat much wheat, did not reach the required level of DEE in winter (February).

Daily Energy Expenditure

The basal metabolic rate (BMR) of a mammal with W (kg) body weight can be estimated by Kleiber's [1961] formula, BMR (kcal/ind/day) = $70 \times W^{3/4}$. However, animals require additional energy according to activities, specific dynamic action, thermoregulation, growth and reproduction. Average daily metabolic rate (ADMR) has been measured, usually in a constant ambient temperature under captive conditions. Robbins [1983] has reviewed studies of various mammals and birds, and reports that ADMR is almost twice BMR, that is, ADMR = $140 \times W^{3/4}$. Likewise, daily energy expenditure (DEE), estimated under completely wild conditions, tended to be higher than ADMR or 2–3 times BMR. This is mainly due to growth, reproduction, and thermoregulation [Robbins, 1983]. Therefore, DEE estimated from only the activity budget express lower values than the above estimates, for example, estimates for an ideal deer by Moen [1973] (1.5 × BMR), for a howler (*Alouatta palliata*) and spider monkey (*Ateles geoffroyi*) by Coelho [1976] (1–1.2 × BMR), or for a gelada (*Theropithecus gelada*) and Japanese monkey [Iwamoto, 1978] (1.3 × BMR). Considering the difficulties in estimating energy requirements with methods other than basal and active metabolic rate in the wild, the direct measurement methods using doubly labeled water or heart rate will give the best estimates of DEE. Nagy and Milton [1979], using labeled water, estimated the DEE of wild howler monkeys to be almost two times the BMR. Since DEE data measured by more direct physiological methods are not available for Japanese macaques, the above rough estimate of DEE $= 2 \times$ BMR will be employed to calculate energy expenditures of the subject monkeys in this study. Three females did not have infants during the study periods in May and September. In February, only Siba had a baby born during the preceding summer, and by this time, according to Iwamoto [1982], the infant was almost nutritionally independent of its mother. It was not necessary to consider additional energy for reproduction in any of the subject females. The body weights of Satsuki and Nasi were both 9 kg, while Siba was 8 kg. These give DEE estimates of 727.5 and 666.0 kcal/ind/day, respectively. Individuals that could not consume enough energy to reach the above levels were Nasi and Siba in February and Satsuki in September (Fig. 3). The latter might have few energy problems because many kinds of fruits were available during this period. However, in February, it is obvious that the energy deficit for Nasi and Siba would cause problems in the maintenance of body weight. Mori [1979b] reported that most females decreased in weight during the winter even during provisioning periods.

DISCUSSION AND CONCLUSIONS

In winter, monkeys did not increase their food intake to compensate for the poor digestibility of leaves, which formed their main diet item. This is not because of the low availability of leaves in the forest, since the island is covered by dense evergreen vegetation. The reason may be that the monkeys are limited by the amount of food they need to process to reach the DEE level. Digestive volume [Chivers and Hladik, 1980] and food passage rate [Milton, 1984] are limiting factors, thus determining a maximum bulk of foods a monkey can process in a day. There is, however, another reason for bulk limitation. Leaves contain various kinds of secondary compounds [McKey, 1978, 1979] to prevent herbivory. Since an animal must keep the intake of each compound under the toxic level, it has to limit also the consumption of particular species with specific toxins [Westoby, 1974]. This limitation especially occurs during winter when the staple food, evergreen tree leaves, high in terpenes, are consumed [see also Fa, 1984, for the Barbary macaque (*Macaca sylvanus*)]. In May and September, females that could not feed or feed only a little on wheat consumed more natural foods than the top-ranking female. Nevertheless, on a yearly basis, female Japanese

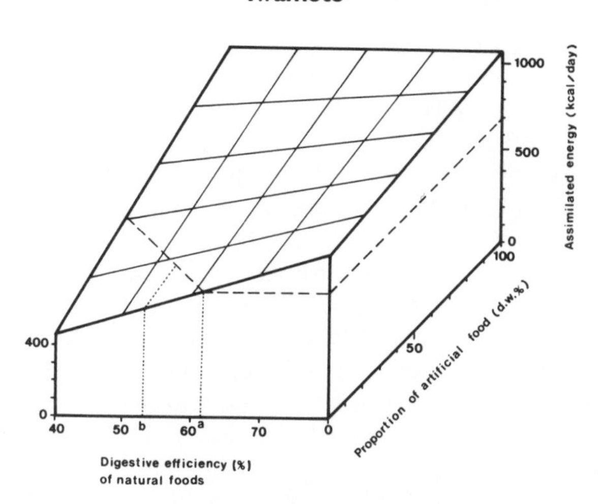

Fig. 5. Three-dimensional plot of quality of natural foods (expressed by digestive efficiency), enhanced level of artificial food (wheat), and assimilated energy. In this model, it was assumed that a monkey could not take more than 300 g dw in a day. The assimilated energy of 700 kcal (broken line) is equivalent to DEE of a monkey with 8.5-kg body weight. If no wheat was provided, a monkey of the above body weight had to collect natural foods with more than 61.4% digestive efficiency (a) as a whole. When the monkey diet was enhanced by wheat at only the 15% level, the required digestive efficiency for natural foods would be lowered to 53.2% (b). See text for details.

macaques in Koshima may be limited to taking less than about 300 g dw per day, though this figure may vary to some extent (Fig. 2). The estimated DEE and assimilated energy showed similar values of around 700 kcal/day. Assuming that a monkey needs to assimilate this energy amount (equivalent to the DEE of a monkey of 8.5 kg body weight), but limited to 300 g dw of food per day, it is possible to calculate the minimum quantity of natural foods (given by its digestive efficiency) it has to collect. Using a caloric content of 3.8 kcal/g dw for an ideal natural food (an average figure for all seasons), the digestive efficiency required for this food should be 61.4%. This efficiency will be given by a food item containing 20.2% crude fiber, using the equation above. The main foods during winter, *M. thunbergii* and *D. teijsmanni,* contained 30.61% and 26.8% crude fiber, respectively. Given this, these foods can not fulfill the required DEE level, which might explain the reported decrease in body weight during winter [Mori, 1979b]. The same calculation will give the required digestive efficiencies of natural foods under the various food enhancement levels (Fig. 5). Here, the values of digestive efficiency and caloric content of wheat were employed. When a monkey eats only wheat, it can obtain 1070.1 kcal/day. However, when it eats no wheat, natural foods with more than 61% digestive efficiency alone can afford

Plate I. Female Japanese macaque feeding on fruits of a wild grape, species *Vitis ficifolia*.

can afford barely 700 kcal/day. The lower-ranking female Nasi fed on wheat at a proportion of 15.1% of its total food dry weight intake, which allows for natural foods with a digestive efficiency of 53% to be consumed (Fig. 5). This is equivalent to 25.9% crude fiber content, a value close to that of *D. teijsmanni*. Thus, provisioning during the most severe season, even if restricted, can greatly improve the monkey's nutritional condition by supplying additional energy in ready use. This effect will be more pronounced for troops during winter in snowy areas, where monkeys are forced to eat mainly bark and winter buds containing much indigestible materials. Wada [1964] measured crude fiber contents of barks to be 43–88% (mean 61.2%) on a caloric basis. This is about three times higher than evergreen leaves. Although there are no experimental studies on the digestibility of barks, these foods probably have enough nutritional content only to slow the rate of body weight loss. Because provisioned foods are shared in a biased manner, as shown in this study, higher-ranking females can obtain additional benefits other than body weight gains. These would be important in lowering mortality rate and increasing reproductive success. Such an improvement in physical condition may cause high fertility and pregnancy rates, lower

Plate II. Koshima troops feeding on wheat scattered on the Ohdomari seashore.

primiparous age and low infant mortality as pointed out by Sugiyama and Ohsawa [1982]. However, provisioning of the Koshima monkey troop has been continued in a fixed manner and in a small feeding ground (Plate II). It can be expected, therefore, that different patterns of food sharing among individuals would appear if the size of the scattering area, manner of food scattering and type of foods changed [Fa, 1986]. The problem for the future is to clarify food-sharing patterns among different-ranking individuals for the various natural as well as provisioned foods.

SUMMARY

The effects of provisioning on the food consumption of individual monkeys were examined for a free-ranging troop of Japanese macaques in Koshima Islet. In nutritional terms, a prominent feature of provisioned foods when compared with natural ones was their high digestibility because of their lower crude fiber content. The three adult females studied showed different proportions of provisioned foods present in their daily food intake. This was related to their dominance ranking order. The amount of provisioned food on which the animals fed influenced the composition of the natural food items consumed. The high ranking female, which fed more on provisioned foods, could decrease her dependence on low quality natural foods such as evergreen leaves throughout the year and obtained enough assimilated energy even in winter. However, the two low ranking females suffered from an energy deficiency. Low ranking females could not compensate for this energy deficit by increasing the amount of evergreen leaf intake, probably because of bulk limitations due to digestive capacity or toxin content in natural foods. These results may explain the differential reproductive success among females in provisioned troops of Japanese macaques.

ACKNOWLEDGMENTS

This work was supported in part by a Grant-in-Aid for Special Project Research on Biological Aspect of Optimal Strategy and Social Structure from the Japan Ministry of Education, Science, and Culture.

REFERENCES

Casimir MJ (1975): Feeding ecology and nutrition of an eastern gorilla group in the Kahuzi Region (Republique de Zaire). Folia Primatol (Basel) 24:81–136.

Chivers DJ, Hladik CM (1980): Morphology of the gastrointestinal tract in primates: Comparisons with other mammals in relation to diet. J Morphol 166:337–386.

Coelho AM (1976): Resource availability and population density in primates: A socio-bioenergetic analysis of the energy budget of Guatemalan howler and spider monkeys. Primates 17:63–80.

Fa, JE (1984): Habitat selection and habitat distribution in Barbary macaques (*Macaca sylvanus*). Int J Primatol 5(3):273–286.

Fa JE (1986): "Use of Time and Resources by Provisioned Troops of Monkeys: Social Behaviour, Time and Energy in the Barbary Macaque (*Macaca sylvanus* L.) at Gibraltar." Contrib. to Primatol N° 23. Basel-Karger.

Goodall AG (1977): Feeding and ranging behaviour of a mountain gorilla group (*Gorilla gorilla beringei*) in the Tshibinda-Kahuzi region (Zaire). In Clutton-Brock TH (ed): "Primate Ecology." London: Academic Press, pp 449–479.

Iwamoto T (1974): A bioeconomic study on a provisioned troop of Japanese monkeys (*Macaca fuscata fuscata*) at Koshima Islet, Miyazaki. Primates 15:241–262.

Iwamoto T (1978): Food availability as a limiting factor on population density of the Japanese monkey and gelada baboon. In Chivers DJ, Herbert J (eds): "Recent Advances in Primatology." London: Academic Press, pp 287–303.

Iwamoto T (1982): Food and nutritional condition of free ranging Japanese monkeys on Koshima Islet during winter. Primates 23:153–170.

Japan Diet Association Corporate (1975): "Standard Table of Food Composition." Tokyo: Daiichi-Shuppan, pp 1–182.

Kay RF, Covert HH (1984): Anatomy and behaviour of extinct primates. In Chivers DJ, Wood BA, Bilsborough A (eds): "Food Acquisition and Processing in Primates." New York: Plenum, pp 467–508.

Kleiber M (1961): "The Fire of Life." New York: John Wiley & Sons, Inc., pp 177–216.

Koyama N, Norikoshi K, Mano T (1975): Population dynamics of Japanese monkeys at Arashiyama. In Kondo S, Kawai M, Ehara A (eds): "Contemporary Primatology." Basel: Karger, pp 411–417.

Masui K, Nishimura A, Ohsawa H, Sugiyama Y (1973): Population study of Japanese monkeys at Takasakiyama. J Anthropol Soc Jpn 81:236–248.

Mckey D (1978): Soils, Vegetation, and seed-eating by black colobus monkeys. In Montgomery GG (ed): "The Ecology of Arboreal Folivores." Washington: Smithsonian Institution Press, pp 423–438.

Mckey D (1979): The distribution of secondary compounds within plants. In Rosenthal GA, Janzen DH (eds): "Herbivores." New York: Academic Press, pp 56–133.

Milton K (1984): The role of food-processing factors in primate food choice. In Rodman PS, Cant JGH (eds): "Adaptations for Foraging in Nonhuman Primates." New York: Columbia University Press, pp 249–279.

Mitchell HH (1964): "Comparative Nutrition of Man and Domestic Animals." New York: Academic Press.

Moen AN (1973): "Wildlife Ecology." San Francisco: W. H. Freeman and Company, pp 333–364.

Mori A (1979a): An experiment on the relation between the feeding speed and the caloric intake through leaf eating in Japanese monkeys. Primates 20:185–195.

Mori A (1979b): Analysis of population changes by measurement of body weight in the Koshima troop of Japanese monkeys. Primates 20:371–397.

Nagy KA, Milton K (1979): Energy metabolism and food consumption by wild howler monkeys (*Alouatta palliata*). Ecology 60:475–480.

Robbins CT (1983): "Wildlife Feeding and Nutrition." New York: Academic Press, pp 99–147.

Sugiyama Y, Ohsawa H (1974): Population dynamics of Japanese macaques at Ryozenyama, Suzuka Mts. I. General view. (in Japanese) Jpn J Ecol 24:50–59.

Sugiyama Y, Ohsawa H (1982): Population dynamics of Japanese monkeys with special reference to the effect of artificial feeding. Folia Primatol (Basel) 39:238–263.

Van Soest PJ (1982): "Nutritional Ecology of the Ruminant." Portland: Durham and Downey, Inc., pp 1–374.

Wada K (1964): Some observations on the monkeys in a snowy district of Japan. (in Japanese) Seiri Seitai 12:151–174.

Westoby M (1974): An analysis of diet selection by large generalist herbivores. Am Nat 108:290–304.

Ecology and Behavior of Food-Enhanced Primate Groups, pages 95–111

5

Feeding Behavior and Activity Patterns of Rhesus Monkeys (*Macaca mulatta*) at Tughlaqabad, India

Iqbal Malik and Charles H. Southwick

Department of Biology, Institute of Home Economics, University of Delhi, N.D.S.E., New Delhi 110049, India (I.M.), and Department of Environmental, Population, and Organismic Biology, University of Colorado, Boulder, Colorado 80309 (C.H.S.)

INTRODUCTION

Few studies on the daily activity patterns of rhesus monkeys (*Macaca mulatta*) in relation to their feeding behavior have been carried out in India. Mukherjee (1969) studied the activity patterns of two rhesus groups living in a roadside habitat of northern Uttar Pradesh, but made only limited observations of their feeding behavior. Lindburg (1975, 1976) and Neville (1968) studied the food habits of rhesus groups around Dehra Dun and Haldwani feeding primarily on natural forest vegetation, whereas Siddiqi and Southwick (this volume, chapter 6) observed the food habits of rural groups in agricultural habitats north of Aligarh. There was no account of daily activity patterns of these groups, however.

The historical site of Tughlaqabad offers a diversity of habitats containing forest patches, agricultural fields, pastures, and a public archaeological site in which to study the relationships of feeding and activity in rhesus monkeys (Fig. 1). Though not a confined colony, the Tughlaqabad rhesus have all the advantages of a protected population. Good visibility provides an opportunity to record data on spatial relations, movements, and social behavior. Information can be obtained on home ranges, night sleeping quarters, grooming, play, locomotion, intergroup relations, and the utilization of available resources.

Many primate studies have shown food to be a key ecological variable, influencing both social behavior and population dynamics. Field and laboratory studies have demonstrated that a large proportion of aggressive interactions occur as a result of competition for food (Chalmers, 1968; Southwick, 1970; Zimmerman et al., 1973). Play, which requires "surplus" energy, decreases as the amount of available food decreases (S.A. Altmann,

Plate 1. A scissors and knife peddler on a bicycle at Tughlaqabad giving rice to a small group of monkeys.

1959; Loizos, 1967; Loy, 1970; Southwick, 1970). Increased availability of food produces a decrease in day range because the group does not have to travel far to secure sufficient food (Altmann and Altmann, 1970; DeVore and Hall, 1965; Crook, 1966). Hall (1963), Crook (1970), and Rowell (1967) postulated that groups that spend less time foraging, spend more time on social activities, especially grooming.

With a continuously abundant food supply, an increase in population size has been noted among the provisioned colonies of Japanese macaques (*Macaca fuscata*) at Takasakiyama (Itani et al., 1963; Itani, 1975) and the rhesus macaques at Cayo Santiago (Koford, 1965; Rawlins and Kessler, 1986; Sade et al., 1985). Conversely, in a food limited population of toque macaques (*Macaca sinica*), the survivorship of infants and juveniles was reduced, and the population remained stable (Dittus, 1975, 1977). The diversity of food consumed by each primate species has not been evaluated, due in part to great differences in observational opportunities to tally the number of plant and animal species eaten by the group under study, and the difficulty of obtaining comparative data.

The extent to which animals select a particular food can be estimated by dividing the amount consumed by the availability of that food in the environment (Clutton-Brock and Harvey, 1976). Since the amounts actually consumed are rarely known in field studies, several investigators have calculated selection ratios for particular foods by dividing the proportion of time spent feeding on a given type of natural vegetation by some measure of the relative abundance of the vegetation, or the relative abundance of the canopy cover that type provides (Clutton-Brock and Harvey, 1976; Struhsaker and Oates, 1975).

Nutritional analyses of diet and energy costs of activities have been done in only a few studies (Coelho et al., 1976; Milton, 1980). Detailed accounts of feeding behavior for macaques have been published only for *Macaca sinica* by Hladik and Hladik (1972).

In regard to time and activity budgets, Marriott (1978) reported that the rhesus monkeys of Kathmandu spent 10.5% of their daytime activity in feeding. Her main interest was the type of food eaten, amounts consumed, nutritional contents, and relative amounts supplied by people compared with natural vegetation. Taylor (1975) observed that the temple monkeys of Kathmandu obtained 68% of their overall diet from worshippers and the remaining 32% from natural sources. Teas (1978) found in Kathmandu rhesus that feeding was the single most time-consuming behavior in the fall season and the second most predominant activity in the summer. Feeding behavior of the temple monkeys of Nepal has also been studied by Bajracharya (1979).

This study attempted to analyze feeding behavior in relation to overall activity patterns on a seasonal basis. Since weather and various ecosystem components change seasonally and from year to year, this work was done over 3 years, from July 1980 through August 1983. During this time, the rhesus population of Tughlaqabad expanded from two groups of 120 monkeys to five groups of 286 (Malik et al., 1984, 1985). This was an average annual increase of 21.4%, virtually a record in the population literature of rhesus monkeys.

STUDY AREA

Tughlaqabad is an ancient city site and fourteenth century fort situated on the southern edge of New Delhi at 30°25′ north latitude and 78°76′ east longitude. The home range of rhesus monkey groups extended throughout the fort and surrounding areas, covering approximately 5 sq km (2.5 × 2.0 km) (Fig. 1). The fort rises 20 to 30 meters above the surrounding plain. The outer walls of the fort form a polygon with a circumference of nearly 5 km. Flat and fertile land surrounding the fort contains croplands, pasture, two forested areas, and encroaching suburban development. A paved road lined

Plate II. Tughlaq's tomb, looking south from the roadway through the Tughlaqabad area.

by trees passes through the area in an approximate east-west direction, between the fort and Tughlaq's tomb (Plates I and II).

The fort constitutes about a quarter of the total area, two forest plantations occupy another quarter, and the surrounding open croplands and pasture make up the remaining half.

Tughlaqabad has a subtropical climate with marked seasonal changes. During May and June, the peak of summer, daytime temperatures often reach 40° to 45°C; in December and January temperatures fall to 7° to 9°C. Monsoon rains occur from the end of June or early July until mid-September, with an annual average of 567 mm. Winter and spring rains occur sporadically and are usually light.

The natural vegetation inside the fort is xerophytic, primarily grasses and arid forbs and shrubs. Outside the fort, vegetation is more mesophytic, and better ground water supports trees and crops, primarily wheat and pulses (legumes). The main trees present are Indian jujube (*Zizyphus jujuba*), neem or margosa (*Azadirachta indica*), sheesham or sissoo (*Dalbergia sissoo*), oak (*Quercus incana*), acacia (*Acacia arabica*), pipal (*Ficus religiosa*), and date palm (*Phoenix dactylifera*). The dominant fauna includes rhesus monkeys, cattle, domestic buffalo, donkeys, goats, dogs, jackals, mongoose, lizards, and a great variety of birds, both migratory and resident. Peacocks,

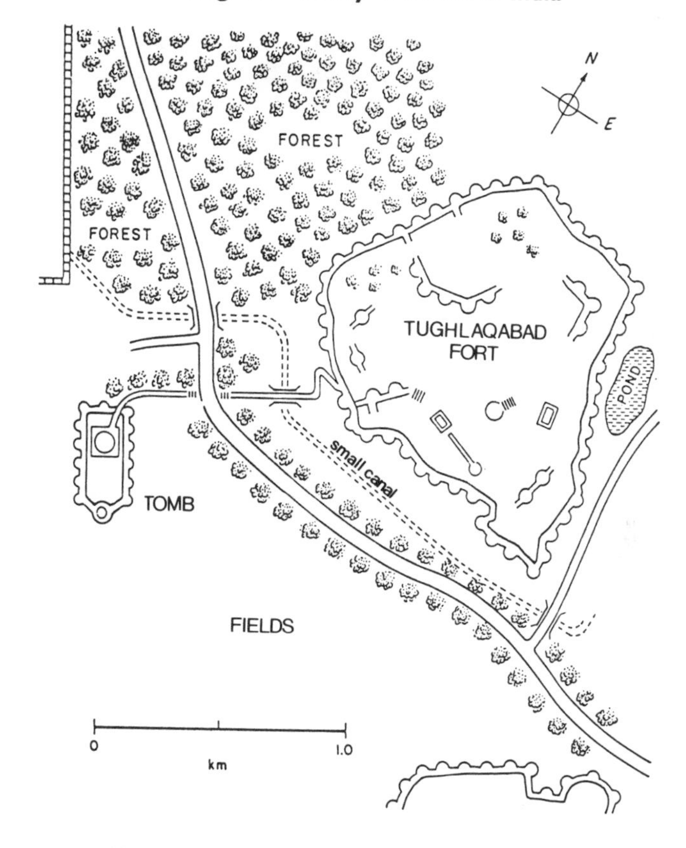

Fig. 1. Map of the Tughlaqabad archaeological site.

partridge, pigeons, crows, sparrows, vultures, mynahs, and kites are all common.

METHODS

Direct observations during 5,800 contact hours were made on the Tughlaqabad rhesus over the 3-year study period. Monkeys were not trapped, marked, artificially fed by the researchers, or otherwise disturbed by the research program. Many monkeys were individually recognizable by unique characteristics. Censuses were conducted when the animals progressed across a roadway, canal, or through some restricted area such as an opening in the wall of the fort. Behavioral observations were based on time sampling methods discussed by Altmann (1974), especially Goodenough's Time

Sampling Method (1928), and Smith's Scan technique (1968). All occurrences of some behaviors were recorded (Rowell, 1967). For average percentage time spent in primary activities, scans of all monkeys in view were made every 5 minutes. Observation periods were randomly distributed throughout the daylight hours from 0600 until 1900.

Although the entire population of Tughlaqabad varied from 120 monkeys in March of 1980 to 286 in July of 1983, the group on which this study is based had an average of 61 individuals throughout the median year of 1982. Its composition in July of 1981 was 12 adult males, 17 adult females, 19 subadults and juveniles, and 13 infants. Data were taken from all individuals except infants.

RESULTS

Rhesus invariably became active at least half an hour before sunrise. During winter mornings, they descended from trees between 0600 and 0645, and sat huddled in groups (up to 18 per huddle) until between 0700 and 0830 when the sun rose fully. At full sunrise, the monkeys slowly and randomly moved out of their huddles. Huddles would temporarily break before this only if food was offered to them by passers-by. Some rhesus separated from these huddles and moved to locations where their surroundings could be surveyed; others groomed or played with intermittent feeding and local movements. From 0900 until 1200, social activities and resting were interspersed with feeding and foraging for food. The peak of feeding activity normally occurred between 0830 and 1030.

The warmest part of the day, 1200 to 1500, was spent resting and grooming in the open while infants and juveniles indulged in active play bouts between resting. Between 1500 and 1730 a second period of intense feeding occurred, primarily from human sources, since rhesus were not likely to raid cultivated fields at this time. The monkeys were back on their sleeping sites or at least near them within a half hour of sunset between 1800 and 1900.

In summer, as daylength increased and morning temperatures were higher, foraging began earlier and there were no huddles. A midday rest time increased in duration from 1100 to 1600, and usually occurred in the cooler shady recesses of Tughlaq's tomb or in the forest. The daily activities were quite predictable so far as behavioral sequences and time allotted to specific behaviors were concerned, but the location of various activities often changed within the home range (Malik, 1986).

Annual Daily Activity Budgets

In our study group, the percentages of time spent in resting averaged 30.0%; locomotion, 18.6%; grooming, 11.5%; fed by humans, 9.5%; play

TABLE I. Average Percentage Time and Distribution of Waking Hours by Activity in Two Seasons

	Winter		Summer		Annual	
Activities	% Time	Waking hr/day	% Time	Waking hr/day	% Time	Waking hr/day
Rest	23.20	2.79	36.74	4.40	30.02	3.60
Locomotion	22.00	2.64	15.14	1.81	18.57	2.22
Groom	12.98	1.55	10.05	1.20	11.51	1.38
Fed by humans	9.00	1.08	10.00	1.20	9.50	1.14
Play	9.76	1.17	8.79	1.05	9.28	1.11
Eating natural vegetation	7.30	0.88	8.49	1.02	7.99	0.96
Drink	2.00	0.24	4.71	0.56	3.35	0.42
Others	13.66		6.08			

9.3%; feeding on natural vegetation, 8.0%; and drinking, 3.3% (Table I). During summers, the average amount of time spent resting (4.4 hr) and drinking water (0.6 hr) was approximately twice that spent resting and drinking during winters (2.8 and 0.2, respectively). The order is reversed in the case of locomotion, however; namely, 2.6 hr in winters and 1.8 hr in summers. There were only slight seasonal changes in amount and percentages of time spent on grooming, feeding by humans, play, and foraging.

Feeding

Feeding played one of the most important roles in determining the daily routine. Although the rhesus of Tughlaqabad did not have to spend long hours foraging (an average of only 2.3 hr daily, constituting 17.5% of daytime activity), nonetheless, their feeding behavior affected many other activities. Feeding usually took precedence over all other activities. This priority was illustrated by the observation of a mating pair who terminated a copulatory bout to obtain food from a visitor. Feeding often interrupted resting, locomotory behavior, grooming, or play, and it frequently brought monkeys down from the trees earlier than usual or broke up early morning huddles.

The Tughlaqabad area provides monkeys with a wide range of foods. Crops are grown in adjacent fields. The monkeys have three sources of foodstuffs: 1) food provided by humans, 2) agricultural crops, and 3) natural vegetation in the terrain. The food provided by humans is fairly consistent, almost ritualistic, but it comes in greatest abundance on Tuesdays and

Saturdays or Sundays, which are Hindu holy days. On each of these three days, an average of 27 people fed the monkeys on an average of 19 occasions. The mean quantity of food given per day was approximately 50 kg: 37 kg of bananas, 9.5 kg of peanuts and grams (a leguminous bean like a soybean), and 2.5 kg of chapaties. With other foods occasionally supplied, such as rice and various vegetables and fruits, this usually represented a surplus of more than the animals would take. The average amounts of food not consumed on Tuesdays, Saturdays, and Sundays were 4 kg of bananas and 2 kg of peanuts and grams. On Mondays, Wednesdays, Thursdays, and Fridays, provisioning by humans was only about 50% of the amounts on Tuesdays and Saturdays. On these days, more time was spent foraging in the croplands surrounding the fort and tomb, and more time feeding on natural vegetation.

The Tughlaqabad rhesus were observed to consume 45 different species of food plants, but only 23 of these constituted a significant intake. Of these 9 were leaves, pods, and fruits of trees, and 14 were agricultural crops (Table II).

Natural vegetation was consumed in proportion to its general abundance and availability in the environment. This differed seasonally to some extent, but on an annual basis, it followed directly the abundance of the trees in sample plots. The numbers of each species in the primary foraging areas of the rhesus were: *Dalbergia sissoo,* 974; *Ficus religiosa,* 428; *Acacia arabica,* 298; *Zizyphus jujuba,* 145; *Azadirachta indica* 140; *Phoenix dactylifera,* 79; and *Quercus incana,* 5. Other species of natural vegetation consumed were grasses and forbs, taken only to a minor extent. The natural vegetation of Tughlaqabad was certainly not as diverse as the forests available to the monkeys in Lindburg's study, where rhesus were observed to eat portions of more than 100 species (Lindburg, 1975, 1976). It was, however, more diverse than the tree species available to the rural monkeys at Chhatari, where only 4 species of native trees were available in addition to mango and guava (this volume, chapter 6).

The first preference of all age groups was food given by humans. Adult rhesus relished bananas, but also had a high preference for grams, peanuts, and chapatis. Juveniles preferred soft foods like bananas and chapatis, and ate grams and peanuts only when soft foods were not available. The Tughlaqabad rhesus spent a yearly average of 59% of their feeding time on food provided by humans (Table III), and 41% of their feeding time on natural vegetation. This differs from the Chhatari group of Siddiqi and Southwick (this volume, chapter 6) where 83% of the feeding time was spent on food from people, 10% on agricultural crops, and only 7% on natural vegetation. In the Sumera Fall group near Aligarh, however, only 29% of feeding time was spent on food provided by people, 17.5% on agricultural

TABLE II. Natural and Agricultural Foods in the Terrain

Local name	Botanical name	Part eaten	Energy in kilocalories*
		On trees	
Babul (Desi kikar)	*Acacia arabica*	Leaves and pods	
Date Palm (Khajoor)	*Phoenix dactylifera*	Fruit	144 fresh
Gum Tree (Kikar)	*Acacia arabica*	Leaves and pods	
Indian Jujube (Ber)	*Zizyphus jujuba*	All but the seed	158
Margosa (Neem)	*Azadirachta indica*	Tender leaves	158
Pepal	*Ficus religiosa*	Figs	110
Oak	*Quercus incana*	Fruit	
Siras	—	Leaves and pods	—
Sissoo (Sheesham)	*Dalbergia sissoo*	Leaves and pods	—
		In fields	
Brinjal	*Solanum melongena*	Leaves and fruits	40 and 24
Cabbage	*Brassica oleracea*	Leaves	27
Cauliflower	*Brassica oleracea*	Leaves and stalk	66
Carrot	*Dancus carota*	Leaves	77
Chari (Jowar)	*Sorghum vulgare*	Leaves	349
Lima beans	*Vigna catjang*	Leaves	290
Maize	*Zea mays*	Grain and leaves	125
Masoor, or radish	*Raphanus sativus*	Leaves	28
Methi	*Medicago falcata*	Leaves	—
Mustard	*Brassica acampestris*	Leaves	34
Peas	*Pisum sativum*	Leaves and pods	315
Spinach	*Spinacia oleracca*	Leaves	26
Turnip	*Brassica rapa*	Leaves	67
Wheat	*Triticum aestivum*	Grain, stalk, and leaves	341

Quantitative food requirements are usually estimated in terms of heat units calories. A physiological calorie (also called kilocalorie and abbreviated kcal*) is the amount of heat necessary to raise the temperature of one kilogram of water by one degree centigrade and this heat unit is different from the physical heat unit which is one-thousandth of the physiological calorie. This is an amount of food having an energy-producing value of one large calorie.

*Also known as the large calorie; figures given are kcal/100 g dw.

crops, and 53% on natural vegetation. In Lindburg's studies on the food habits of rhesus in the forests around Dehra Dun, virtually 100% of the diet was based on natural vegetation (Lindburg, 1975, 1976). Thus rhesus groups modify their feeding behavior markedly depending upon specific habitat and environmental conditions. When the Tughlaqabad rhesus did not have enough food provided by people, they foraged upon agricultural crops or natural foods. The food provided by nature and by crops varied seasonally: 1) in January and February when crops have been sown and trees bear fruits, rhesus have access to both young shoots in the fields and fruits of trees,

TABLE III. Distribution of Time by Activity in Different Seasons

	Seasons	Eat natural vegetation	Fed by humans	Drink
Average percentage time spent per day	Winter	7.30	9.00	2.00
	Summer	8.49	10.00	4.71
	Annual	7.99	9.50	3.35
Average hours per day	Winter	0.88	1.08	0.24
	Summer	1.02	1.20	0.56
	Annual	0.96	1.38	0.42
Percentage of total feeding time		41%	59%	

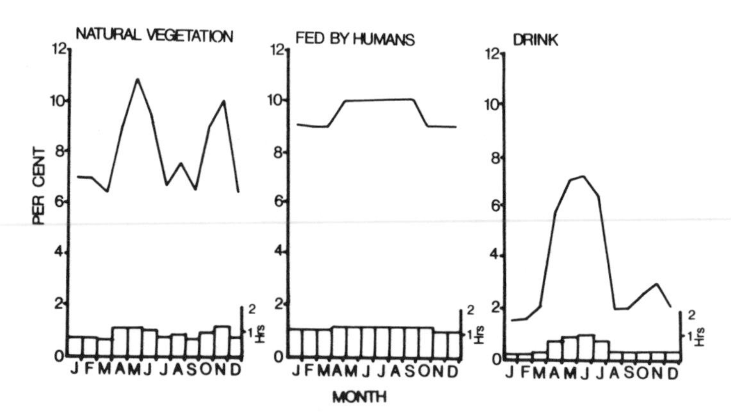

Fig. 2. Seasonal patterns of feeding on natural vegetation, feeding by humans, and drinking. Percent of daily time budgets shown as monthly averages.

2) in May and November, trees bear no fruit, but crops have been harvested, and rhesus can glean some food from harvest spillage. In these latter months, they have to spend more time foraging; e.g., 11% of their total time budget in May (the peak of the dry season) compared with only 6.5% in March, September and December (Fig. 2). Thus, the dietary pattern is variable and adaptable at different times of the year.

The quantity and caloric value of food consumed by different sex and age class individuals were estimated by focal animal observations on 4 monkeys. All food items and their rates of consumption eaten by these animals were recorded continuously from 0530 until 1930 on a Tuesday. Table IV shows that the adult male consumed in one day a total of 243 gm containing a total of 2,196 calories, whereas an adult female consumed 173 gm containing 1,449 calories in one day. A juvenile male consumed only 102 gm containing 1,235 calories, and a juvenile female consumed approximately 290 gm

TABLE IV. Focal Animal Study of Individual Food Intake in the Tughlaqabad Rhesus

Age and sex class	Bananas (gms)	Grams, peanuts, and other seeds (gms)	Chapaties (gms)	Leaves, shoots and herbs (gms)	Wild fruit (gms)	Total calories consumed
Adult male	12	225	2	—	4	2,196
Adult female	7	40	1	100	25	1,449
Juvenile male	6	35	1	50	10	1,235
Juvenile female	4	50	1	200	35	1,115

containing only 1,115 calories. The low calorie intake of the juvenile female occurred because a sizable part of her food intake consisted of leaves, shoots, and herbs, items of low caloric value.

Subordinate animals approached food only after dominants had their pick, especially at times when they were fed by people. Hence the adult male took most of the bananas and the juveniles received the least. The lowest caloric values and poorest quality of food was consumed by the juveniles, especially the juvenile females.

Drinking

For drinking, rhesus frequented the village pond, tubewell, and canal, though they readily drank from rainwater puddles whenever available. At times of acute scarcity, they explored their entire territory for new sources of water, even if it was dirty water collected from construction work. On one day in April (dry season) they moved to Adilabad (a village east of the fort and pond), and drank water from a pit at a construction site.

When foraging for water, one leader took the entire troop, but when ample water was available in their core area, intermittent drinking occurred among all members of the group. In times of limited water, adults took priority over subadults and juveniles if they happened to reach a water hole at the same time. If subadults and juveniles arrived first, adults chased them away with threats and baring of teeth.

During winters (especially January and February), most of the monkeys' water requirements seemed to be met from leaves and succulent fruits. Time spent drinking was reduced to less than 2% (Fig. 2). In the dry season (April and May), when natural water sources dried up, the monkeys spent more time (4.7%) looking for drinking water.

Rhesus were observed drinking at all hours of the day, as early as 0600 shortly after waking. There seemed to be no consistent relationship between troop spacing or numbers and the availability of water. At times, large groups of over 100 animals gathered at one place and spaced their drinking over a long period, interrupted by bouts of playing and eating. On numerous other occasions, only a few or even single animals came to the water source for drinking. To drink, rhesus lean on their forelegs, dip their mouths in the water, and suck through their lips for 2–3 seconds, often lifting their heads sharply to look around, and then dip their mouths again if they need more. This is the only method observed while the rhesus drink water. Hands were never used to facilitate drinking water except to clear the surface with rapid hand swipes before drinking. Sometimes as many as 10 to 15 rhesus have been observed drinking at the same time.

Adults were not observed playing with water, although juveniles and infants were often seen jumping, splashing, and swimming in ponds and puddles.

Grooming

During winter, grooming was common but occurred only after feeding when the sun was sufficiently high to provide enough warmth for the monkeys to bask and groom. In the summer season, grooming was also frequent and occurred throughout the day. Grooming did not take priority over feeding, however.

The peak of grooming occurred in the afternoon and was observed right until the monkeys slept. There were some seasonal differences. In winter, peak grooming time was 1200 to 1500, and in summer, the peak extended from 1100 to 1600. However, grooming was slightly more frequent in winter because less time was spent sleeping during midday. Thus, the percentage of time spent on grooming was 13.0% (1.55 hr/day) on average in winter and 10.0% (1.2 hr/day) in summer (Table I).

Locomotion

Rhesus have ample protection from wind in the fort; there are trees for shade and the tomb with a cool interior. Locomotory behavior is fairly predictable in all seasons. In the winter, on cool cloudy days, the monkeys have been observed locomoting at speeds up to 100 meters per minute covering the extremes of their home ranges. Bouts of grooming, feeding, and relaxation occurred between locomotion. Monkeys normally moved less on hot summer days, but occasionally they did cover the extremes of their home ranges in search of water. In summer, locomotion was done only in the mornings and evenings, and that for specific purposes of food and water acquisition. Summer afternoons were spent in cool, shady places. In winters, time spent on locomotion was 22% of the daily activity budget (2.64 hr/day), compared with 15.1% (1.8 hr/day) on the average in summers.

DISCUSSION AND CONCLUSIONS

The rhesus monkeys of Tughlaqabad spent 17.5% of their daytime hours in feeding or on active foraging for food and 3.3% for water. The time spent foraging on a particular day depended upon the availability of food from visitors. When food was provided in abundance by people, the monkeys spent less time foraging and more time on other activities, especially resting and grooming. Southwick (1962) found that provisioned monkeys in an Aligarh temple spent only 10% of their time feeding. Conversely, Altmann (1962) found in his study of rhesus on Cayo Santiago that up to 80% of their time was spent on foraging despite the provision of monkey chow in feeding bins. At Maroth in western India, Ohja (1982) found that feeding time was almost the same as that of the Tughlaqabad monkeys.

The activity patterns of the rhesus of Tughlaqabad were strongly influenced by the human population. This supports the theory of Shukla et al. (1982) that primate activity patterns are based on the components of the ecosystem. At Tughlaqabad, human activities are a prominent component of the ecosystem.

Tughlaqabad monkeys rejected all types of nonvegetarian food except eggs, which they occasionally stole from food baskets brought by picnickers. Koford (1963) reported that the Cayo Santiago rhesus frequented bird's nests, but there are no other reports of rhesus eating eggs. Lindburg (1971) noted that rhesus in the forests of Dehra Dun ate termites, grasshoppers, ants, and beetles. Tughlaqabad monkeys were not observed eating insects, but juveniles were seen eating earth in small quantities on at least seven different occasions. Blanford (1888–91), Roonwal (1956), Mandal (1964), Mukherjee and Gupta (1965), Lindburg (1971), Puget (1971), Krishnan (1972), and Teas et al. (1980) have all reported rhesus eating earth, though they did not mention any specific age class. During the present study, the only individuals eating earth were juveniles less than 2 years of age. This was not limited to any particular month or season, but was observed at different times of the year. On one occasion a monkey was observed eating bird droppings, but no record of such a habit has been reported by others.

During the summer months, rhesus have been observed drinking water as early as 0600, shortly after waking, followed by drinking at least 4 or 5 times a day from rain puddles or nala in the fields. Mukherjee (1969) observed rhesus near Bareilly to drink stagnant water 2 or 3 times daily from roadside ditches. On the other hand, rhesus in the mangrove forest studied by Mandal (1964) were never seen drinking water. According to Mukherjee and Gupta (1965) rhesus in mangrove obtain water by licking dew from leaves, and by eating succulent leaves. The rhesus of Tughlaqabad met some of their water requirements in a similar way, especially in winter.

In Tughlaqabad the diet of the monkeys comprised only 45 plant species, considerably less than the 100 plus species eaten by forest monkeys at Dehra Dun in Lindburg's studies (Lindburg, 1975, 1976). None of the Tughlaqabad monkeys was seen to eat insects, perhaps because of the abundance and variety of food from visitors.

The eating of earth by juveniles at all times of the year is a phenomenon not easily explained. With food in abundance, it is not likely that hunger per se is the explanation. Since earth eating is observed in young children, and only young juveniles were seen eating earth, this could perhaps be due to a playful mood, or to some special nutritional need.

The animals under study were healthy and well fed. Only 2 cases of sick individuals were observed in the entire study, a very unusual occurrence in the rhesus of India. The Tughlaqabad monkeys spent 30.0% of their time resting, whereas Shukla (1979) reported that the 2 Maroth groups in Rajasthan spent 9.3% and 16.1% of their time resting, and Southwick et al. (1982) reported that rhesus in Nepal spent only 8% of their time resting. The greater amount of rest in the Tughlaqabad monkeys is probably related to their more abundant food supply; less time is needed for foraging. As further evidence of this, the rhesus in Tughlaqabad spent 18.6% of their daily time budget in locomotion, as compared with 25% in the Nepal rhesus.

In conclusion, the Tughlaqabad rhesus spent relatively less time foraging and locomoting than rhesus in other locations. This permitted more time for other activities, especially resting and grooming. In a vernacular sense, the Tughlaqabad rhesus enjoyed a relatively high degree of behavioral leisure. Food provisioning by the human population, coupled with an abundant and moderately varied source of food from agricultural crops and natural vegetation, contributed to their good health, rapid population growth, and modified activity patterns.

REFERENCES

Altmann J (1974): Observational study of behaviour: Sampling methods. Behaviour 49:227–267.

Altmann SA (1959): Field observations on a howling monkey society. J Mamm 40:317–330.

Altmann SA (1962): A field study of the sociobiology of rhesus monkey. Ann NY Acad Sci 102:338–435.

Altmann SA, Altmann J (1970): "Baboon Ecology." Bibl Primatol No. 12. Basel: S. Karger and Univ Chicago Press.

Bajracharya AN (1979): "Feeding behaviour of the rhesus monkey in Swayambhu." Ph.D. Dissertation, Trichandra campus, Tribhuvan University.

Blanford WW (1888–91): "The Fauna of British India including Ceylon and Burma. Mammalia," London: Taylor and Francis.

Chalmers NR (1968): The social behaviour of free living mangabeys in Uganda. Folia Primatol 8:263–31.

Clutton-Brock TH, Harvey PA (1976): A statistical analysis of some aspects of primate ecology and social organization. Paper presented at the International Primatological Society Meeting, Cambridge, England.

Coelho AM, Bramblett CA, Quick LB, Bramblett SS (1976): Resource availability and population density in primates: A socio-bioenergetic analysis of energy budgets of Guatemalan howler and spider monkeys. Primates 17:63–80.

Crook JH (1966): Gelada baboon herd structure and movements. A comparative report. Symp Zool Soc London 18:237–258.

Crook JH (1970): The social ecology of primates. In Crook JH (ed): "Social Behaviour in Birds and Mammals." New York: Academic Press, pp 103–166.

Devore I, Hall KRL (1965): Baboon ecology. In DeVore I (ed): "Primate Behavior: Field Studies of Monkeys and Apes." New York: Holt Rinehart and Winston Inc., pp 20–52.

Dittus WPJ (1975): Population dynamics of the toque macaque (*Macaca sinica*). In Tuttle RH (ed): "Socioecology and Psychology of Primates." The Hague: Mouton, pp 125–151.

Dittus WPJ (1977): The social regulation of population density and age distribution in the toque monkey. Behaviour 63:281–322.

Gabow SL (1973): Dominance order reversal between two groups of free-ranging rhesus monkeys. Primates 14:215–223.

Goodenough FL (1928): Measuring behaviour traits by means of repeated short samples. J Jur Res 12:230–235.

Hall KRL (1963): Observational learning in monkeys and apes. Br J Psychol 54:201–226.

Hladik CM, Hladik A (1972): Disponsibilites alimentaires et domaines vitaux des primates a Ceylon. Terre et Vie 26:149–215.

Itani J (1975): Twenty Years with Mt. Takasaki monkeys. In Bermant G, Lindburg DG (eds): "Primate Utilization and Conservation." New York: Wiley, pp 101–126.

Itani J, Tokuda K, Furuya Y, Kano K, Shin Y (1963): The social construction of natural troops of Japanese monkeys in Takasakiyama. Primates 4:1–42.

Jolly A (1972): Troop continuity and troop spacing in *Propithecus verreauxi* and *Lemur catta* at Berenty, Madagascar. Folia Primatol 17:335–362.

Koford CB (1963): Group relations in an island colony of rhesus monkeys. In Southwick CH (ed): "Primate Social Behavior." Princeton, NJ: D. Van Nostrand Co., pp 136–152.

Koford CB (1965): Population dynamics of rhesus monkeys on Cayo Santiago. In DeVore I (ed): "Primate Behavior: Field Studies of Monkeys and Apes." New York: Holt, Rinehart and Winston, pp 160–174.

Krishnan M (1972): An ecological survey of the larger mammals of Peninsular India, part I. J Bombay Nat Hist Soc 68:503–555.

Lindburg DG (1971): The rhesus monkeys in North India: An ecological and behavioral study. In Rosenblum LA (ed): "Primate Behavior; Developments in Field and Laboratory Research." New York: Academic Press, pp 1–106.

Lindburg DG (1975): Feeding behaviour and diet of rhesus monkeys in a Siwalik Forest in northern India. Unpubl. ms. 44 pp.

Lindburg DG (1976): Dietary habits of rhesus monkeys (*Macaca mulatta* Zimmerman) living in Indian forests. J Bombay Nat Hist Soc 73:261–269.

Loizos C (1967): Play behaviour in higher primates: A review. In Morris D (ed): "Primate Ethology." Chicago: Aldine, pp 179–219.

Loy J (1970): Behavioral responses of free-ranging rhesus monkeys to food shortage. Am J Phys Anthropol 33:263–272.

Malik I (1986): Time budgets and activity patterns in free-ranging rhesus monkeys. In Else JG, Lee PC (eds): "Primate Ecology and Conservation." Cambridge: Cambridge University Press, pp 105–123.

Malik I, Seth PK, Southwick CH (1984): Population growth of free-ranging rhesus monkeys at Tughlaqabad. Am J Primatol 7:311–321.

Malik I, Seth PK, Southwick CH (1985): Group fission in free-ranging rhesus monkeys of Tughlaqabad Northern India. Int J Primatol 6; No. 4:411–421.

Mandal AK (1964): The behaviour of the rhesus monkeys (*Macaca mulatta*) in the Sunderbans. J Bengal Nat Hist Soc 33:153–165.

Marriott B (1978): A preliminary report on the feeding behaviour of rhesus monkeys (*Macaca mulatta*) in Kathmandu, Nepal. Ann Nepal Nat Conserv Soc 2:68–72.

Milton K (1980): "The Foraging Strategy of Howler Monkeys." New York: Columbia University Press, 165 pp.

Mukherjee AK, Gupta S (1965): Habits of the rhesus macaques, *Macaca mulatta* in the Sunderbans, 24 Parganas, West Bengal. J Bombay Nat Hist Soc 62:145–146.

Mukherjee RP (1969): A field study on the behaviour of two roadside groups of rhesus macaque (*Macaca mulatta*) in northern Utter Pradesh. J Bombay Nat Hist Soc 66:47–56.

Neville MK (1968): Behaviour of rhesus monkeys in a town of northern India. Am J Phys Anthropol 29:131.

Ohja PR (1982): Population trends of rhesus monkey (*Macaca mulatta*) at Maroth Primate Research Center. Paper read in Symposium on National Primate Programme organized by Primatological Society of India.

Oppenheimer JR: Effects of intra and interspecific competition and habitual structure on use of time and space by Hanuman Langur (*Presbytis entellus*). Baltimore, Maryland: Johns Hopkins University unpublished m.s.

Pirta RS (1984): Cooperative behaviour in rhesus monkeys (*Macaca mulatta*) living in urban and forest areas. In Roonwal ML, Mohnot SM, Rathore NS (eds): "Current Primate Researches." Jodhpur: University of Jodhpur, pp 271–283.

Puget A (1971): Observations sur le macaque rhesus *Macaca mulatta* en Afganistan. Mammalia 35:199–203.

Rawlins RG, Kessler MJ (eds) (1986): "The Cayo Santiago Macaques: History, Behavior, and Biology." Albany: State University of New York Press, 256 pp.

Roonwal ML (1956): Macaque monkey eating mushrooms. J Bombay Nat Hist Soc: 54–171.

Rowell TE (1967): A quantitative comparison of the behaviour of a wild and a caged baboon group. Anim Behav 15:499–509.

Rowell TE (1972): Female reproductive cycles and social behaviour in primates. Adv Study Behav 4:69–105.

Sade DS, Chepko-Sade BD, Schneider JM, Roberts SS, Richtsmeier JT (1985): "Basic Demographic Observations on Free-Ranging Rhesus Monkeys." Vol. 1. New Haven, CT.: HRAF Press. 150 pp.

Shrestha J, Malla YK, Majupuria TC (1980): Rhesus monkey: In Majpuria TC (ed): "Wild is Beautiful," introduction to the fauna and wildlife of Nepal." pp. 388–399.

Shukla AK (1979): Activity patterns in the rhesus monkeys (*Macaca mulatta*). Research Proceedings Dept of Anthropology, University of Delhi.

Shukla AK, Seth PK, Seth S (1982): The ecology of free-ranging rhesus monkeys (*Macaca mulatta*) in an arid forest of India. In Symposium on National Primate Programme, by Primatological Society of India. Abstract.

Smith CC (1968): The adaptive nature of social organization in the genus of tree squirrels *Tamiasciurus*. Ecol Monogr 38:31–63.

Southwick CH (1962): Patterns of inter-group social behaviour in primates, with special reference to rhesus and howling monkeys. Ann NY Acad Sci 102:436–454.

Southwick CH (1970): Genetic and environmental variables affecting animal aggression. In

Southwick CH (ed): ''Animal Aggression.'' New York: Van Nostrand Reinhold, pp 213–229.

Struhsaker TT, Oates JF (1975); Comparison of the behaviour and ecology of red colobus and black and white colobus monkeys in Uganda: A Summary. In Tuttle RH (ed): ''Socioecology and Psychology of Primates.'' The Hague: Mouton, pp 103–123.

Taylor HG (1975): Cited from Majpuria TC: 1979–80.

Teas J (1978): ''Ecology and Behavior of Rhesus Monkeys in Kathmandu, Nepal.'' Ph.D. Dissertation. Johns Hopkins University.

Teas J, Richie T, Taylor H, Southwick C (1980): Population patterns and behavioral ecology of rhesus monkeys in Nepal. In Lindburg DG: ''The Macaques: Studies in Ecology, Behavior and Evolution.'' New York: Van Nostrand Reinhold, pp 247–262.

Vessey SH (1968): Interactions between free-ranging groups of rhesus monkeys. Folia Primatol 8:228–239.

Zimmerman RR, Wise LA, Strobel DA (1973): Dominance measurements of low and high protein reared rhesus macaques. Behav Biol 9:77–84.

Ecology and Behavior of Food-Enhanced Primate Groups, pages 113–123

6

Food Habits of Rhesus Monkeys (*Macaca mulatta*) in the North Indian Plains

M. Farooq Siddiqi and Charles H. Southwick

Department of Geography, Aligarh Muslim University, Aligarh, Uttar Pradesh, India (M.F.S.), and Department of Environmental, Population, and Organismic Biology, University of Colorado, Boulder, Colorado, 80309 (C.H.S.) [1]

INTRODUCTION

Rhesus macaques are among the most adaptable of all nonhuman primates. They have an immense geographic range, which extends from Afghanistan through India and Nepal to eastern China, and a broad ecological range, which includes subalpine habitats at elevations of 3,500 m, through many kinds of agricultural areas to tropical mangrove at sea level. They occur in urban locations and temple sites in close commensal relationships with human populations, and they also occur in remote forests where they avoid human contact. Their ecological relationships and behavior differ considerably in these various habitats, but few comparative studies of these differences have been done.

This fieldwork was undertaken to compare the dietary patterns of rhesus groups in different habitats within an agricultural area. The main purpose of the research was to study not only how the food habits of rhesus vary in different environments, but also to determine the extent to which rhesus monkeys in agricultural areas depend upon crops and direct human food sources.

Although rhesus monkeys have been studied extensively in both laboratory and field conditions, their food habits in India have been neglected except for the studies of D.G. Lindburg [1971, 1975] and M.K. Neville [1968] dealing with the behavioral ecology of forest groups around Dehra Dun and Haldwani. Their work concentrated on rhesus groups feeding primarily on natural forest vegetation. Marriott (this volume, chapter 7) has recently

[1]This chapter is adapted from a paper entitled ''Feeding Behaviour of rhesus monkeys (*Macaca mulatta*) in the north Indian plains,'' which appeared in the Proceeding of the Zoological Society of Calcutta, 31:53–61, 1980.

analyzed the food habits of Nepalese rhesus living in a temple and parkland habitat. We decided, therefore, to investigate the feeding behavior of rhesus groups in rural agricultural areas typical of the Gangetic plain.

Two rhesus groups in Aligarh District, approximately 130 k southeast of Delhi, were selected for comparative study. Both groups were commensal with people, and both were free-ranging. One group, Chhatari B, lived in a rural schoolyard at a crossroads surrounded by agricultural fields; the other, Sumera Fall Jungle group, lived in a small forest patch bordering a canal, also surrounded by agricultural fields.

GROUPS AND STUDY SITES

Chhatari B was a large group of approximately 70 monkeys at the time of the study inhabiting a rural schoolyard and grove of trees at a crossroads 22 kms north of Aligarh. This group grew substantially after this study to over 140 monkeys, and was then forced from the schoolyard to live along the roadside 0.5 km from the school [Southwick et al., 1986], but this disturbance had not occurred at the time of these observations.

All passenger buses between Aligarh and Anupshahr stopped at the schoolyard crossroads, and both passengers and numerous pedestrians fed the monkeys with gram nuts, peanuts, maize, wheat, and other foods (Plate I). Within the general home range of the group are many trees including sheesham, peepul, banyan, neem, jamun, and mango. The croplands around the school and roads are devoted primarily to wheat, barley, gram, sarson, sugarcane, millet, bajra, and pulses. This group, and an adjacent smaller one, has been the subject of previous behavioral studies [Southwick et al., 1974], and long-term population work [Southwick et al., 1980].

The Sumera Fall Jungle group consisted of 21 individual rhesus inhabiting a small forest patch near a canal and along a rural road used only by pedestrians. The area is bounded on one side by the Upper Ganga Canal and on the other by a smaller distributary canal. It is 10 km south of the Chhatari group. The jungle is composed of trees like sheesham, sirsi, amaltash, peepul, bamboo, and jamun, and many shrubs and bushes of babul, tiksee, heens, ber, kakronda, and akila. The botanical names of these are provided in Table I. The jungle patch is rectangular in shape, and approximately 5 ha in size. Some part of the jungle near the guest house has been cleared and converted into a plant nursery devoted to flowering plants and food crops, including potatoes, carrots, methi, sarson, wheat, barley, and rizka. There are also guava, mango, and jamun trees in the guest house compound, which are visited by members of this group. Surrounding the canals, jungle patch, and guest house compound are agricultural fields similar to those at Chhatari.

Plate I. The Chhatari group feeding on the road after people in a bullock cart have tossed food on the roadway.

METHODS OF OBSERVATION

Data were collected during two winter seasons (December–January) of 1970 and 1971, just after the general breeding season of the rhesus (September–November). This nonbreeding season was selected since rhesus groups are behaviorally and spatially more stable at this time than during the breeding season [Wilson and Boelkins, 1970; Lindburg, 1971] and therefore devote more consistent attention to feeding without frequent or excessive behavioral disruption.

The total observation period for this study extended over 152 hr in the case of Chhatari-do-Raho and 141 hr for the Sumera Fall Jungle. Observations were made throughout the day without any break from 0700 in the morning to 1800 in the evening, except on two occasions when they had to be abandoned at 1600 owing to heavy rains. For each feeding instance, records were made of 1) the time of the day, 2) total number of individuals involved in feeding, and 3) the nature and type of food being eaten—i.e., whether it

TABLE I. Local and Scientific Names of Common Plants and Trees in Study Areas

Local name	English name	Botanical name
Akola	—	*Mimosa rubicaulis*
Amaltash	Indian labyrinth	*Cassia fistula*
Arhar	Pigeon-pea	*Cajanus cajan*
Babul	Gum-tree	*Acacia arabica*
Bel	Wood-apple	*Aegale marmelos*
Ber	Wild jujube	*Zizyphus oenoplia* and *Zizyphus nummularia*
Heens	—	*Capparis sepiaria*
Jamun	Jamol	*Eugenia jambolana*
Kakronda	—	*Carissa carandas*
Masur	Lentil	*Lens culinaris*
Methi	Fenugreek	*Trigonella foenumgraecum*
Neem	Neem	*Melia azadirachta*
Peepul	Bo-tree	*Ficus religiosa*
Rizka	Lucerne	*Mudicago sativa*
Sarson	Mustard	*Brassica campestris*
Sheesham	Sissoo	*Dalbergia sissoo*
Sirsi	Siris	*Albizzia lebbek*
Tiksee	—	*Peristophe bicalyculata*

was from natural vegetation, agricultural crops, or was given by people. The authors tried to note the quantity of actual food taken but could not do so in case of natural vegetation and agricultural crops. Similarly, the actual duration of each feeding instance was recorded when possible, but could not be recorded in all cases because different individuals involved in feeding at any one instance varied considerably in the length of their individual feeding bouts. A feeding observation or feeding bout consisted of an accurately observed instance of one or more monkeys eating identifiable food. Usually a number of monkeys were involved, and each monkey was counted as one feeding observation or bout. The group was scanned once every 5 min, and the number of feeding bouts was tallied for each type of food being eaten. Feeding on natural vegetation and agricultural crops was usually of a long duration (if not chased by rakhwallas or village guards, usually children or elderly villagers, in the case of agricultural crops) as compared to feeding on food provided by the people.

RESULTS AND DISCUSSION

The monkeys took every type of food given by people in the form of peanuts, potatoes, bananas, guava, sugarcane, wheat, barley, gram, maize, and various types of local food preparations such as chapatis. In the case of natural vegetation they devoted themselves to fresh young leaves and buds of amaltash, kakronda, heens, akola, mango, bel, sirsi, jamun, and guava; leaves and fruits of ber; fresh leaves and beans of sheesham and babul; leaves and flowers of tiksee; fruits of peepul tree locally known as *peepli;* and grass. Scientific names are given in Table I. The important items of diet from the agricultural crops were young leaves of wheat, barley, gram, sarson, potato, carrot, methi, masur, and rizka plants; beans and leaves of arhar; and leaves and flowers of various nursery plants.

During the whole observation study there was no instance of monkeys feeding on insects as was noted by Lindburg [1975] in his studies of forest monkeys. It may be due to the fact that during the period of this study all the plants had young, fresh leaves and there was in no way any dearth of food to compel them to resort to insect eating. Insect eating may occur, of course, at other seasons. There was one occasion of a monkey killing a crow but it was not eaten. There was another attempt at a different time of catching a bird but it escaped.

In the Chhatari-do-Raho group, out of the total number of monkey-feeding observations made, 83% involved food from people, 10% was obtained from agricultural crops, and only 7% from natural vegetation (Fig. 1, Table II).

Of the food purposely fed to monkeys by people at Chhatari, 70% consisted of groundnuts (peanuts) or grams (a leguminous bean). Bread, wheat, and gur were also commonly provided and eaten (Table III).

Of agricultural crops consumed at Chhatari during the winter season, 84% was wheat and 12% was barley. Of natural vegetation consumed, 54% consisted of peepul, 20% mango, and 15% grass (Table III).

Thus, according to these observational estimates, the Chhatari group diet during the winter season consisted of 58% groundnuts or grams obtained directly from people. This estimate is obtained as the multiple of 0.83×0.70.

In contrast, the Sumera Fall Jungle group was observed to obtain 53% of their feeding from natural vegetation, 18% from agricultural crops, and 29% directly from people (Fig. 1, Table II).

Within each major category, the actual food source utilized differed remarkably from that of Chhatari. In the natural vegetation that dominated the diet of the Sumera group, peepul and mango were unimportant, and the three most important species were babul, sheesham, and ber. These three constituted 57% of the natural vegetation eaten during the winter season, and along with tiksee and grass, these five items accounted for 80% of the natural vegetation eaten, and 45% of the total diet.

Of the food provided by people for the Sumera group 42% was bread, 21% wheat, and 12% gram. Of agricultural crops, 26% was methi, 15% rizka, 13% wheat, and 10% each for arhar and carrots. The Sumera group certainly

TABLE II. Basic Food Sources of Two Rhesus Groups in North Central India

Group/source of food	No. of observed feeding bouts	Feeding time (%)
Chhatari-Do-Raha		
Natural vegetation	957	6.9
Agricultural crops	1,428	10.2
Food given by people	11,561	82.9
Total	13,946	100.0
Sumera Fall Jungle		
Natural vegetation	1,300	53.4
Agricultural crops	426	17.5
Food given by people	707	29.1
Total	2,433	100.0

showed a more diverse diet in the winter season than the Chhatari group, in terms of both the sources of the food, and the species utilized under each source (Table IV).

These data clearly emphasize the outstanding role of the local environment in determining the food habits of rhesus monkeys. The Chhatari group, which lives near a busy crossroad, depends more than 80% on local people for its food requirements, whereas the Sumera group, inhabiting a small jungle patch, relied on local people for less than 30% of its food. The Sumera group was observed to feed more than 50% of the time on natural vegetation. Another indication of this local adjustment to food supplies was provided by the work of Lindburg [1975], who found that forest-dwelling monkeys around Dehra Dun derived their food almost completely from natural vegetation, consuming more than 100 species of natural vegetation.

The daily feeding activity in both groups showed a very similar temporal pattern (Fig. 2). Two prominent periods of feeding activity occurred interspersed with a period of rest or low activity. Seth and Seth [1985] have observed similar patterns in rhesus groups in urban habitats, arid zones, and forests. Most of the feeding takes place between 0800 and 1200 in the morning and between 1400 to 1700 in the afternoon. Rhesus in Chhatari start feeding earlier than the other group, and the heaviest feeding takes place from 0800 to 1100 in the morning. Peak feeding hours in the morning for Sumera group are between 0900 and 1100. However, during the afternoon, the highest feeding activity takes place between 1500 and 1700 in both groups.

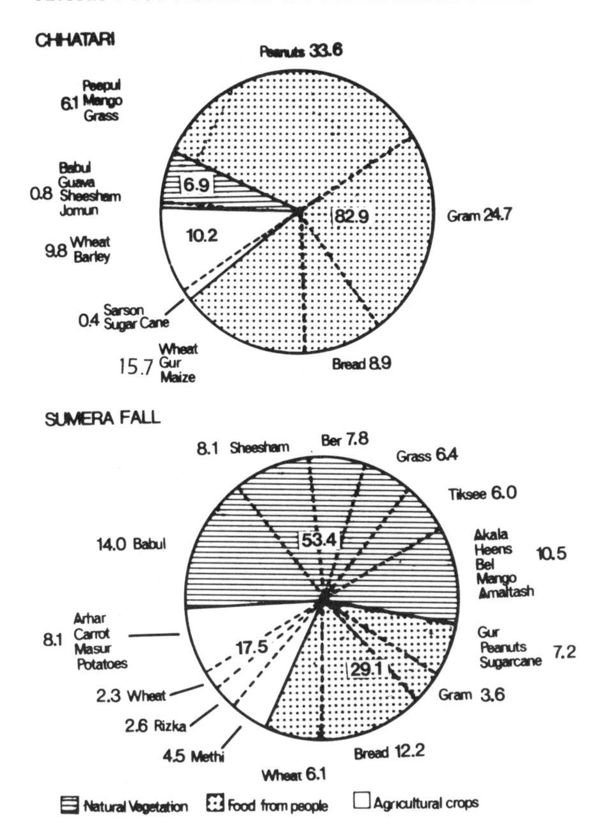

Fig. 1. Observed composition of diet of the Chhatari and Sumera Fall rhesus groups. Percentages refer to total observed food intake.

RESEARCH NEEDS

This work has barely scratched the surface of what should be done in the study of rhesus monkey food habits. By showing the remarkable differences in food habits between two groups living in essentially the same general environment, and comparing these with the totally different patterns shown by forest groups as studied by Lindburg [1971, 1975], and Neville [1968], our work raises questions as to how great the differences in food habits of rhesus living at greater ecological extremes may be. It would be interesting to compare feeding behavior of rhesus in pine forests of northern India and Nepal, mangrove forest of the Sundarbans in Bengal and Bangladesh, arid environments of Rajasthan, temperate forests of central China, etc.

There are also questions of seasonality. How do food habits change in any

TABLE III. Chhatari-do-Raha Rhesus Group: Species Composition of Food Ingested in Each Major Category as Recorded in Field Observations

Natural vegetation[a]			Agricultural crops			Food from people		
Food item	No. of feeding bouts	%	Food item	No. of feeding bouts	%	Food item	No. of feeding bouts	%
Peepul	516	53.9	Wheat	1,195	83.7	Peanuts	4,688	40.5
Mango	189	19.7	Barley	171	12.0	Gram	3,441	29.8
Grass	142	14.8	Sarson	42	2.9	Bread	1,241	10.7
Guava	40	4.2	Sugarcane	20	1.4	Wheat	945	8.2
Sheesham	37	3.9				Gur	539	4.7
Jamun	32	3.3				Guava	206	1.8
Babul	1	0.1				Maize	168	1.4
						Bananas	82	0.7
						Potatoes	23	0.2
						Peas	13	0.1
						Sweet potatoes	8	0.1
						Bajra	4	0.03
						Carrots	4	0.03
						Mixed grains	48	0.4
						Misc. sweet preparations	151	1.3
Total	957	100.0		1,428	100.0		11,561	100.0

[a]Primarily young leaves, buds, and shoots, but fruits of peepul, sheesham, and babul were also eaten.

of these habitats in relation to season? Throughout most of the rhesus macaque's range, seasonal differences are prominent in terms of temperature, rainfall, plant growth, and fruiting season.

There are also questions of nutrient values in terms of calories, protein contents, and vitamin and mineral values. Marriott (this volume, chapter 7) has been investigating this in her rhesus studies in Nepal, and is showing the potential and importance of this field.

Finally, there are questions of demography and food habits. How do dietary patterns affect reproductive behavior, birth rates, and survivorship? Certainly food intake is a major influence on general health and physiology, and will thereby affect many aspects of population ecology, but we know little or nothing about these influences in natural populations. It would be difficult to study these matters in wild populations, however, due to the complexity of uncontrolled variables. Many factors other than diet affect reproductive and mortality patterns in natural populations, and it would require a complicated set of experimental situations to investigate the specific role of food.

TABLE IV. Sumera Fall Jungle Rhesus Group: Species Composition of Food Ingested in Each Major Category as Recorded in Field Observations

Natural vegetation[a]			Agricultural crops			Food from people		
Food item	No. of feeding bouts	%	Food item	No. of feeding bouts	%	Food item	No. of feeding bouts	%
Babul	355	27.3	Methi	110	25.8	Bread	297	42.0
Sheesham	196	15.1	Rizka	64	15.0	Wheat	148	20.9
Ber	190	14.6	Wheat	56	13.1	Gram	87	12.3
Grass	156	12.0	Arhar	44	10.3	Gur	76	10.8
Tiksee	147	11.3	Carrot	44	10.3	Peanuts	21	3.0
Akola	94	7.2	Masur	21	4.9	Sugarcane	11	1.6
Heens	83	6.4	Potatoes	29	6.8	Mixed grains	37	5.2
Bel	30	2.3	Gram	10	2.4	Misc. sweet preparations	30	4.2
Mango	21	1.6	Coriander	10	2.4			
Amaltash	15	1.2	Barley	8	1.9			
Peepul	7	0.5	Sarson	7	1.6			
Kakronda	5	0.4	Podeena	2	0.5			
Jamun	1	0.1	Nursery plants	21	4.9			
Total	1,300	100.0		426	100.0		707	100.0

[a]Primarily young leaves, buds, and shoots, but fruits of peepul, sheesham, and babul were also eaten.

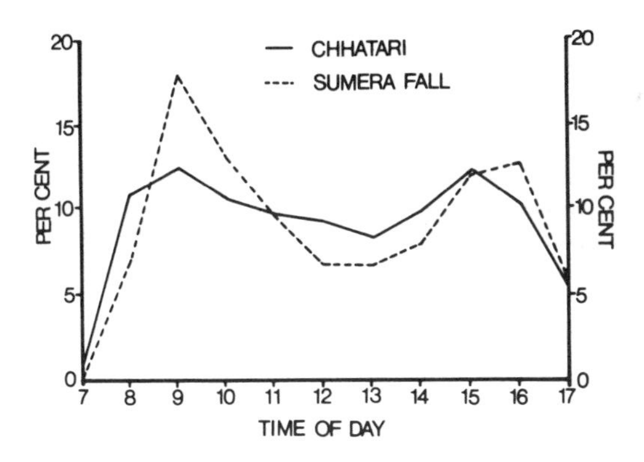

Fig. 2. Diurnal pattern of feeding activity of the Chhatari and Sumera Fall rhesus groups.

POSTLUDE—SUBSEQUENT HISTORY OF CHHATARI AND SUMERA GROUPS

The two groups in this study of food habits had remarkably different histories since this study. The Chhatari group doubled in size to over 140 individuals and was then forcibly expelled from its school yard and grove of trees in 1978. Throughout the mid 1970s it showed remarkably high birth rates, averaging 90% per annum, and low mortality rates. Subsequent to its expulsion, it stopped expanding, actually declined to 120 individuals, owing largely to striking increases in infant and juvenile mortality. In 1984 and 1985, however, it resumed its pattern of growth and in late 1986 numbered 175 individuals. On the roadside, with a less favorable habitat and considerably more human disturbance, the Chhatari group retains its basic dietary pattern, possibly depending even more on human sources for food.

The Sumera jungle group, although ostensibly living in a better habitat than Chhatari, has had a much less favorable history since our study. From 21 individuals, it declined more or less steadily to only 3 remaining adults in 1985. These were two adult males and one adult female, so the group appears headed for extinction. Throughout the 1970s, the group had a satisfactory birth rate, but infants, juveniles, and adults were dying or disappearing. We cannot say what role diet may have played in this. All indications at the time of this study were that the food intake was adequate, but the small size of their forest patch (5 ha) plus inadequate supplemental feeding from people may have contributed to poor survivorship, excessive dispersal, or both. This group may also have been subjected to more trapping, for Chhatari had a semiprotected status while it remained within the schoolyard grove of trees. On the other hand, Sumera Fall had a better habitat from which to escape trapping in its jungle patch. In any case, trapping for export ceased for both groups in April of 1978, so if the ecological setting of Sumera had been adequate, it should have recovered. At the time of the export ban on trapping, the Sumera group numbered four individuals, including two adult females and one juvenile. These histories suggest that diet may have played a role in the success of Chhatari and the demise of Sumera, but this is entirely speculative. This illustrates how little we know about the role of diet in the population ecology of natural primate groups.

SUMMARY

Rhesus groups living in slightly different habitats in an agricultural area of the Gangetic plain show remarkable differences in the food habits. The Chhatari group, occupying a rural schoolyard and busy crossroad, took 93% of their observed food intake from human sources, either via direct feeding by people or from adjacent field crops. In contrast, the Sumera Fall group, occupying a small forest patch along the upper Gangetic canal, took 53% of their food from natural vegetation. In both groups, the temporal pattern of daily feeding activity was similar. The peak of feeding occurred between

0800 and 1100 in the morning, and 1500 and 1700 in the afternoon with an intervening period of rest and reduced feeding.

There are many opportunities for further research in the feeding behavior and dietary intakes of natural rhesus groups in different habitats and different seasons. We know very little about the full range of dietary adaptability of this species.

Also unknown are the demographic consequences of different feeding patterns. Subsequent to this study, the Chhatari group has more than doubled despite a loss of its habitat and displacement to a less favored location. In contrast, the Sumera Fall group has declined to the point of near extinction. Diet may have played a role in the demographic factors leading to these population changes, but this is unknown.

REFERENCES

Lindburg DG (1971): The rhesus monkey in north India: An ecological and behavioral study. In Rosenblum LA (ed): "Primate Behavior: Developments in Field and Laboratory Research." New York: Academic Press., 1–106.

Lindburg DG (1975): Feeding behavior and diet of rhesus monkeys (*Macaca mulatta*) in a Siwalik forest in north India, upubul. ms. 44 pp.

Neville MK (1968): Ecology and activity of Himalayan foothill rhesus monkeys. Ecology, 49:110–123.

Seth PK, Seth S (1985): Ecology and feeding behavior of free ranging rhesus monkeys in India. Indian Anthropol 15:51–62.

Southwick CH, Siddiqi MF (1977): Demographic characteristics of semi-protected rhesus groups in India. Yearbook Phys Anthropol, 20:242–252.

Southwick CH, Siddiqi MF, Farooqui MY, Pal BC (1974): Xenophobia among free-ranging rhesus groups in India. In Halloway RK (ed): "Primate Aggression, Territoriality and Xenophilbia: A Comparative Perspective." New York: Academic Press: pp 185–209.

Southwick CH, Siddiqi MF, Johnson R (1986): Demographic effects of home range displacement on rhesus monkeys at Chhatari, northern India. Am J Primatol 10:433.

Southwick CH, Richie T, Taylor H, Teas HJ, Siddiqi MF (1980): Rhesus monkey populations in India and Nepal: Patterns of growth, decline, and natural regulation, Chap. 7. In Cohen MN, Malpass, RS, Klein HG (eds): "Biosocial Mechanisms of Population Regulation." New Haven: Yale University Press, pp 151–170.

Wilson AP, Boelkins RC (1970): Evidence for seasonal variations in aggressive behavior in *Macaca mulatta*. Anim Behav 18:719–724.

Ecology and Behavior of Food-Enhanced Primate Groups, pages 125–149

7

Time Budgets of Rhesus Monkeys *(Macaca mulatta)* in a Forest Habitat in Nepal and on Cayo Santiago

Bernadette M. Marriott

Caribbean Primate Research Center and Department of Obstetrics and Gynecology, University of Puerto Rico, Medical Sciences Campus, San Juan, Puerto Rico 00936

INTRODUCTION

How an organism divides its time during its waking hours reflects the social and environmental variables with which it must contend for survival on a daily, seasonal, and lifetime basis. Time budgets can thus be considered estimates of the way in which animals expend energy. Some studies of primate activity rhythms have described the presence of consistent diurnal variation across wild and captive or enclosed populations of the same species [Clutton-Brock, 1977; Bernstein, 1970a,b, 1975]. Additionally, diurnal similarities in time budgets such as early-morning and late-day peaks in eating have been reported for species with different social organizations and habitats, such as lemurs [Sussman, 1977], spider monkeys [Carpenter, 1935], patas monkeys [Hall, 1965], colobus monkeys [Strusaker, 1975], and chimpanzees [Wrangham, 1977]. It has been suggested that, considered together, the factors involved in diurnal variation in activity patterns may represent universal adaptations to temperature variations [Chivers, 1975; Bernstein and Mason, 1963; Harrison, 1985], digestive/energy needs [Clutton-Brock, 1974], or social facilitation of ingestion [Kummer, 1971]; or they may reflect an underlying basic arousal mechanism [Maxim et al., 1976]. One difficulty in comparing inter- and intraspecific time budgets is that these data are sensitive to differences in methodology [cf. Marsh, 1981]. In two studies in which the methodology has been carefully controlled, however, significant differences in time budgets between intraspecific populations were demonstrated [Oates, 1977; Marsh, 1981]. There is currently no consensus about the functional significance of interspecific similarities and intraspecific differences in diurnal rhythms of primate activities. The degree to which activity rhythms are species specific and governed by underlying neurological/perceptual mechanisms is unknown.

The activities of social groups of rhesus monkeys (*Macaca mulatta*) have been studied extensively in their natural habitats, as introduced on island habitats and in large enclosures [cf. Southwick et al., 1982; Teas et al., 1982; Baulu and Redmond, 1980; Drickamer, 1973; Bernstein and Mason, 1963; Post and Baulu, 1978]. Data have been collected on diurnal variation in activities of rhesus monkeys using different methods and behavioral definitions that are not directly comparable. There are no studies that have directly compared the activity rhythms of wild rhesus monkeys with a captive provisioned group to assess the impact of provisioning or other environmental factors on time budgets.

As the human population increases worldwide, many species of nonhuman primates that have ranged freely in natural forests may have to be restricted to smaller reserves and as a result provisioned to maintain population survival. Our knowledge of the effects of provisioning on overall behavior rhythms is limited. Throughout the Indian subcontinent, rhesus monkeys are fed local foods that serve either as their main dietary source or as supplements [Siddiqi and Southwick, 1977]. In order to examine food enhancement in rhesus monkey populations in a more controlled setting than that typically found in Indian villages, data were collected on the time budgets of the food-enhanced population of monkeys on Cayo Santiago in Puerto Rico for comparison with parallel data on a forest-dwelling troop near Kathmandu, Nepal, which received no regular provisioning.

In considering the impact of provisioning on nonhuman primate groups, two aspects were of interest. First, do the conditioned stimuli that accompany food presentation lead to pronounced behavioral changes that are reflected in time budgets? Second, do food-enhanced primate groups exhibit any diurnal food-type selection patterns that may be similar to those of their nonprovisioned counterparts?

The purpose of this chapter is to compare the time budgets of wild rhesus monkeys in Nepal with provisioned rhesus monkeys on Cayo Santiago. The emphasis of the approach has been to focus on animal states of behavior rather than brief events [Altmann, 1974], including records of food types consumed during the observation day.

MATERIALS AND METHODS

Study Areas

Nepal. The home range of the study troop consists of two forest patches and a connecting parkland situated 4.1 km northeast of Kathmandu, Nepal

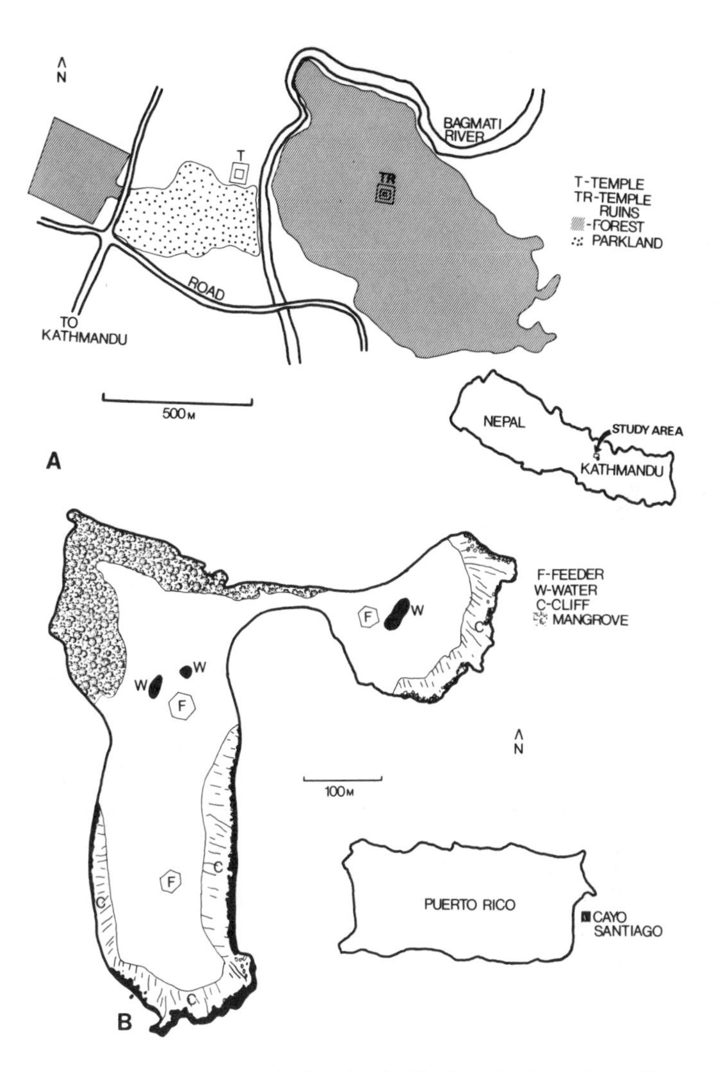

Fig. 1. **(A):** Gaushalla study site, Nepal. **(B):** Cayo Santiago, Puerto Rico.

(N) (Fig. 1A). This 70-ha area, located at 27° 42′ N latitude and 85° 21′ E longitude, elevation 1,320–1,340 m above sea level, is physically isolated from other forests by surrounding agricultural and low-density residential development. The parkland is contiguous with the grounds of a major Hindu temple. The forest has not been systematically surveyed botanically but is tentatively classified as a mixed broad-leaved forest [Malla, 1967]. A walled forest patch (7 ha) and the central parkland (3.6 ha) compose the troop's core area. The walled forest is at the western end of the troop's home range, and it is used exclusively by the study troop, which sleeps there nightly. The parkland and larger forest patch are also inhabited by six additional troops, an undetermined number of all-male bands and solitary males [Teas et al., 1982; Teas, 1983]. The forest patches are densely wooded, with an undulating canopy rising to 18 m. In contrast, the parkland has a flat bilevel canopy rising to 11 m on plateau areas or 46 m in lowland areas with low-density ground cover and no shrub layer.

Cayo Santiago. Approximately 1 km off the southeastern coast of Puerto Rico at 18° 09′ N latitude and 65° 44′ W longitude is Cayo Santiago (CS) (Fig. 1B). This 15.2-ha hilly island is patchily wooded primarily with introduced vegetation. Throughout most of the island the wind-buffeted trees are short, with an average canopy height of 10 m. There is little ground cover and only a sparse shrub layer. The island is composed of two arms (cays) connected by a narrow sand peninsula. There are six naturally formed social groups inhabiting Cayo Santiago. During the course of the observations, one social group that solely inhabited the smaller of the two cays was studied. This "small cay" is approximately 3.5 ha in non-cliff surface area.

Subjects

Nepal. Activity samples were collected during October 1979 on a single, well-habituated troop of rhesus monkeys: the Gaushalla troop. This troop was chosen from the six troops that jointly inhabit the home range because the troop confined its day range to the forested or parkland habitats and had little contact with people. The individuals of this troop were not regularly provisioned, but they occasionally stole cultivated produce from people on their way to the temple or were fed dried corn or rice while passing through the parkland. Water was available to the animals in natural ground pools and from moist vegetation. Troop counts were recorded daily when the monkeys first left their sleep site in the early morning. Average counts over the study period yielded a total troop size of 82.4 ± 9.5 animals. For the purpose of this study animals were coded by the age/sex class designations as modified from Southwick et al. [1965]: adult male, subadult male, lactating adult female, pregnant adult female, cycling adult female, juvenile, yearling, and infant. Adults were determined by body size and presence of red sex skin. Juveniles were distinguished from adults and yearlings by body size. For the purposes of analysis data on the one subadult male was combined with that of the juvenile category. All adults in the study troop were individually recognizable to the observers by distinctive facial markings, permanent fur patterns, and scars. Infants were identified by sex and were recognized by association with their mothers. Only those few juveniles with distinctive

TABLE I. Troop Compositions and Densities

	Cayo Santiago (CS)	Nepal (N)	Ratio (CS:N)
Adult males	32	12	2.7 : 1
Pregnant females	0	0	1 : 1
Lactating females	21	15	1.4 : 1
Cycling females	6	12	1 : 2
Total adult females	27	27	1 : 1
Adult males: females	1.2 : 1	1 : 2.3	
Juveniles	36	18	2 : 1
Yearlings	17	12	1.4 : 1
Infants	21	15	1.4 : 1
Total	133	84	1.6 : 1
Home range (ha)	15.2	70.0	1 : 4.6
Core area (ha)	3.5	15.2	1 : 3.1
Density (HR: animals/ha)	8.8	1.2	7.3 : 1
(CA: animals/ha)	38.0	7.9	4.8 : 1

Approximate age spans for both populations: infants, 0–1 year; yearlings, 1–2 years; juveniles—males, 2–4 years; females, 2 to 3 years; adults—males, 5+ years; females, 4+ years.

bodily features were individually identified. Over a 5-year period from May 1977 through August 1982 the birth peak in the Gaushalla troop consistently occurred during the second week in June.

Cayo Santiago. Group L was chosen as the study group on Cayo Santiago because it was the sole inhabitant of the small cay where the natural vegetation is most dense. The monkeys were provisioned with approximately 0.22 kg of commercial diet per animal per day, which was distributed in several metal hog feeders located inside a fenced corral. Water was available from three gravity-fed fountains dispersed on the cay. Daily census of the entire Cayo Santiago population was maintained by the staff of the Caribbean Primate Research Center. The verified group total for the months of June and July 1981 during the study period was 133 animals. Table I presents a comparison of the troop compositions for Nepal and Puerto Rico.

Comparability of Animal Groups

Consideration was given in data assessment to timing of sample collection within the reproductive cycle of the rhesus monkeys in each population. Temperature, rainfall, and animal density were also considered in assessing these data sets for comparability.

In Nepal, the activities of the troop individuals were collected for at least two months each year from June 1977 through August 1982. A data set collected in October 1979 in Nepal was selected to match the timing within the reproductive season and duration of the sample collected on group L monkeys on Cayo Santiago during June and July 1981. On Cayo Santiago the median birthdate of the animals in group L occurred 4 months prior to the beginning of the observations. Observations conducted in Nepal in October 1979 were 4 months after the median birthdate for the Gaushalla troop of June 15.

Yearly variability in temperature and rainfall have been shown to correlate with cyclical changes in activity patterns in nonhuman primates in natural environments [cf. Lindburg, 1977; Richard, 1977; Harrison, 1985]. There was considerable climatic variation between the two study sites during the observation periods. The average maximum temperature in Nepal for the month of October 1979 was 24.9°C. with a minimum of 13.4°C. [Shrestha, 1982]; on Cayo Santiago the average maximum temperature was 32.4°C and the minimum was 24.8°C. The monthly accumulation of rainfall during the study period on Cayo Santiago was approximately 3.5 times the total rainfall in October 1979 in Nepal: CS, 12.7 cm; N, 3.6 cm (daily records of Cayo Santiago) Nepal [Shrestha, 1982]. Although the total accumulation was greater on Cayo Santiago, the number of rainy days was roughly equivalent: CS, 12 days; N, 11 days. In general, the observations in Nepal were collected during a cooler, drier period of the year than on Cayo Santiago.

Density is another factor that can influence the behavior and presumably time budgets of nonhuman primates [cf. Alexander and Roth, 1971; Wilson, 1972; Erwin and Erwin, 1976; Nieuwenhuijsen and de Waal, 1982]. Table II presents the densities of the monkeys in Gaushalla forest, on Cayo Santiago, and selected densities calculated from reports of rhesus monkeys in the literature. Presentation of density estimates on the basis of home range allows comparison with the literature, however, since both group L and the Gaushalla troop encounter other troops within their home ranges, more appropriate comparison figures are their core area-based densities. From this table it is evident that the home range-based densities of group L and the Gaushalla forest troop, and the core area-based density of the Gaushalla troop are comparable with town troop densities in India, whereas the core-area based density of group L is more comparable to the densities of temple troops in Nepal and India. Considering core area alone, the density of group L on the small cay is 4.8 times the density of the Gaushalla troop in its core area.

Sampling Methods

The sampling method was identical in both locations. Instantaneous scan samples [Altmann, 1974] of the total number of individuals who were engaged in each of 29 mutually exclusive behaviors were recorded at 20-min intervals. (See Table III.) With this method, the activity in progress was

TABLE II. Troop Sizes, Home Ranges, and Densities for Rhesus Monkeys in Free-Ranging and Natural Conditions

Source[a]	Location/ identifier[b]	Troop size	Home range (km^2)	Density[c]
Lindburg [1971]	North India(f)	115(tot.pop.)	2.59	44.4
Lindburg [1977]	Asaror,India(f)	85	15	5.7
Neville [1968]	Haldwani,India(f)A	15 to 16	3.11	5.2
	Haldwani,India(f)B	20	1.04	19.2
	Haldwani,India(f)C	54+	1.04+	51.9
	Haldwani,India(f)E	38+	1.55+	25.5
	Haldwani,India(t)H	25	0.05	500.0
	Haldwani,India(t)J	11	0.05+	220.0
Oppenheimer[a]	Dacca,Bangaladesh(t)	18	.001	18,000.0
Southwick [1972]	Calcutta,India(tm)	62	.007	8,857.1
Teas[a]	Kathmandu,Nepal(tm)	341(tot.pop.)	.1	3,410.0
This study	Cayo Santiago (HR)	133	.15	886.7
This study	Gaushalla (HR)	84	.7	120.0
This study	Cayo Santiago (CA)	133	.035	3800.0
This study	Gaushalla (CA)	84	.1	840.0

[a]Complete source information: Oppenheimer et al. (unpubl); Teas et al. [1975, 1981].
[b]Abbreviations: f = forest troop; t = town troop; tm = temple troop; HR = home range; CA = core area. For Neville [1968] troop identities were designated by letters.
[c]Animals/km^2 (hypothetical).

scored as closely as possible at the instant the target animal was observed. Care was taken in both locations not to record changes in behavior resulting from the appearance of the observer but rather the behavior of the animals when first spotted.

Struhsaker [1975] has argued that the observer needs to observe an animal for a longer time period and should base the selection of a minimum time/target animal on the bout length of behaviors in question. He suggested a 5-sec observation per target animal per scan to avoid overestimating activities that may merely represent pauses in more sustained activities. As Marsh [1981] has described, there are tradeoffs with both methods.

In a comparison of observations of red colobus monkeys (*Colobus badius rufomitratus*) Marsh [1981] used both instantaneous and 5-sec scan sampling. He reported that the major difference between these methods during a year of observations was a lower estimate of ingestion and a higher estimate of locomotion using the instantaneous scan method. Marsh's interpretation of this difference was that eating was often preceded by other short-duration activities. Instantaneous sampling was the method chosen here because it eliminated the extra time dimension when working with large animal groups and reduced error owing to observer decisions as well as variance from a point-sample technique.

TABLE III. Definitions of Activities Scored in Instantaneous Scan Samples in Both Locations

Activity	Definition
Nurse	Infant holding nipple in mouth
Drink	Obtaining water from any source
Rest	In relaxed posture with eyes closed and scored as either:
Individual	Not in body contact with another while resting
Social	In contact with another monkey while resting
Groom	Searching through the fur or rubbing, licking or scratching the fur and scored as either:
Individual	Grooming oneself
Social	Grooming another
Eat	Actual ingestion of food; does not include chewing, licking or food selection or preparation and scored as one of the following categories of food types: mature leaves, young leaves, fruit, stems, flowers, seeds, roots, grass, soil, insects, commercial diet, cultivated plants, other, unknown
Play	Nonaggressive energetic activity such as wrestling, chasing, tumbling, leaping, etc.; may include one or more animals
Investigate	Close olfactory, visual, or oral inspection of an object or individual
Threat	Directed open-mouth display, which may include threat bark, head bounce, or flattened ears
Chase	One or more monkeys pursue each other aggressively
Attack	Actual physical contact usually involving a hit, bite, pull, or push
Fight	A combination of threats, chases, and attacks among animals in which no submissive behavior is seen
Sexual	Any occurrence of heterosexual mounting, thrusting, or sexual presentation
Locomote	Any movement of an animal from one place to another that does not include short movements that are a part of feeding, play, investigation, or agonism
Look	Scanning the environment while stationary
Other	Any behaviors that do not readily fit into the above stated categories
Unknown	Common usage

In both populations it was typical for groups of animals to be simultaneously engaged in the same activity, and each group could be observed from a distance without disturbance. For social behaviors, both the actor and recipient of a behavior were scored as participating in that behavior; for example, in a threat, both the animal threatening and the recipient were both scored as engaged in a threat encounter. In both locations the observers walked around and through the study group in a counterclockwise direction following the general movement and location of the animals. No individuals were scored more than once per scan. The time necessary to complete a scan varied with activity of the animals and group spread. Observers were to allocate no more than 15 min to each scan and were to return to the approximate starting point within the animal group to begin the next scan at the 20-min interval.

Observations in Nepal were collected from earliest to latest daily visibility (0630–1730) by rotating teams of two observers. During the scans one person dictated the age/sex class and activity to the other, who recorded the data. All observers were skilled in recording scan sampling data and had an independently sampled minimum interobserver reliability [Lehner, 1979] of 92% on age/sex class identification and behavior category. Scans were considered acceptable for data summary if 60% of the total of the project average for the daily troop counts or 50 animals were observed. A total of 340 scans were analyzed.

On Cayo Santiago two observers, with an interobserver reliability of 98% on age/sex class determination and 94% on independently scored behaviors, collected instantaneous scan data on group L from 0730 to 1730. The observations thus began approximately 1 hr after clear visibility in this location. One observer both observed and recorded the scans. Because of the larger group size and large number of peripheral males, 50 animals or 37% of the group total was the criterion for scan inclusion in final analyses (113 scans).

Data Analyses

Each acceptable scan was converted to a proportion by dividing the number of animals observed in each activity by the total number of animals observed in that scan. Activities were summed across hours of the day and averages taken to illustrate diurnal rhythms. Hourly sums were added across the day and divided by the total number of scans to generate the average time distribution of the animals in each location. For the purpose of analysis, activities which occurred at frequencies less than 1% and/or were of short bout duration—thereby approaching events rather than states of behavior [Altmann, 1974]—were grouped into major categories. Ingestion of leaves, fruit, stems, flowers, seeds, roots, and grass were grouped into the category of vegetation. Threats, chases, attacks, and fights were termed agonistic behaviors. Nurse, investigate, sexual behavior, and others were grouped as "other."

Wilcoxon's matched pairs signed ranks test [Siegel, 1956; Sokal and Rohlf, 1981] was utilized to examine the differences in overall time budgets between the two study populations. Evaluation of the contribution of the six age/sex classes to the scan sample across the hours of the day for each group was accomplished using the Wilcoxon test. Associations of diurnal variation in activity patterns within groups across the ten 1-hr observation time blocks were analyzed using the Spearman rank correlation coefficient [Siegel, 1956]. The significance level for both tests was set at $P<.05$.

TABLE IV. Group Time Budgets for Group L on Cayo Santiago and the Gaushalla Troop in Nepal

Behavior	Group L	Gaushalla troop
Look**	30.0	22.0
Drink**	2.5	0.2
Total rest*	12.9	5.0
Individual rest	5.6	1.8
Social rest	7.3	3.2
Total groom	12.3	9.8
Individual groom	0.7	3.2
Social groom	11.6	6.6
Eat**	10.8	15.6
Play**	5.3	3.7
Agonistic encounters*	0.8	1.6
Locomote**	23.1	39.4
Other	2.3	2.7

This table shows mean percentages of time spent in each behavior by the study groups.
*P<.05.
**P<.01 (Wilcoxon test [Siegel, 1956]).

RESULTS

Table IV shows the time budget for both groups calculated by summing the proportions across the ten hourly time blocks and dividing by the total number of scans. As can be seen from this table there were significant differences in the amount of time spent in the major activity categories by the two groups. Both groups spent over 50% of their time engaged in the combined activities of locomotion and looking. As might be hypothesized for a nonprovisioned troop, the Nepalese monkeys devoted significantly more time to feeding, moving, and agonistic encounters than did the Puerto Rican monkeys (n = 10, T = 0, 0, 5; $P<.01$, .01, .02). Group L exhibited significantly more looking, resting, playing, and drinking than did the Gaushalla troop (n = 10; T = 0, 4, 2, 0; $P<.01$, .02, .01, .01).

As has been previously reported [cf. Marsh, 1981; Jolly, 1972], the age/sex composition of troops and of behavior scan samples can significantly affect the resultant composite time budgets and lead to possible errors in interpretation of intra- and interspecific comparisons. Differences in overall behaviors seen in this study could have been due to the age/sex composition disparity of the respective groups. To evaluate this possibility the composition of each group was converted to percentages of the total (see Table V). From inspection it is evident that the only real differences in age/sex composition of the two troops were the higher percentage of adult males and the much lower percentage of nonlactating females in the group L monkeys

TABLE V. A Comparison of the Distribution of the Age/Sex Classes in Each Location by Percentage Within the Population and the Sample

	Age/sex class					
Location	Adult male	Lactating female	Cycling female	Juvenile	Yearling	Infant
Nepal						
Troop	14.29	17.85	14.29	21.43	14.29	17.85
Sample	13.64**	19.70**	13.64**	19.70**	12.12**	21.20**
Cayo Santiago						
Troop	24.06	15.79	4.51	27.07	12.78	15.79
Sample	22.50	20.00**	1.25**	22.50**	13.75	20.00**

**P<.01 (Wilcoxon test [Siegel, 1956]).

as compared with the Gaushalla troop. If one assumes that the largest source of variation when comparing aggregate time budgets is likely to result from comparison of adult classes with the young animals [Marsh, 1981; Marriott and Sultana, in prep.], then the composition of these two groups should not affect the overall time budgets since the sum of the adult and juvenile percentages for both groups is very similar.

Table V also presents the average frequency of each class in the activity samples as summed across the ten 1-hr time blocks. Using the troop percentage as an expected value and the means of the sample classes for each 1-hr block, Wilcoxon tests demonstrated significant differences between the expected and observed values as illustrated. For all but the yearling sample the direction of the differences was consistent across the two groups: mothers and infants were overrepresented in the samples, while males, nonlactating females, and juveniles were underrepresented. Yearlings were overrepresented in the Puerto Rican sample and underrepresented in the Nepalese sample. The effect of this bias can be estimated by calculating corrected activity estimates for activities that largely reflect the contributions of limited age/sex classes [Marsh, 1981]. New or expected percentage for these activities can be estimated as follows: [after Marsh, 1981]

$$\text{Expected } \% = \text{observed } \% \times \frac{\text{Proportion of age/sex class in group}}{\text{Proportion of age/sex class in sample}}$$

Since play is a behavior demonstrated primarily by young animals, and agonistic behavior is primarily an adult behavior, these two activities were used to evaluate the potential effects of sample bias. For play estimates, infant, yearling, and juvenile data were used; for agonistic estimates the data from lactating females, cycling females, and males were summed.

GAUSHALLA TROOP (observed play = 3.7% ; observed agon. = 1.6%)

$$\text{Expected \% time play} = 3.7 \times \frac{.1785}{.2120} + \frac{.2143}{.1970} + \frac{.1429}{.1212} = 3.74$$

$$\text{Expected \% time agon.} = 1.6 \times \frac{.1429}{.1364} + \frac{.1785}{.1970} + \frac{.1429}{.1364} = 1.58$$

GROUP L (observed play = 5.3% ; observed agon. = 0.8%)

$$\text{Expected \% time play} = 5.3 \times \frac{.1579}{.2000} + \frac{.2707}{.2250} + \frac{.1278}{.1375} = 5.24$$

$$\text{Expected \% time agon.} = 0.8 \times \frac{.2406}{.2250} + \frac{.1579}{.2000} + \frac{.0451}{.0125} = 0.81$$

From this comparison it can be seen that although the sample distribution of the age/sex classes differed statistically from the distribution within the troops, the differences should not be large enough to lead to substantial bias in the overall time budgets.

Figures 2 and 3 illustrate the diurnal rhythm of activities, respectively, for group L and the Gaushalla troop. In these figures the 1-hr time periods have been combined into 3-hr time blocks. Blocking did not mask any trends in diurnal activity rhythms for either group. Overall it is evident that there are smaller diurnal shifts in behavior in Nepal than on Cayo Santiago. In the Gaushalla troop there is a bimodal increase in eating in the early morning and late afternoon, which correlates negatively ($r_s = -0.746$; $P<.01$) with a bimodal decrease in resting in the same time periods. Although total grooming parallels resting with a midday increase, there is no significant correlation. Grooming peaks daily in Nepal from 0930 to 1030. Eating occurs most actively from 0630 to 0839 in the morning and from 1530 to 1730 in the afternoon. After a major movement from the sleeping area and an initial rapid foraging bout, there is a marked drop in locomotion, which corresponds with the grooming peak from 0930 to 1030, followed throughout the rest of the day by repeated cycles of resting and foraging bouts. The Gaushalla troop moves the greatest distances from 1530 to 1730 each day during the final intensive foraging bouts and mass movement to the sleep site. In general, with the exception of increased grooming in the morning, there is little distinct diurnal variation in the activity patterns of the Gaushalla troop. Distribution of the data indicates that this is due to low variability in activity distribution throughout the day. Very consistently the troop members are divided roughly with 15% feeding, 20% engaged in vigilance, 10% resting or grooming, and 40% locomoting. Throughout the day in Nepal, there are always some animals engaged in the major activities. Although the troop moves as a unit from place to place, once in a new location they do not all engage in one activity type. Even during movements, some animals may continue to forage as they move or stop to eat along the movement path.

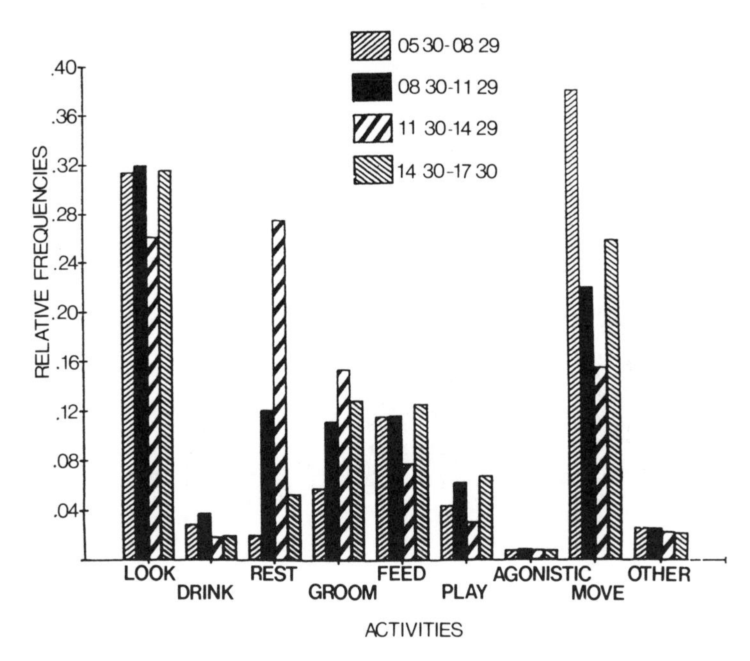

Fig. 2. Diurnal rhythm of activities for group L on Cayo Santiago.

These results are in contrast with the more distinct activity rhythms of the monkeys on Cayo Santiago. The bimodal peak in eating negatively correlates with resting ($r_s = -0.9394$; $P<.01$) and the peak eating period has a longer duration from 0730 to 1030 and 1430 to 1730. The distinctive difference lies in the amount of synchrony. From 1130 to 1330 the percentage of animals eating drops to 7% from the earlier and later peaks of 14.5%, and there is little eating activity between the morning and afternoon ingestion peaks. The Nepalese monkeys, in contrast, exhibit more continuous foraging throughout the day; an average of 15.7 ± 2.57% across the ten 1-hr time periods. On Cayo Santiago resting significantly correlates with grooming ($r_s = 0.6303$; $P<.05$) with an increase from 1230 to 1530, whereas play behavior negatively correlates with resting ($r_s = -0.6970$; $P<.05$). Monkeys on Cayo Santiago are highly synchronized in their movement patterns, correlating with increased vigilance ($r_s = 0.8424$; $P<.01$).

Table VI presents the distribution of food types utilized by the two populations on the basis of percent total ingestion activity. At any given time it is equally likely that monkeys will be eating vegetative matter or commercial diet on Cayo Santiago. Rhesus in Nepal concentrate on natural vegetation. Geophagy (soil eating) exists in both locations, and bouts of this

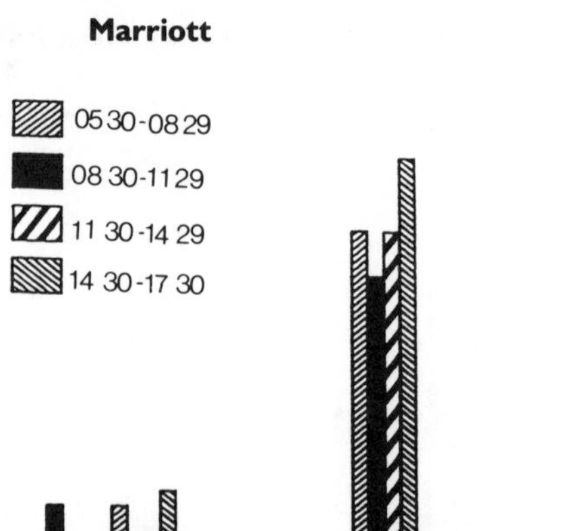

Fig. 3. Diurnal rhythm of activities for the Gaushalla troop in Nepal.

TABLE VI. Distribution of Food Types Consumed by the Two Study Groups on the Basis of Percent Total Eating Time

Food type	Location	
	Cayo Santiago	Nepal
Vegetation	46.65	68.95
Soil	0.09	0.00
Insects	0.15	0.25
Commercial/cultivated	49.30	10.75
Other	0.93	15.81
Unknown	2.88	4.24

activity in Nepal are highly synchronized because they only occur when the troop moves to a soil site area. Geophagy is of such low frequency in Nepal that the representation using scan sampling averaged less than 0.1%. On Cayo Santiago the soil sites are located near the feeding corral and in several other highly visible locations on the small cay, eliminating the need for individuals specifically to travel to encounter edible soils. The overall frequency of geophagy is higher on Cayo Santiago, and it temporally correlates with consumption of commercial diet [Sultana and Marriott, 1982].

Insectivory is not well represented by the scan sampling technique since it

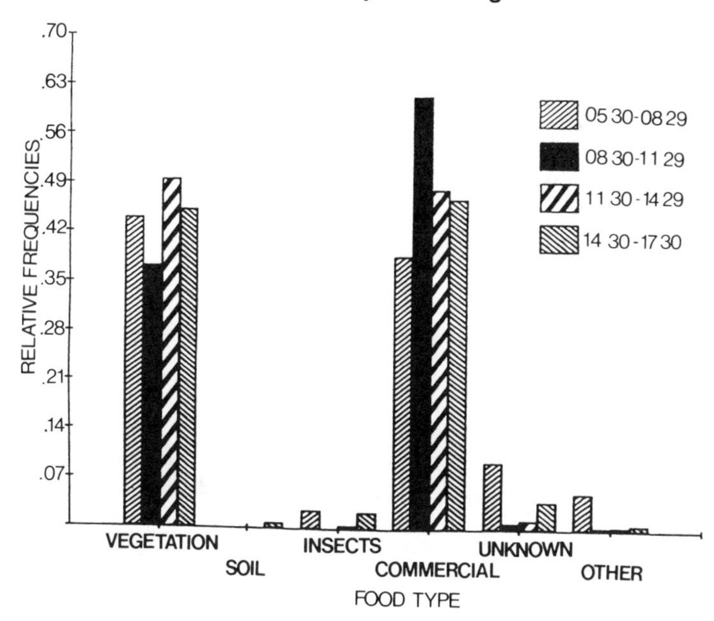

Fig. 4. Diurnal selection of food types by group L on Cayo Santiago.

usually involves solitary stalking behavior by the target individual. Typically the target animal moves away from the group, goes into denser vegetation and carefully visually inspects leaf surfaces. At this time the animal is less tolerant of observation and moves in a slow, stealthy manner. We have observed this behavior in both populations and termed it "insect-foraging mode." When scoring insect feeding in an activity scan, an observer clearly could discern that the target animal was ingesting an insect. These estimates of insectivory, which is more an event than a state, must be considered dependent upon the seasonal fluctuations in insect availability. The percentages presented here are considered underestimates of the contribution of insects to total food intake of rhesus monkeys in both locations.

Figures 4 and 5, respectively, depict the diurnal trends in food type selection for the two populations. On Cayo Santiago, commercial diet intake increases in late morning, corresponding with distribution of fresh commercial diet to the food hoppers. Selection of vegetative parts shows no related pattern with commercial diet consumption across the ten 1-hr time blocks. In Nepal, vegetative feeding peaks in late morning. This increase reflects a dramatic rise in fruit consumption from 0830 to 1130. A comparison of fruit versus other vegetative intake for the two populations is shown in Figure 6. Note the similarity between the two populations in the timing of their

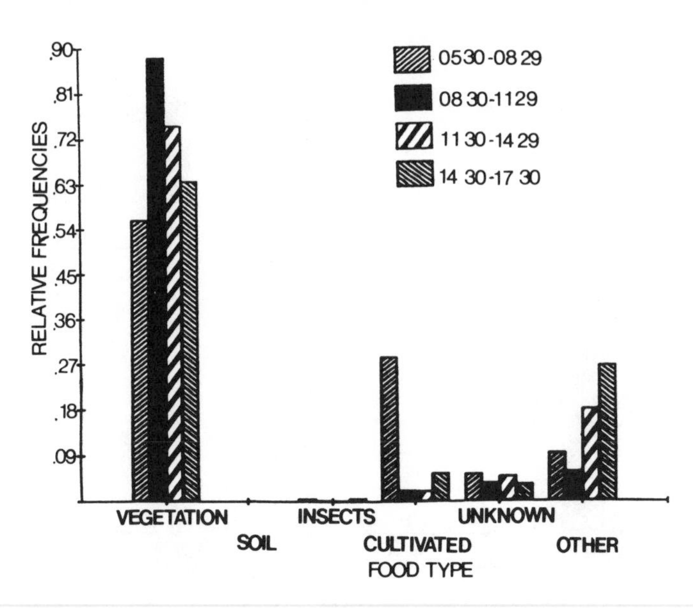

Fig. 5. Diurnal selection of food types by the Gaushalla troop in Nepal.

consumption of vegetative parts: while the pattern of fruit ingestion differs, ingestion of other vegetation (over 90% of which represents young leaves) approximately doubled for both populations during the last time period of the day: N, 4.4% to 7.2%; CS, 1.0% to 4.0%.

DISCUSSION

Comparison of activity budgets across populations of the same species often presents methodological difficulties, including differences in definition of activities and type of sampling [cf. Marsh, 1981]. In this study, methodology was constant and significant differences between the activity budgets of the two populations were demonstrated. Since age/sex class distribution within the two populations was not shown to affect the overall activities, these differences can be examined in terms of environmental variables.

An important difference between the two populations is that the monkeys on Cayo Santiago are provisioned daily. This provisioning removes the necessity of locating, preparing, and ingesting many needed nutrients and energy from the natural vegetation. In terms of the overall time budgets of the two populations, the major activities are exhibited in roughly the same proportions, with over half of the time being accounted for by locomotion

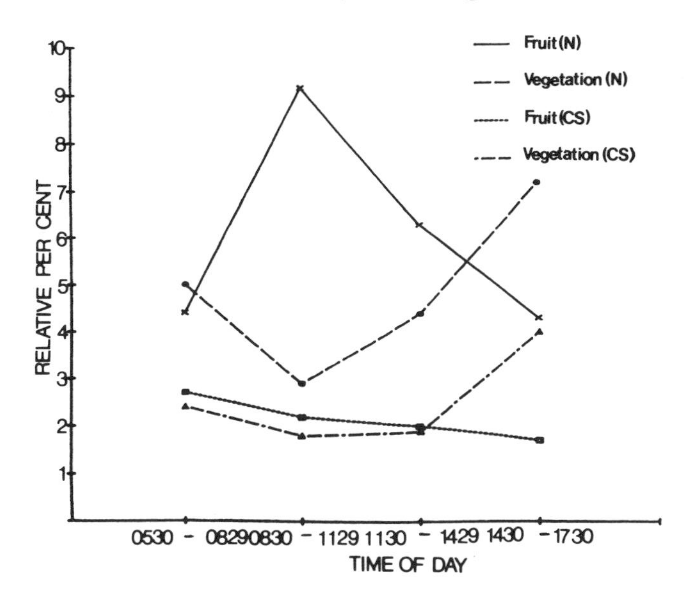

Fig. 6. Diurnal fluctuation in selection of fruit and vegetation by group L and the Gaushalla troop.

and vigilance. The significant differences in activities can logically be interpreted in terms of the provisioning variable. As one might expect, the provisioned animals on Cayo Santiago spent less time in locomotion and ingestion while exhibiting more resting, playing, and drinking time than the wild monkeys.

It should be noted that unlike some comparative studies of captive and wild monkeys [cf. Rowell, 1967; Gartlan, 1968; Southwick, 1969], in this study the provisioned rhesus monkeys of Cayo Santiago exhibited fewer aggressive encounters than the wild monkeys in Nepal. Since the data samples were matched within the yearly reproductive cycle, the twofold difference in aggression may be attributable to higher levels of intra- or intertroop aggression in the wild monkeys. Separate studies in both populations have indicated an extremely low level of aggression related to food acquisition and/or ingestion [Roemer and Marriott, 1982; Marriott and Lowney, in prep.]; therefore, the difference in time spent in food acquisition patterns does not account for the difference in aggression. Similarly, the results seen here are contrary to studies that have correlated increased density with increased aggression [Southwick, 1967; Alexander and Roth, 1971; Elton and Anderson, 1977] since the density of group L on the small cay is 4.8 times as great as that of the animals in Gaushalla Forest and the aggression levels are reversed. These results also do not support the idea that provisioning may increase intragroup aggression [cf. Southwick et al., 1976]. Caution must be exercised, however, when interpreting any scan-sampling-based data on aggression because activities such as threat, attack, chase, and fight are very brief behaviors that resemble events rather than states.

The dryness of the commercial pellets and reduced intake of fluid from a diet that includes fewer fresh vegetative parts can readily account for the significantly higher frequency of drinking by the Cayo Santiago monkeys. Additionally, the monkeys have often been seen carrying the commercial pellets to the drinking fountains and soaking them as they eat. By contrast, drinking is relatively rare in the wild monkeys. Water is obtained primarily from the high moisture content of the consumed vegetation and through licking dew and rain from leaf surfaces.

Cayo Santiago monkeys rest more than twice as much as the Nepal troop; however, the relationship of social resting to individual resting is the same in both populations. During the adult rest periods of both rhesus monkey populations it is typical to find the infants and juveniles playing. The increased resting is readily correlated temporally with the increase in playing on Cayo Santiago. These results corroborate those of Loy [1970], who found a decrease in play and resting behavior with an increase in foraging during a temporary halt in provisioning on Cayo Santiago. From Loy's work one would therefore expect less play and resting in nonprovisioned troops.

Several reports have described decreased frequencies of foraging in intraspecific comparisons where one population inhabited a more marginal habitat [Oates, 1977; Marsh, 1981], providing an alternate basis for predicting the incidence of foraging. Estimates of food abundance were done [Roemer and Marriott, 1982; Marriott and Lowney, in prep.] at both study sites, but these results cannot be compared with intraspecific comparisons, where both populations are not provisioned since the major volume consumed by the rhesus monkeys on Cayo Santiago is commercial diet—a more readily bioavailable dietary source than most vegetative matter. The difference in overall time spent eating between the two populations most likely reflects a combination of 1) the rate of digestion of the major dietary food items; 2) the quality/bioavailability of nutrients per food source; 3) abundance and type of plant secondary compounds in foods; and 4) general energy intake/expenditure requirements at both locations.

Although group L's core area is very small, the group moves in a habitual pathway around the island similar to their wild counterparts. Both populations can predictably be found by observers in a specific location in their home range based upon the time of day and the season. Location maps that were plotted concurrently with the activity scans for both populations indicated clear diurnal movement patterns that were temporarily synchronized on a daily basis [Marriott and Sultana, in prep.]. Although the commercial diet was readily available to group L throughout the day, the animals did not remain in the corral feeding area but moved through the corral as a group twice daily as part of their habitual day range.

In addition to differences in overall time budgets provisioning appears to affect the synchrony of activities within the troop individuals. There are larger diurnal shifts in the major activities exhibited by the Cayo Santiago monkeys when compared with the Gaushalla troop. More individuals in group L are engaged in the same behavior at any given time than in the Gaushalla troop. Ready access to food in a large feeding arena results in group L monkeys' eating, moving, resting, and socializing more as a unit than the Gaushalla troop, whose individuals continually forage whenever the opportunity arises.

The warmer, wetter climate during the observations on Cayo Santiago as compared to Nepal could also contribute to both the increase in overall resting behavior of group L and hence increased group synchronous behavior. In particular, the heightened temperature could lead to longer daily rests, with peaks in eating as have been described for rhesus monkeys in India during the hotter months of the year [Lindburg, 1971, 1977] and in captivity [Bernstein and Mason, 1963]. It is less likely that the level of rainfall was an important variable [cf. Harrison, 1985]. Rhesus monkeys in both populations generally continue to eat, move, and interact in all but the heaviest downpours, and the total number of rainy days in both sites was approximately the same.

Both on Cayo Santiago and in Nepal there was a significant increase in eating during the early morning and late afternoon. Data were carefully checked to avoid masking other diurnal eating peaks before the daily data sets were combined [Chalmers, 1968]. Similar bimodal peaks in eating have been described for a number of nonhuman primate species [cf. Clutton-Brock, 1977]. Bimodal peaks in the daily time budgets of food intake have been reported for rhesus monkeys in natural habitats [Lindburg, 1977; Siddiqi and Southwick, 1977; Seth and Seth, 1985] and in enclosures [Post and Baulu, 1978]. Some authors have suggested that bimodal peaks in feeding are a result of daily or seasonal temperature variation which leads to a lengthened midday rest period during hotter, dryer weather, with increased time spent in food ingestion during the cooler early-morning and late-afternoon hours [cf. Chivers, 1974]. The studies that have investigated rhesus monkeys food intake have commented on or included data that described a consistent seasonal bimodal food intake pattern. The exact timing of the peaks within the pattern varied somewhat with the seasons [cf. Lindburg, 1977]; however, across all seasons and in a wide variety of forest types the bimodal distribution has been evidenced [cf. Seth and Seth, 1985]. The fact that this pattern has also been reported for provisioned monkeys in India [Siddiqi and Southwick, 1977], monkeys on Cayo Santiago, and in enclosures [Post and Baulu, 1978] lends support to the hypothesis that the functional significance of this type of diurnal rhythm is due to physiological digestive/energy requirements rather than climatic variables [Clutton-Brock, 1974]. In the morning the animals eat to obtain energy for the day, then rest or participate in less active behaviors while the food is digested. The late-afternoon peak in eating provides energy for the longest inactive period of the 24-hr day—the night rest. This is also the time when more energy may be required for thermoregulation. Increased attempts to standardize methodology of data collection and comparative studies [Marsh, 1981; Oates, 1977] may lead to the conclusion that a diurnal bimodal peak in food intake is a more universal primate behavior pattern.

Of particular interest is the pattern exhibited by both populations in the diurnal rhythm of vegetative (primarily young leaf) ingestion. Ingestion of leaf matter doubled during the last time periods of the day in both locations. Diurnal variation in food item selection has been reported for a number of nonhuman primates species in natural habitats [cf. Clutton-Brock, 1977; Raemaekers, 1978]. This behavior may be beneficial because leaves are digested more slowly than fruit and therefore will provide for a slow energy release throughout the long nighttime period of inactivity [Clutton-Brock, 1977; Raemaekers, 1978]. An additional consideration is that the sugar content of leaves is higher later in the day, and for some species the level of secondary compounds may also be lower at that time [Raemaekers, 1978]. Young leaves are selected over mature leaves by most nonhuman primate species owing to lower content of tannins, higher digestibility, and higher nutrient content—particularly proteins [McKey et al., 1978, 1981; Milton, 1979; Waterman and Choo, 1981]. It is noteworthy that the Cayo Santiago rhesus monkeys, despite provisioning with commercial diet, continue to exhibit a late-day peak in young leaf ingestion that parallels their wild counterparts.

An increase in fruit eating in the late morning in the Gaushalla troop was observed, but it cannot readily be attributed to the need for a rapid energy source as described for several nonhuman primate species [Clutton-Brock, 1977; Raemaekers, 1978]. The most likely explanation is that during the late morning in October 1979, the habitual day route of the Gaushalla troop led through an area containing trees with fruit. There was no evidence of rapid movement to a particular fruiting source for the initial daily eating bout, but rather rapid movement to a densely forested part of the day range that contained fruits, flowers, and many species of leafy or flowering ground covers. Diversity of food types consumed was hence greatest in the early morning.

TABLE VII. A Comparison of Group Time Budgets by Percentages of Time Spent in Each Activity for Rhesus Monkeys in Different Habitat Types

	Habitat type[a]				
Activity	Enclosed	Town	Temple	Island	Forest
Look	—	31.7	23	30.0	22.0
Drink	—	1.2	—	2.5	0.2
Rest	39	2.9	1	12.9	5.0
Groom	24	12.2	8	12.3	9.8
Eat	—	15.8	36	10.8	15.6
Play	—	4.8	1	5.3	3.7
Agonism	2	1.8	2	0.8	1.6
Locomotion	10	29.6	28	23.1	39.4
Other	—	—	—	2.3	2.7
Density[b]	100,000+	18,000	1,967	887	120

[a]Source from left to right: Bernstein and Mason [1963]; Oppenheimer et al. (unpubl.); Teas et al. [1975]; Cayo Santiago, this study; Nepal, this study.
[b]Approximate densities based on an extrapolated estimate of No. of animals per km^2.

Comparison with Other Studies

Table VII presents comparative data on time budgets of rhesus monkeys from the literature with the present study. Only time budgets generated using instantaneous scan sampling were included; however, the time between scans and behavioral definitions varied somewhat. Approximate densities on the basis of number of animals per km^2 are included at the bottom of the table. As can be seen from this table, the monkeys in the enclosed area studied by Bernstein and Mason [1963] were the least active, whereas the temple monkeys spent the most time eating. Locomotion, vigilance, and feeding composed the principal activities for the free-ranging monkeys in all locations, but the percentage and rank orders of percentages within a study varied across habitats. Animals in all locations spent little time in aggressive encounters. There are no evident relationships among the distribution of the behaviors in the time budgets and the approximate densities. In general, the variation that exists in these overall time budgets can be explained in terms of habitat type and food source. The troops with the least predictable food source spent the largest amount of their time engaged in eating or food acquisition and least time resting. The major conclusion from this comparison is that the provisioned monkeys of Cayo Santiago exhibit overall time budgets that are more similar to those of provisioned wild monkeys than to rhesus monkeys in enclosures.

CONCLUSIONS

Provisioning of the rhesus monkeys on Cayo Santiago can be viewed as a contributing factor to significant differences exhibited in daily time budgets when compared with a seasonally matched sample of activity patterns for a wild monkey troop. It is important, however, that climatic factors and, particularly, differences in temperature be considered. The major differences in time budgets recorded between the two populations were that the Cayo Santiago monkeys spent significantly more time resting and playing, and less time eating, moving, and engaged in aggressive encounters than did their wild counterparts. The behaviors of group L on Cayo Santiago were more internally synchronized, with significantly more group members exhibiting the same behaviors at any given time of day than in the Nepal troop. Of particular interest was that both populations evidenced bimodal eating peaks in the early morning and late afternoon and an increase in young leaf consumption in the late afternoon. These data on bimodal food intake parallel data reported for other nonhuman primate species and rhesus monkeys in natural habitats. Together these eating patterns can be explained in terms of energy intake needs and the requirement for digestive functioning. Specifically, the late-day peak in leaf consumption can be attributed to the longer time course for digestion of leaf matter when compared with other plant parts and the subsequent slow release of energy during the night rest period. Through comparison with other reports of time budgets for rhesus monkeys, there is no evidence for an overall species-specific time budgeting pattern; rather, there is evidence for daily time distributions that reflect the variables of the particular habitat and troop.

The continued diurnal selection of young leaves as a late-afternoon food source and habitual movement through a day range illustrates two perhaps functionally basic behavior patterns that have not been affected by provisioning. In general these results indicate that the containment and provisioning of omnivores such as the rhesus monkey in seminaturalistic habitats at high population densities may lead to statistically significant but logically different changes in overall time distributions for the animals. It is important that these differences be taken into consideration when studies of monkeys in captive/enclosed conditions and on island colonies are interpreted.

ACKNOWLEDGMENTS

I am indebted to Dr. S.B. Malla, Director General, Department of Medicinal Plants, His Majesty's Government, Nepal, and the staff of the National Herbarium, Nepal, for their help and support through all phases of my work, and to EARTHWATCH and The Center for Field Studies for financial support in Nepal. Through the EARTHWATCH program, 12 people volunteered their time in 1979 to assist with data collection. I am grateful for their time and dedication. The work in Puerto Rico was supported by a John Jay Hopkins Fund Award 0047.26.2210 to B. Marriott. I wish to thank especially Carmen Sultana, who assisted with collection and preliminary analysis of the data from Cayo Santiago, and Robert Marriott for general logistical support during all phases of my work. I appreciate the support of former Scientist-in-Charge of Cayo Santiago, Richard Rawlins, and the staff of the Caribbean Primate Research Center. Cayo Santiago was supported by NIH contracts RR-7-2115 and grant RR-01293 to the University of Puerto Rico during this study. A special thank-you goes to Dr. Jane Teas, who first introduced me to the monkeys of the Kathmandu Valley, and to Dr. J.R. Oppenheimer for the use of his unpublished data and encouragement over the years.

REFERENCES

Alexander BK, Roth FM (1971): The effects of acute crowding on aggressive behavior of Japanese monkeys. Behaviour 39:73–89.

Altmann J (1974): Observational study of behavior: sampling methods. Behaviour 49:227–243.

Baulu J, Redmond DE (1980): Social and nonsocial behaviours of sex- and age-matched enclosed and free-ranging rhesus monkeys (*Macaca mulatta*). Folia Primatol (Basel) 34:239–258.

Bernstein IS (1970a): Activity patterns in pigtail monkey groups. Folia Primatol (Basel) 12:187–198.

Bernstein IS (1970b): Daily activity cycles and weather influences on a pigtail monkey group. Folia Primatol (Basel) 18:390–415.

Bernstein IS (1975): Activity patterns in a gelada monkey group. Folia Primatol (Basel) 23:50–71.

Bernstein IS, Mason WA (1963): Activity patterns of rhesus monkeys in a social group. Anim Behav 11(4):455–460.

Carpenter CR (1935): Behavior of red spider monkeys in Panama. J Mammal 16:171–180.

Chalmers NR (1968): Group composition, ecology and daily activities of free living mangabeys in Uganda. Folia Primatol (Basel) 8:247–262.

Chivers DJ (1974): "The Siamang in Malaya: A Field Study of a Primate in a Tropical Rain Forest." Basel: Karger, Contrib Primatol, Vol 4, pp 1–335.

Chivers DJ (1975): Daily patterns of ranging and feeding in siamang. In Konds S, Kawai M, Ehara A (eds): "Contemporary Primatology." Basel: Karger, pp 362–372.

Clutton-Brock TH (1974): Activity patterns of red colobus (*Colobus badius*). Folia Primatol (Basel) 21:161–187.

Clutton-Brock TH (1977): Some aspects of intraspecific variation in feeding and ranging behaviour in primates. In Clutton-Brock TH (ed): "Primate Ecology: Studies of Feeding and Ranging Behaviour in Lemurs, Monkeys and Apes." New York: Academic Press, Ch 18, pp 539–556.

Drickamer LC (1973): Semi-natural and enclosed groups of *Macaca mulatta:* A behavioral comparison. Am J Phys Anthropol 39(2):249–254.

Elton RH, Anderson BV (1977): The social behavior of a group of baboons (*Papio anubis*) under artificial crowding. Primates 18:225–234.

Erwin N, Erwin J (1976): Social density and aggression in captive groups of pigtail monkeys (*Macaca nemestrina*). Appl Anim Ethol 2:265–269.

Gartlan JS (1968): Stucture and function in a primate society. Folia Primatol (Basel) 8:89–120.

Hall KRL (1965): Behaviour and ecology of the wild patas monkey, *Erythrocebus patas*, in Uganda. J Zool 148:15–87.

Harrison MJS (1985): Time budget of the green monkey *Cercopithecus sabaeus:* Some optimal strategies. Int J Primatol 6(4):351–376.

Jolly A (1972): "The Evolution of Primate Behaviour." New York: Macmillan.

Kummer H (1971): "Primate Societies." Chicago: Aldine-Atherton.

Lehner PN (1979): "Handbook of Ethological Methods." New York: Garland STPM Press.

Lindburg DG (1971): The rhesus monkey in North India: an ecological and behavioral study. In Rosenblum LA (ed): "Primate Behavior: Developments in Field and Laboratory Research," Vol. 2. New York: Academic Press, pp 1–106.

Lindburg DG (1977): Feeding behavior and diet of rhesus monkeys (*Macaca mulatta*) in a Siwalik forest in North India. In Clutton-Brock TH (ed): "Primate Ecology: Studies of Feeding and Ranging Behaviour in Lemurs, Monkeys and Apes." New York: Academic Press, Ch 8, pp 223–249.

Loy J (1970): Behavioral responses of free-ranging rhesus monkeys to food shortage. Am J Phys Anthropol 33:262–272.

Malla SB (1967): "Notes on Flora of Rajnikunj (Gokarna Forest)." Kathmandu, Nepal: His Majesty's Government, Ministry of Forests, Bulletin No. 1.

Marriott BM, Lowney PA (in prep.): Seasonal variation in food and nutrient intake in a forest dwelling troop of rhesus monkeys (*Macaca mulatta*) in Nepal.

Marriott BM, Sultana C (in prep.): Age and sex differences in activity patterns of free-ranging rhesus monkeys (*Macaca mulatta*) on Cayo Santiago.

Marsh C (1981): Time budget of Tana River red colobus. Folia Primatol (Basel) 35:30–50.

Maxim PE, Bowden DM, Sackett GP (1976): Ultradian rhythms of solitary and social behavior in rhesus monkeys. Physiol Behav 17:337–344.

McKey DB, Waterman PG, Mbi CN, Gartlan JS, Struhsaker TT (1978): Phenolic content of vegetation in two African rain forests: ecological implications. Science 202:61–64.

McKey DB, Gartlan JS, Waterman PG and Choo GM (1981): Food selection by black colobus monkeys (*Colobus satanus*) in relation to plant chemistry. Biol J Linn Soc 16:115–146.

Milton K (1979): Factors influencing leaf choice by howler monkeys. A test of some hypotheses of food selection by generalist herbivores. Am Nat 114:362–378.

Neville MK (1968): The ecology and activity of Himalayan foothill rhesus monkeys. Ecology 49:110–123.

Nieuwenhuijsen K, de Waal FBM (1982): Effects of spatial crowding on social behavior in a chimpanzee colony. Zoo Biol 1:5–28.

Oates JF (1977): The guereza and its food. In Clutton-Brock TH (ed): "Primate Ecology: Studies of Feeding and Ranging Behaviour in Lemurs, Monkeys and Apes." New York: Academic Press, Ch 10, pp 275–321.

Oppenheimer JR, Akondo AW, Husain KZ (unpubl. man.): Urban rhesus monkeys (*Macaca mulatta*): Activity patterns and agonistic interactions.

Post W, Baulu J (1978): Time budgets of *Macaca mulatta*. Primates 19:125–140.

Raemaekers J (1978): Changes through the day in the food choice of wild gibbons. Folia Primatol (Basel) 30:194–205.

Richard A (1977): The feeding behaviour of *Propithecus verreauxi*. In Clutton-Brock TH (ed): "Primate Ecology: Studies of Feeding and Ranging Behaviour in Lemurs, Monkeys and Apes." New York: Academic Press, Ch 3, pp 72–96.

Roemer J, Marriott BM (1982): Feeding patterns of rhesus monkeys (*Macaca mulatta*) on Cayo Santiago Island, Puerto Rico. Int J Primatol 3(3):327.

Rowell TE (1967): A quantitative comparison of the behaviour of a wild and caged baboon group. Anim Behav 15:499–505.
Seth PK, Seth S (1985): Ecology and feeding behaviour of the free-ranging rhesus monkeys in India. Indian Anthropol 15(1):51–62.
Shrestha PK (1982): "A Ten Year Report of Metereological Service of His Majesty's Government, Nepal: The Kathmandu Valley." Kathmandu, Nepal: His Majesty's Government Press.
Siddiqi MF, Southwick CH (1977): Feeding behaviour of rhesus monkeys (*Macaca mulatta*) in the North Indian plains. Paper presented at the Annual Anim Behav Soc Meeting, Penn State University.
Siegel S (1956): "Nonparametric Statistics for the Behavioral Sciences." New York: McGraw-Hill Book Company, Inc.
Sokal RR, Rohlf FJ (1981): "Biometry. The Principles and Practice of Statistics in Biological Research." San Francisco: WH Freeman and Co.
Southwick CH (1967): An experimental study of intragroup agonistic behavior in rhesus monkeys (*Macaca mulatta*). Behaviour 28:182–209.
Southwick CH (1969): Aggressive behaviour of rhesus monkeys in natural and captive groups. In Garattini S, Sigg EB (eds): "Aggressive Behavior." Amsterdam: Excerpta Medica, pp 32–43.
Southwick CH (1972): Aggression among nonhuman primates. In "Module in Anthropology," No. 23, Reading, Massachusetts: Addison-Wesley, pp 1–23.
Southwick CH, Beg MA, Siddiqi MF (1965): Rhesus monkeys in northern India. In DeVore I (ed): "Primate Behavior. Field Studies of Monkeys and Apes." New York: Holt, Rinehart and Winston, pp 111–160.
Southwick CH, Siddiqi MF, Farooqui MY, Pal BC (1976): Effects of artificial feeding on aggressive behaviour of rhesus monkeys in India. Anim Behav 24:11–15.
Southwick CH, Teas J, Richie T, Taylor H (1982): Ecology and behavior of rhesus monkeys (*Macaca mulatta*) in Nepal. Nat Geog Soc Res Repts 14:619–630.
Struhsaker TT (1975): "The Red Colobus Monkey." Chicago: University of Chicago Press.
Sultana C, Marriott BM (1982): Geophagia and related behavior of rhesus monkeys (*Macaca mulatta*) on Cayo Santiago Island, Puerto Rico. Int J Primatol 3(3):338.
Sussman RW (1977): Feeding behaviour of *Lemur catta* and *Lemur fulvus*. In Clutton-Brock (ed): "Primate Ecology: Studies of Feeding and Ranging Behaviour in Lemurs, Monkeys and Apes." New York: Academic Press, pp 1–36.
Teas J (1983): Ecological considerations important in the interpretation of census data on free-ranging rhesus monkeys in Nepal. In Seth PK (ed): "Perspectives in Primate Biology." New Delhi, India: Today & Tomorrow's Printers and Publishers, pp 211–225.
Teas J, Taylor H, Richie T, Southwick CH (1975): Nepal rhesus studies: Ecological influences on the behavior of rhesus monkeys in Kathmandu Valley. Progress report to the National Geographic Society, unpubl.
Teas J, Richie TL, Taylor HG, Siddiqi MF, Southwick CH (1981): Natural regulation of rhesus monkey populations in Kathmandu, Nepal. Folia Primatol (Basel) 35:117–123.
Teas J, Feldman HA, Richie TL, Taylor HG, Southwick CH (1982): Aggressive behavior in the free-ranging rhesus monkeys of Kathmandu, Nepal. Aggressive Behav 8:63–77.
Waterman PG, Choo GM (1981): The effects of digestibility-reducing compounds in leaves on feed selection by some Colobinae. Malays Appl Biol 10(2):147–162.
Wilson CC (1972): Spatial factors and the behavior of nonhuman primates. Folia Primatol (Basel) 18(3 to 4):256–275.
Wrangham RW (1977): Feeding behaviour of chimpanzees in Gombe National Park, Tanzania. In Clutton-Brock TH (ed): "Primate Ecology: Studies of Feeding and Ranging Behaviour in Lemurs, Monkeys and Apes." New York: Academic Press, Ch 17, pp 504–538.

SECTION II: DEMOGRAPHY AND LIFE HISTORY PATTERNS

Ecology and Behavior of Food-Enhanced Primate Groups, pages 153–166

8

Effects of Supplementary Feeding on Maturation and Fertility in Primate Groups

James Loy

Department of Sociology and Anthropology, University of Rhode Island, Kingston, Rhode Island, 02881

INTRODUCTION

Growth and reproduction are two major areas of primate research and are the subject of scores of publications annually. Many studies utilize nonhuman primates as substitutes or "models" for humans in order to gain insight into human behavioral or biomedical problems. Other studies are designed to provide information about primate evolution. Regardless of an investigator's objectives, an understanding of the variables that influence the behavior and biology of the research population is an important prerequisite to success.

Nutrition is obviously a factor that can significantly affect both growth and reproduction. For example, among several human populations, age at menarche has declined over the past 100 years, while height at all ages has shown an increase. Improved nutrition is usually cited as the probable cause for these trends [Harrison et al., 1977]. Nonhuman primates living on a natural diet supplemented by food from humans or, in the most extreme case, subsisting entirely on provisioned food, might be expected to grow faster and larger, and begin reproduction earlier than unsupplemented animals. Such differences between "fed" and "unfed" animals may not be obstacles to research; indeed, they may be positive elements for some studies. Helpful or harmful, however, the behavioral and biological effects of supplementary feeding must be recognized and dealt with by investigators during the selection of populations for study and the interpretation of research results.

This chapter reviews the effects of supplementary feeding on maturation and reproduction in primates. It emphasizes data on macaques, because populations of these monkeys have been provisioned for decades in Japan and on the Caribbean island of Cayo Santiago, and have lived as commensals of humans in Asia for hundreds of years. The best information on provisioning effects is from studies of Japanese macaques *(Macaca fuscata)* and rhesus monkeys *(Macaca mulatta)*. Japanese monkey groups were first intentionally provisioned by scientists in 1952 at Koshima and 1953 at Takasakiyama [Sugiyama, 1965]. Rhesus monkeys were established as provisioned populations on the Puerto Rican islands of Cayo Santiago in 1938 [Carpenter 1942, 1972] and La Cueva and Guayacan (near the town of La Parguera) in 1961 [Vandenbergh, 1967]. Furthermore, in parts of Asia, and especially in India, many (perhaps most) groups of wild macaques are given some portion of their diet by humans [Southwick et al., 1976; Southwick and Siddiqi, 1985]. Primate species other than macaques are mentioned in the chapter, but most of the data have been derived from rhesus and Japanese monkeys.

As far as possible, only data from studies of wild, or at least free-ranging, animals have been used in order to minimize the effects of the numerous nonnutritional variables that can influence the maturation and fertility of captive animals [Sadleir, 1969]. In a few instances, however, instructive comparisons with captive groups have been made.

MATERIALS AND METHODS

In this study, maturation has been considered in terms of the onset of an animal's reproductive career, and age at first parturition used as the marker of female maturation. This measure is commonly reported in the literature, probably because it is relatively easy to observe in the field. Other indicators of female maturation, such as menarche, the initial development of the sex skin, or the start of the adolescent growth spurt, were not chosen because they are mentioned in only a few studies of wild primates.

Sexual maturation is more difficult to detect in males than in females, and there seems to be no consensus as to which indicators are most useful. The beginnings of spermatogenesis and intravaginal ejaculation were taken as the primary markers of male maturation; other measures, such as age at testicular descent and the onset of adult testosterone production, were used in a supplementary fashion.

Two measures of fertility were employed: birth rate and interbirth interval. Birth rate is defined as the percentage of reproductively adult (i.e., postmenarchial) females who produce live young during a specified time period (usually the annual birth season because most data came from seasonally breeding macaques). Interbirth interval is the period from one birth to the next in the same female.

Like most reviews, this one suffers from a degree of incomparability among the relevant reports. The cited studies vary greatly in length; some lasted only a few months, others a year, and others covered several years. Some studies concentrated on a single primate group, while others focused on several groups at once. In some cases, ages for the animals were known

TABLE I. Age at First Birth in Japanese Monkeys

Group/population	Mean age at first birth (yr)	Source
A. Substantially provisioned groups[a]		
Koshima	6.2	Mori [1979]
Arashiyama B	5.7	Takahata [1980]
Mt. Ryozen	5.2	Sugiyama and Ohsawa [1982]
Ohirayama	5.1	Nigi [1976]
Okinoshima	4.6	Nomura et al. [1972]
Miyajima	5.0	Nigi [1976]
B. Unprovisioned (or only lightly provisioned) groups[b]		
Mt. Ryozen	6.7	Sugiyama and Ohsawa [1982]
Koshima[c]	6.1	Mori [1979]
Takagoyama I	5.3	Hiraiwa [1981]

[a]Mean of group means = 5.2 years.
[b]Mean of group means = 6.0 years.
[c]Koshima value is an average of the data from 1952 to 1963 and 1972 to 1977.

with precision; in others, ages were estimated from body size and other anatomical features. Most importantly, many reports were unclear on the level of supplementary feeding experienced by the animals. When such information was given, it was usually the frequency of feeding rather than the amount of food provided. Therefore, for comparisons in this study, a rough classification scheme based on the frequency of supplementary feeding was constructed. "Substantially provisioned" groups or populations were given food at least every few days. "Lightly provisioned" groups were fed less than once a week. "Unprovisioned" groups received no supplementary food.

Despite the uneven nature of the data, there was sufficient information to compare groups receiving different levels of provisioning on a few measures of maturation and fertility. All comparisons of feeding effects were intraspecific in order to avoid the complicating factor of species differences in biology and behavior. When there were adequate samples, differences between provisioning categories were tested for significance using nonparametric statistical procedures. For all statistical tests, the alpha level was 0.05.

RESULTS

Female Maturation: Age at First Birth

Information on age at first birth was found for six substantially provisioned groups of Japanese monkeys and three unprovisioned or lightly provisioned groups (Table I). The overall mean age of females from substantially provisioned groups at the birth of their first infant was 5.2 years (range of

4.6–6.2 years), while that of females from unfed or lightly fed groups was 6.0 years (range of 5.3–6.7 years). The difference in mean ages did not achieve statistical significance (Mann-Whitney U-test, $P = 0.083$).

Despite the nonsignificant statistical difference, the conclusion that provisioning is unrelated to age at first birth among female Japanese macaques seems questionable for two reasons. First, the probability value of 0.083 is a "near miss" based on the analysis of small samples. A slight change in the Table I samples could move the "*P-value*" to the alpha level. Second, there are two studies of Japanese monkeys in which the animals were observed during both a period of substantial provisioning and a period with little or no supplementary feeding. Both groups showed feeding-related shifts in age at first birth. At Mt. Ryozen in central Japan, the monkeys of Ryozen group A were fed from 1965 to 1973. The mean female age at first birth during provisioning was 5.2 years. In contrast, the mean age at first birth in a subsequent period without provisioning at Mt. Ryozen (1974–1980) rose to 6.7 years [Sugiyama and Ohsawa, 1982]. Similarly, during substantial provisioning at Koshima in southern Japan (1964–1971), mean age at first birth was 6.2 years. This figure rose to 6.8 years during a later period (1972–1977) with very restricted supplementary feeding [Mori, 1979]. It should be noted that the lowest mean age at first birth at Koshima (mean age of 5.3 years) was from the 1952–1963 period of extremely light feeding—less than once per week—that preceded the start of intensive provisioning [Mori, 1979]. The validity of this low figure is questionable. Jolly [1985] suggested that it may be attributable to the investigators' knowing only the youngest females' ages at the start of the Koshima project. Another contributing factor could be inaccurate age estimations at this early stage in the study of Japanese monkeys [see Altmann et al., 1981, for a discussion of shifting age estimations as a function of observers' increased field experience].

There is very little information concerning the effects of provisioning on age at first birth among rhesus monkeys. This is doubly unfortunate because such information might have helped interpret the Japanese monkey data. Comparative data for rhesus monkeys are scanty mainly because of a lack of longitudinal studies of maturation and reproduction in the animals' natural habitat. Excellent information exists for the fully provisioned colonies at Cayo Santiago and La Parguera, where the mean age at first birth is about 4 years [Cayo Santiago: Korford, 1965, 1966; M. J. Kessler pers. comm.; La Parguera: Vandenbergh, 1973; Drickamer, 1974]. At both Puerto Rican colonies a few 2-year-old females conceive each year [Koford, 1965, 1966; Drickamer, 1974], an occurrence not reported for the wild. Furthermore, females from mid-ranking and high-ranking matrilineal genealogies tend to begin reproduction at an earlier age than females from low-ranking genealogies [Drickamer, 1974; Sade et al., 1976]. Although rank-related differences in nutrition might be suspected to be the basis of intergenealogy differences in age at first birth, Sade et al. [1976] doubted that this was the case on Cayo Santiago. Further research is needed on this point since it bears directly on the question of whether nutrition can affect reproduction even within substantially provisioned populations.

Rhesus monkeys housed in large, outdoor corrals at the Yerkes Primate Center Field Station typically give birth to their first infant between 3 and 4 years of age [Hadidian and Bernstein, 1979; Wilson et al., 1984], a figure in close agreement with the Puerto Rican data.

In summary, no conclusions can be drawn at present concerning feeding effects on female age at first birth in rhesus monkeys. Longitudinal studies of maturation and reproduction in unprovisioned rhesus groups are needed for comparative analysis. Supplementary feeding may lower the age at first birth among Japanese monkeys, although this possibility must be verified by additional data.

Male Maturation

For both Japanese macaques and rhesus monkeys, good data are available on age at male sexual maturity among substantially provisioned, free-ranging groups and captive animals. In the provisioned *M. fuscata* group of Arashiyama B, males appear to start ejaculating at about 4.5 years of age [Takahata, 1980]. The same general age at first ejaculation was reported for males in the provisioned Arashiyama West group [4.5–5.5 years: Wolfe, 1978] and the corral-housed group at the Oregon Primate Center [4.5 years: Hanby et al., 1971]. These findings also agree with the start of spermatogenesis in Japanese monkey males. In the provisioned Takasakiyama population, spermatogenesis and adult testosterone production start during the breeding season when males are four years old [Nigi et al., 1980]. In the laboratory, *M. fuscata* males show the first appearance of sperm in the seminiferous tubules at 3–4 years of age [Matsubayashi and Mochizuki, 1982].

Among the Cayo Santiago rhesus monkeys, males' testes descend at about 3.5 years of age, and these young males actively series-mount females [Loy, 1971]. Estimated age at first ejaculation among wild-caught juvenile males tested in the laboratory was 3–5 years [Michael and Wilson, 1973]. In an all-juvenile group corral-housed on La Cueva, 2.5-year-old males series-mounted females, but did not ejaculate [Loy and Loy, 1974]. Spermatogenesis begins at about 3.5 years of age on Cayo Santiago [Conaway and Sade, 1965], and at about the same time plasma testosterone levels begin to approach adult values in both corral-housed males [Rose et al., 1978] and laboratory-housed males [Resko, 1967]. A male 3.25 years of age is reported to have impregnated a female in the Yale University laboratory colony [Catchpole and van Wagenen, 1975].

These data suggest that Japanese monkey males mature at about 4.5 years of age, while rhesus males mature about a year earlier. Unfortunately, it is impossible at present to determine the effects of supplementary feeding on male maturation in either species due to a lack of data on unprovisioned groups. Future studies should correct this lack of information.

Fertility: Birth Rate

Considerable data are available concerning birth rates among both rhesus and Japanese macaques. Both species are seasonal breeders, and birth rates indicate the percentage of adult females producing young during a particular year's birth season. The birth season ranges from 3 to 6 months in length among Japanese monkeys [Kawai et al., 1967], and from 4 to 8 months in rhesus monkeys [Roonwal and Mohnot, 1977; Koford, 1966; Rawlins et al., 1984; Malik et al., 1984].

Birth rates from ten free-ranging groups of Japanese macaques are shown in Table II, with the groups classed according to their level of supplementary feeding. Groups that were not fed, or were only lightly fed, showed an average birth rate of 39.1% (range of 24.7–54.0%). In contrast, groups that received substantial provisioning showed an average birth rate of 59.3% (range of 51.3–63.3%). The difference between the two classes of groups was statistically significant (Mann-Whitney U-test, $P = 0.005$).

Birth rates in eight free-ranging groups or populations of rhesus monkeys are presented in Table III. Seven of the groups/populations were substantially provisioned, and for these animals, birth rates averaged 82% (range of 73–90.3%). The single population receiving light supplementary feeding (Kathmandu) showed a much lower birth rate of 63%. These data were not subjected to statistical testing because the sample of lightly provisioned groups was so small. Nonetheless, it is interesting that the available rhesus data are consistent with the trend observed in Japanese macaques—the rhesus group receiving the smallest amount of supplementary feeding had the lowest birth rate.

Fertility: Interbirth Interval

Very few reports of interbirth intervals among free-ranging macaques were found in the literature. This was rather surprising given the number of studies that have been done on reproduction in these monkeys. It appears, however, that interbirth interval is used as a measure of fertility much less often than birth rate, possibly because 2 or more years of observations are required for the interval measure, whereas 1 year of study will yield information, albeit limited, on birth rate.

TABLE II. Birth Rates in Japanese Monkeys

Group/population	Birth rate (%)	Source
A. Substantially provisioned groups[a]		
Arashiyama B	60.2	Takahata [1980]
Mt. Ryozen	59.3	Sugiyama and Ohsawa [1982]
Takasakiyama	63.3	Ohsawa et al. [1977][b]
Koshima[c]	62.5	Mori [1979]
Shiga Heights A	51.3	Suzuki et al. [1975][b]
B. Unprovisioned (or only lightly provisioned) groups[d]		
Shiga Heights B_2	35.2	Suzuki et al. [1975][e]
Kojiba	40.0	Takahata [1980]
Kawaradake[f]	52.1	Ikeda [1982]
Mt. Ryozen	33.6	Sugiyama and Ohsawa [1982]
Koshima[g]	34.0	Mori [1979]
Shiga Heights C	24.7	Suzuki et al. [1975][e]
Takagoyama I	54.0	Hiraiwa [1981]

[a]Mean of group means = 59.3%.
[b]As cited by Sugiyama and Ohsawa [1982].
[c]Birth rate for multiparous females five years old and older.
[d]Mean of group means = 39.1%.
[e]As cited by Sugiyama and Ohsawa [1982].
[f]Mean birthrate in unprovisioned group before capture.
[g]Average for multiparous females five and older from 1952 to 1963 and 1972 to 1977.

TABLE III. Birth Rates in Rhesus Monkeys

Group/population	Birth rate (%)	Source
A. Substantially provisioned groups[a]		
Morgan Island, SC[b]	74.5	D.M. Taub pers. comm.
La Parguera	73.0	Drickamer [1974]
Cayo Santiago	80.3	Rawlins et al. [1984]
Tughlaqabad	82.4	Malik et al. [1984]
Aligarh	83.3	Malik et al. [1984]
Chhatari	90.2	Southwick and Siddiqi [1976]
Forest Research Inst. (Dehra Dun)	90.3	Lindburg [1971]
B. Lightly provisioned groups/populations		
Kathmandu	63.0	Teas et al. [1980]

[a]Mean overall birth rate = 82.0%.
[b]MI value is the mean birth rate between 1981 and 1985 for "core" females 4 years old and older.

TABLE IV. Interbirth Intervals in Macaques

Group/population	Level of provisioning	Interbirth interval (yr)	Source
A. Japanese monkeys (*M. fuscata*)			
Arashiyama B	Substantial	2.2	Koyama et al. [1975]
Ohirayama	Substantial	2.0	Tanaka et al. [1970]
Takasakiyama	Substantial	Some females give birth in successive years	Ohsawa et al. [1977][a]
Koshima	Light	Strong repression of births in successive years	Mori [1979]
B. Rhesus monkeys (*M. mulatta*)			
Cayo Santiago	Substantial	~1.2	Sade et al. [1985]
Yerkes Primate Center[b]	Full	1.0	Hadidian and Bernstein [1979]

[a]As cited by Mori [1979].
[b]Captive in outdoor corrals.

The data on interbirth intervals are summarized in Table IV. Under conditions of substantial provisioning at Takasakiyama, some Japanese monkey females give birth in successive years [Ohsawa et al., 1977, as cited by Mori, 1979]. Such successive parturitions were strongly repressed at Koshima after supplementary feeding was drastically reduced [Mori, 1979]. Even when substantially provisioned, however, females in some *M. fuscata* groups (e.g., Arashiyama B and Ohirayama) give birth only every 2 years.

Heavily provisioned rhesus females on Cayo Santiago usually have an infant each year [Sade et al., 1985]. The genealogies presented in Sade et al. [1985] indicate an approximate interbirth interval of 1.2 years. This figure agrees closely with the data from the captive rhesus at the Yerkes Primate Center. At Yerkes, rhesus females living in heterosexual social groups in outdoor corrals have an average interbirth interval of 1 year [Hadidian and Bernstein, 1979]. Unfortunately, it is impossible to compare validly the interbirth intervals shown by fed and unfed macaques with the available data. Mori's [1979] suggestion that provisioned animals have shorter intervals between births remains unverified; indeed, it appears to be contradicted by the data from Arashiyama B and Ohirayama. Studies of interbirth intervals in completely unprovisioned macaques are needed to clear up this question.

DISCUSSION

The effects of nutrition on maturation and reproduction are generally known, although specific causal pathways may not be understood. Undernourished humans and other animals show delayed puberty compared to well-nourished individuals [Frisancho, 1978; Frisch, 1983; Sadleir, 1969]. A degree of overnourishment, in contrast, can lead to early puberty in humans [Frisancho, 1978; Cheek, 1974]. The timing of menarche in female mammals may be related to the attainment of a critical body weight, or more specifically to a critical amount of body fat [Frisch, 1974, 1983; but see Malina, 1978 for an opposing view]. Several other factors, however, such as age and, for some species, reproductive seasonality can also influence the timing of menarche [Resko et al., 1982; Hadidian and Bernstein, 1979; Wilson et al., 1984].

Age at menarche is difficult to determine among free-ranging primates, and therefore it is rarely reported in the literature. The more commonly reported measure of age at first birth is a product of age at menarche, the length of the period of adolescent sterility, and gestation length. Because nutrition should be capable of affecting the timing of first birth by influencing age at menarche and/or the length of the sterile period, the exact operation of nutritional effects may be expected to be complex.

Macaque data are equivocal on the question of supplementary feeding effects on age at first birth, although they suggest the occurrence of the expected reduction in age among substantially provisioned animals. While nothing can be said about rhesus monkeys owing to a lack of data on unprovisioned groups, supplementary feeding may lower age at first birth among Japanese monkeys. Although the mean values for substantially fed and unfed/lightly fed groups in Table I do not quite reach statistical significance, the age changes in the Koshima and Mt. Ryozen groups were in the expected direction (later age at first birth) following cessation of heavy provisioning. At both Koshima and Mt. Ryozen, reduction or cessation of supplementary feeding produced lower body weights in young females compared to periods with heavy provisioning [Mori, 1979; Sugiyama and Ohsawa, 1982], supporting the hypothesis of an association between critical body weight (or some component of body weight, such as fat) and the start of reproduction in females. The probable relationship between body weight (or fat) and female sexual maturation is reinforced by Wolfe's [1979] observations on the Arashiyama West group of Japanese monkeys. Following relocation from Japan to Texas, sexual maturation of many females was delayed 1–2 years, possibly as a result of low body weights.

There are few reports from species other than macaques that address the question of nutritional effects on female age at first birth. Strum and Western's [1982] report on the baboons (*Papio cynocephalus anubis*) of Gilgil, Kenya, however, does discuss this problem. At Gilgil, female age at first birth increased greatly over a 4-year period, apparently as a result of reduced food supply owing to competition from ungulates. This finding parallels the reports of delayed reproduction in female Japanese macaques following cessation of supplementary feeding [Mori, 1979; Sugiyama and Ohsawa, 1982].

An interesting difference in mean age at first birth is evident between Japanese macaques and rhesus monkeys, and it appears to be independent of provisioning effects. Heavily provisioned rhesus females usually produce their first offspring at 3–4 years of age, while substantially fed Japanese macaque females are typically 5 years old before their first birth. This difference in the start of females' reproductive careers is paralleled by maturational differences between rhesus and Japanese monkey males. Male *M. fuscata* appear to reach sexual maturity at 4.5 years of age, while rhesus males mature about 1 year earlier. As for females, this interspecific difference does not correlate with the presence or absence of supplementary feeding. Wolfe's [1979] data from the Arashiyama West group indicate that low body weights are correlated with delayed sexual maturity in macaque males as well as females.

The only statistically significant difference found between substantially provisioned and lightly fed/unfed macaques concerned birth rates in Japanese monkeys (Table II). Fifty-nine percent of females in heavily fed groups gave birth annually as compared with only 39% of females in unfed/lightly fed groups. Data from rhesus monkeys were in apparent agreement, but could not be statistically tested because of an inadequate sample of groups receiving little or no provisioning (Table III). Among the Gilgil baboons, fertility (births per female day) declined along with the amount of available food, supporting the hypothesis of a link between nutrition and reproductive rate [Strum and Western, 1982].

When all birth rates of Japanese monkey groups are compared with those of rhesus monkeys (compare Tables II and III), it is clear that, regardless of the presence or absence of supplementary feeding, rhesus reproduce at a higher rate. The lowest birth rate for a rhesus population (63% at Kathmandu) is equal to the highest rate reported for a Japanese macaque group (63.3% at Takasakiyama).

The macaque data were inconclusive on the question of feeding effects on interbirth intervals (Table IV). Studies of species other than macaques, however, provide some additional evidence that suggests a correlation between good nutrition and short intervals between births. Rudran [1973] found that the Horton Plains, Sri Lanka, population of purple-faced langurs (*Presbytis senex*) had shorter interbirth intervals than the langurs living at a second Sri Lankan site of Polonnaruwa. Food availability was considerably higher at Horton Plains. Furthermore, whereas baboon (*Papio cynocephalus*) infants are usually born at 1-year intervals in well-fed zoo populations, the interval increases to 1.5–2 years in wild, unfed populations such as that of Amboseli [Altmann, 1980].

Rhesus monkeys appear to have shorter interbirth intervals than Japanese macaques (Table IV). It seems clear, therefore, that by virtually every measure, rhesus reproduction exceeds that of *M. fuscata*. Both male and female rhesus monkeys reach sexual maturity earlier than their Japanese macaque counterparts; rhesus groups have higher birth rates than *M. fuscata* groups; and the rhesus interbirth interval is shorter than that of Japanese monkeys. Why these reproductive differences should exist between close congeners is unclear, but this topic has been discussed recently by Wolfe [1986].

CONCLUSIONS

A survey of the literature on supplementary feeding effects on primate maturation and fertility allows only a limited set of conclusions. This is due to a lack of information for most species on both fed and unfed groups, and because most reports on wild primates fail to give detailed information on how much food, if any, the study animals received from humans. The following conclusions, however, seem warranted by the available data.

Among Japanese macaques, substantial supplementary feeding probably contributes to more rapid sexual maturation in females than is found in unfed/lightly fed groups. The same effect may occur in other species, but the point requires further study. The available data do not allow any conclusion concerning provisioning effects on male sexual maturation in any species.

Birth rates are significantly higher in substantially provisioned groups of Japanese macaques than in groups receiving little or no supplementary food. Increased birth rates owing to supplementary feeding cannot be proved for any other species. Similarly, shortened interbirth intervals owing to provisioning cannot be proved for any species given the available data.

In theory, an extra-nutritious diet through supplementary feeding should hasten female and male sexual maturity and increase fertility in primate groups. In fact, provisioning effects can be documented presently for only one measure of fertility, birth rate, and in only one species of nonhuman primate, the Japanese monkey. As shown by comparisons of rhesus and Japanese macaques, variation between species, even congeners, can equal or exceed the apparent effects of provisioning. Further research is required before broad generalizations about the effects of supplementary feeding can be made.

ACKNOWLEDGMENTS

Several people contributed to the completion of this review, and I extend my thanks to them all. John Fa, Paul Winkler, Mary Knezevich, Jean

DeRousseau, Matt Kessler, and R. Rudran all shared information about various populations of wild or provisioned primates. David Boss Taub provided unpublished data on the Morgan Island rhesus monkeys. Peter August and Breck Peters critically reviewed the manuscript, and Betty Jones typed the paper.

REFERENCES

Altmann J (1980): "Baboon Mothers and Infants." Cambridge: Harvard University Press.

Altmann J, Altmann S, Hausfater G (1981): Physical maturation and age estimates of yellow baboons, *Papio cynocephalus*, in Amboseli National Park, Kenya. Am J Primatol 1:389–399.

Carpenter CR (1942): Sexual behavior of free ranging rhesus monkeys (*Macaca mulatta*): Specimens, procedures, and behavioral characteristics of estrus. J Comp Psychol 33:113–142.

Carpenter CR (1972): Breeding colonies of macaques and gibbons on Santiago Island, Puerto Rico. In Beveridge W (ed): "Breeding Primates." Basel: Karger, pp 76–87.

Catchpole HR, van Wagenen G (1975): Reproduction in the rhesus monkey, *Macaca mulatta*. In Bourne GH (ed): "The Rhesus Monkey, Vol. II, Management, Reproduction, and Pathology." New York: Academic Press, pp 117–140.

Cheek DB (1974): Body composition, hormones, nutrition, and adolescent growth. In Grumbach MM, Grave GD, Mayer FE (eds): "Control of the Onset of Puberty." New York: John Wiley and Sons, pp 424–442.

Conaway CH, Sade DS (1965): The seasonal spermatogenic cycle in free ranging rhesus monkeys. Folia Primatol (Basel) 3:1–12.

Drickamer LC (1974): A ten-year summary of reproductive data for free-ranging *Macaca mulatta*. Folia Primatol (Basel) 21:61–80.

Frisancho AR (1978): Nutritional influences on human growth and maturation. Yearbook Phys Anthropol 21:174–191.

Frisch RE (1974): Critical weight at menarche, initiation of the adolescent growth spurt, and control of puberty. In Grumbach MM, Grave GD, Mayer FE (eds): "Control of the Onset of Puberty." New York: John Wiley and Sons, pp 403–423.

Frisch RE (1983): Fatness, puberty, and fertility: The effects of nutrition and physical training on menarche and ovulation. In Brooks-Gunn J, Petersen AC (eds): "Girls at Puberty." New York: Plenum Press, pp 29–49.

Hadidian J, Bernstein IS (1979): Female reproductive cycles and birth data from an Old World monkey colony. Primates 20:429–442.

Hanby JP, Robertson LT, Phoenix CH (1971): The sexual behavior of a confined troop of Japanese macaques. Folia Primatol (Basel) 16:123–143.

Harrison GA, Weiner JS, Tanner JM, Barnicot NA (1977): "Human Biology: An Introduction to Human Evolution, Variation, Growth and Ecology" (2*nd* edition). Oxford: Oxford University Press.

Hiraiwa M (1981): Maternal and alloparental care in a troop of free-ranging Japanese monkeys. Primates 22:309–329.

Ikeda H (1982): Population changes and ranging behaviour of wild Japanese monkeys at Mt. Kawaradake in Kyushu, Japan. Primates 23:338–347.

Jolly A (1985): "The Evolution of Primate Behavior" (2*nd* edition). New York: Macmillan Publishing Company.

Kawai M, Azuma S, Yoshiba K (1967): Ecological studies of reproduction in Japanese monkeys (*Macaca fuscata*). I. Problems of the birth season. Primates 8:35–74.

Koford CB (1965): Population dynamics of rhesus monkeys on Cayo Santiago. In DeVore I (ed): "Primate Behavior: Field Studies of Monkeys and Apes." New York: Holt, Rinehart and Winston, pp 160–174.

Koford CB (1966): Population changes in rhesus monkeys: Cayo Santiago, 1960–1964. Tulane Stud Zool 13:1–7.

Koyama N, Norikoshi K, Mano T (1975): Population dynamics of Japanese monkeys at Arashiyama. In Kondo S, Kawai M, Ehara A (eds): "Contemporary Primatology: 5*th* International Congress of Primatology, Nagoya 1974." Basel: Karger, pp 411–417.

Lindburg DG (1971): The rhesus monkey in north India: an ecological and behavioral study. In Rosenblum LA (ed): "Primate Behavior: Developments in Field and Laboratory Research," Vol. 2. New York: Academic Press, pp 1–106.

Loy J (1971): Estrous behavior of free-ranging rhesus monkeys (*Macaca mulatta*). Primates 12:1–31.

Loy J, Loy K (1974): Behavior of an all-juvenile group of rhesus monkeys. Am J Phys Anthropol 40:83–96.

Malik I, Seth PK, Southwick CH (1984): Population growth of free-ranging rhesus monkeys at Tughlaqabad. Am J Primatol 7:311–321.

Malina RM (1978): Adolescent growth and maturation: Selected aspects of current research. Yearbook Phys Anthropol 21:63–94.

Matsubayashi K, Mochizuki K (1982): Growth of male reproductive organs with observation of their seasonal morphologic changes in the Japanese monkey (*Macaca fuscata*). Jpn J Vet Sci 44:891–902.

Michael RP, Wilson M (1973): Changes in the sexual behaviour of male rhesus monkeys (*M. mulatta*) at puberty. Folia Primatol (Basel) 19:384–403.

Mori A (1979): Analysis of population changes by measurement of body weight in the Koshima troop of Japanese monkeys. Primates 20:371–397.

Nigi H (1976): Some aspects related to conception of the Japanese monkey (*Macaca fuscata*). Primates 17:81–87.

Nigi H, Tiba T, Yamamoto S, Floescheim Y, Ohsawa N (1980): Sexual maturation and seasonal changes in reproductive phenomena of male Japanese monkeys (*Macaca fuscata*) at Takasakiyama. Primates 21:230–240.

Nomura T, Ohsawa N, Tajima Y, Tanaka T, Kotera S, Ando A, Nigi H (1972): Reproduction of Japanese monkeys. In Diczfalusy E, Standley CC (eds): "The Use of Non-Human Primates in Research on Human Reproduction." Stockholm: WHO, pp 473–482.

Ohsawa H, Sugiyama Y, Nishimura A (1977): Population dynamics of Japanese monkeys at Takasakiyama by the marking trace. In Sugiyama Y (ed): "Population dynamics of Japanese monkeys at Takasakiyama." Oita City: Oita, pp 19–36.

Rawlins RG, Kessler MJ, Turnquist JE (1984): Reproductive performance, population dynamics and anthropometrics of the free-ranging Cayo Santiago rhesus macaques. J Med Primatol 13:247–259.

Resko JA (1967): Plasma androgen levels of the rhesus monkey: Effects of age and season. Endocrinology 81:1203–1212.

Resko JA, Goy RW, Robinson JA, Norman RL (1982): The pubescent rhesus monkey: Some characteristics of the menstrual cycle. Biol Reprod 27:354–361.

Roonwal ML, Mohnot SM (1977): "Primates of South Asia: Ecology, Sociobiology, and Behavior." Cambridge: Harvard University Press.

Rose RM, Bernstein IS, Gordon TP, Lindsley JG (1978): Changes in testosterone and behavior during adolescence in the male rhesus monkey. Psychosom Med 40:60–70.

Rudran R (1973): The reproductive cycles of two subspecies of purple-faced langurs (*Presbytis senex*) with relation to environmental factors. Folia Primatol (Basel) 19:41–60.

Sade DS, Chepko-Sade BD, Schneider JM, Roberts SS, Richtsmeier JT (1985): "Basic Demographic Observations on Free-ranging Rhesus Monkeys." New Haven: Human Relations Area Files.

Sade DS, Cushing K, Cushing P, Dunaif J, Figueroa A, Kaplan JR, Lauer C, Rhodes D, Schneider J (1976): Population dynamics in relation to social structure on Cayo Santiago. Yearbook Phys Anthropol 20:253–262.

Sadleir RMFS (1969): "The Ecology of Reproduction in Wild and Domestic Mammals." London: Methuen and Company, Ltd.

Southwick CH, Siddiqi MF (1976): Demographic characteristics of semi-protected rhesus groups in India. Yearbook Phys Anthropol 20:242–252.

Southwick C, Siddiqi MF (1985): The rhesus monkey's fall from grace. Nat Hist 94(2):62–70.

Southwick CH, Siddiqi MF, Farooqui MY, Pal BC (1976): Effects of artificial feeding on aggressive behaviour of rhesus monkeys in India. Anim Behav 24:11–15.

Strum SC, Western JD (1982): Variations in fecundity with age and environment in olive baboons (*Papio anubis*). Am J Primatol 3:61–76.

Sugiyama Y (1965): Short history of the ecological and sociological studies on non-human primates in Japan. Primates 6:457–460.

Sugiyama Y, Ohsawa H (1982): Population dynamics of Japanese monkeys with special reference to the effect of artificial feeding. Folia Primatol (Basel) 39:238–263.

Suzuki A, Wada K, Yoshihiro S, Tokita E, Hara S, Aburada Y (1975): Population dynamics and group movement of Japanese monkeys in Yokoyugawa Valley, Shiga Heights. Physiol Ecol 16:15–23.

Takahata Y (1980): The reproductive biology of a free-ranging troop of Japanese monkeys. Primates 21:303–329.

Tanaka T, Tokuda K, Kotera S (1970): Effects of infant loss on the interbirth interval of Japanese monkeys. Primates 11:113–117.

Teas J, Richie T, Taylor H, Southwick C (1980): Population patterns and behavioral ecology of rhesus monkeys (*Macaca mulatta*) in Nepal. In Lindburg DG (ed): "The Macaques: Studies in Ecology, Behavior and Evolution." New York: Van Nostrand Reinhold Company, pp 247–262.

Vandenbergh JG (1967): The development of social structure in free-ranging rhesus monkeys. Behaviour 29:179–194.

Vandenbergh JG (1973): Environmental influences on breeding in rhesus monkeys. In Phoenix CH (ed): "Symposia of the IV*th* International Congress of Primatology, Vol. 2: Primate Reproductive Behavior." Basel: Karger, pp 1–19.

Wilson ME, Gordon TP, Blank MS, Collins DC (1984): Timing of sexual maturity in female rhesus monkeys (*Macaca mulatta*) housed outdoors. J Reprod Fertil 70:625–633.

Wolfe L (1978): Age and sexual behavior of Japanese macaques (*Macaca fuscata*). Arch Sex Behav 7:55–68.

Wolfe L (1979): Sexual maturation among members of a transported troop of Japanese macaques (*Macaca fuscata*). Primates 20:411–418.

Wolfe L (1986): Reproductive biology of rhesus and Japanese macaques. Primates 27:95–101.

Ecology and Behavior of Food-Enhanced Primate Groups, pages 167–198

9

Dynamics of Provisioned and Unprovisioned Primate Populations

Anna Marie Lyles and Andrew P. Dobson

Department of Biology, Princeton University, Princeton, New Jersey 08544

INTRODUCTION

Other papers in this volume describe the effects of provisioning on the feeding and social behavior of a specific primate species. This chapter is primarily concerned with demographic aspects of primate ecology and will attempt to condense a broad array of behavioral mechanisms into a few simple mathematical functions. Such models allow us to predict how changes in resource availability may affect the short-term population dynamics of primate populations. We feel that this knowledge may prove useful in determining how endangered populations of primates might be managed and whether or not artificial provisioning may be used to help sustain such species. This exercise is also a useful step for developing more detailed models that will further enhance our understanding of the underlying ecological and behavioral mechanisms that ultimately determine primate group size and structure.

The main body of the paper falls into two halves: one concerned with demographic data, and one with mathematical models of primate populations. Initially, demographic data collected from provisioned and unprovisioned primate groups will be compared. Some simple population models will then be developed in which we assume that the basic demographic rates of mortality and fecundity are independent of population size. This model seems most applicable to primate populations that are unnaturally small owing to poaching, or those in large zoos or artificial colonies, where artificial living conditions minimize the effects of resource limitation. Some models for primate populations in which food, habitat requirements, or predation act in some way to regulate population growth are then considered. Potential consequences of provisioning upon the mechanisms that normally regulate population density are illustrated. The main body of the text concludes with an assessment of some stochastic models for primate population dynamics and a final discussion. In each section the amount of mathematics has been kept to a minimum and all significant results have been portrayed graphically. The more important mathematical results are presented in the concluding appendices.

POPULATION BIOLOGY OF PROVISIONED AND UNPROVISIONED TROOPS

This section briefly presents some of the empirical differences between unprovisioned and provisioned primates. Data were collated from several different types of study populations. Though we found limited demographic data for nearly 20 primate species [Lyles and Dobson, manuscript], there were only 6 species with data available for both provisioned and unprovisioned populations. Data were included in the survey only if they came from free-ranging or semi-free-ranging study populations that were either natural, introduced, or, if captive, then housed as natural social groups in large, outdoor enclosures and left relatively undisturbed. Table I presents the data from 43 studies of six primate species.

The three most extensively studied species, vervet monkeys (*Cercopithecus aethiops*), Japanese and rhesus macaques (*Macaca fuscata* and *mulatta*), provide sufficient data for analysis using a Spearman rank correlation. The level of provisioning is positively correlated with the birth rates of all three species at the .05 level of significance. Artificial feeding is also positively associated with infant (first year) survival of vervets and rhesus macaques at this level, but not significantly so for Japanese macaques. The age at first reproduction was negatively correlated with provisioning in the one species (Japanese macaque) with sufficient data for this analysis.

The clearest pattern to emerge from this analysis is that birth rates increase along with infant survival rates when more food is provided (Table I:columns G and H). This is often matched by an increase in adult survival (column J). Thus, in provisioned troops more females are likely to survive to age of first reproduction, both because their mortality rates are lower and because they produce their first offspring sooner than unprovisioned females (columns F and I). An exception to this pattern may occur in species, such as langurs or chimpanzees, in which infanticide is a common source of infant mortality. Social stress may increase with the concentration of animals about an artificial food source, and this could boost infant mortality rates, resulting in a shorter interbirth interval in females that lose their infants. Thus, high birth rates in wild primate troops might be indicative of either high infant mortality or high female fertility. (Chapter 8 examines in more detail the factors affecting fertility in provisioned and nonprovisioned primate troops.)

Many of the studies relied on repeated censuses to estimate demographic parameters. This technique depends upon accurate determination of age-sex

TABLE I. Demographic Parameters

(A) Species and site	(B) Level of Artificial feeding	(C) Type Population	(D) Study years	(E) No. studied (No. groups)	(F) Age 1st reproduct. in years	(G) Birth rate (mean No. per year)	(H) Survival to age 1 year	(I) Survival to 1st repro.	(J) Adult mortality (yearly)	(K) Studies
Cercopithecus aethiops										
Amboseli, Kenya	0	N	'77–80	11–28(3)	—	0.73	0.40	—	—	Cheney et al. [1981]
North Senegal	0	N	'75–76	33–47(1)	—	~0.50	0.65	—	0.15	Galat and Galat-Luong [1977]
Samburu/Isiolo, Kenya	0	N	'77–80	37–40(2)	—	0.58	0.59	—	—	Whitten [1983]
Lolui Island, Uganda	1	N	'63–64	—(1)	—	0.83	—	—	—	Gartlan [1969]
St. Kitts, Caribbean	1	I	'71–73	4–65(18)	—	0.83	—	—	—	McGuire [1974]
Barbados, West Indies	1	I	'79–83	12–19(1)	~4	1.2	~0.68	~0.5	—	Horrocks [1986]
Sepulveda, CA, USA	4	C	'75–83	14–23(2)	4.1	1.0	0.90	0.90	—	Fairbanks and McGuire [1984]
Macaca fuscata										
Mt. Kawaradake, Japan	0	N	'72–74	100(1)	—	0.52	0.64	—	—	Ikeda [1982]
Mt. Kuniwari, Japan	0	N	'75–79	13–47(8)	—	0.44	0.74	0.70	0.12	Maruhashi [1982]
Shiga Heights B, Japan	0	N	—	(1)	—	0.35	0.53	—	—	Suzuki, in Sugiyama and Ohsawa [1982b]
Mt. Ryozen, Japan	0	N	'74–80	20–60(2)	6.7	0.34	0.73	0.57	0.026	Sugiyama and Ohsawa [1982b]
Koshima Islet, Japan	1	N	'52–63	~20–50(1)	5.3	0.46	0.96	—	0.039	Mori [1979]
Koshima Islet, Japan	1	N	'72–77	~105(1)	6.8	0.32	0.31	—	0.081	Mori [1979]
Shiga Heights A, Japan	2–4	N	'62–75	(1)	—	0.51	0.86	—	—	Suzuki, in Sugiyama and Ohsawa [1982b]
Mt. Ryozen, Japan	3	N	'69–73	45–80(1)	5.2	0.59	0.82	0.72	0.0081	Sugiyama and Ohsawa [1982b]

(continued)

TABLE I. Demographic Parameters *(Continued)*

(A) Species and site	(B) Level of Artificial feeding	(C) Type Population	(D) Study years	(E) No. studied (No. groups)	(F) Age 1st reproduct. in years	(G) Birth rate (mean No. per year)	(H) Survival to age 1 year	(I) Survival to 1st repro.	(J) Adult mortality (yearly)	(K) Studies
Macaca fuscata										
Ohirayama, Japan	~3	I	'57–69	—(1)	—	0.58	0.76(6 mon)	—	—	Tanaka et al. [1970]
Koshima Islet, Japan	3	N	'64–71	~50–120(1)	6.2	0.67	0.85	—	0.021	Mori [1979]
Laredo, TX, USA	3	I	'74–79	130–222(1)	5.9	0.18	0.93	0.64	~0.09	Fedigan et al. [1983] Gouzoules et al. [1982]
Arishiyama A, Japan	3–4	N	'54–74	34–158(1)	5–6	0.73	0.90	—	—	Koyama et al. [1975] Fedigan et al. [1983]
Arishiyama B, Japan	3–4	N	'75–77	210–270(1)	—	.53–.68	—	—	—	Takahata [1980]
Takasakiyama, Japan	3–4	N	'50s–75	—(1–3)	5	0.63	0.89	—	—	Itani [1975] Sugiyama and Ohsawa [1982b]
Rome Zoo	4	C	'77–81	27–51(1)	—	0.86	—	—	—	Scucchi [1984]
Macaca mulatta										
Dunga Gali, Pakistan	0	N	'78–79	~290(7)	—	0.38	0.55	—	—	Melnick 1981, in Richard [1985]
Aligarh District, India	1	N	'59–77	0–36(19)	—	0.77	0.82	~0.08	0.17	Southwick et al. [1980]
Kathmandu, Nepal	2	N	'75–78	292–358(2)	—	0.62	0.78	0.41	0.22	Teas et al. [1981]
La Parguera, Puerto Rico	3	I	'62–72	106–364(1)	4.0	0.73	0.83	0.63	—	Drickamer [1974]
Chhatari, Aligarh, India	3–4	N	'59–77	8–132(2)	—	0.90	0.84	~0.3	0.091	Southwick et al. [1980]
Cayo Santiago, Puerto Rico	3–4	I	'59–62	30–130(6)	~4	0.80	0.92	—	0.065	Koford [1965]
Cayo Santiago, Puerto Rico	3–4	I	'76–83	59–306(6)	~4	0.80	0.93	~.80	0.068	Rawlins et al. [1984]

A	B	C	D	E	F	G	H	I	J	K
Macaca mulatta										
Tuglaqabad, India	4	N	'80–83	28–133(5)	—	0.82	>.96	0.96	—	Malik et al. [1984]
CPRC, Davis, USA	4	C	'77–79	(6)	3–4	—	0.81	0.71	—	Smith [1982]
Yerkes, Georgia, USA	4	C	'72–77	—	~3.8	0.84	—	—	—	Wilson et al. [1978]
Macaca sylvanus										
Ain Kahla, Morocco	0	N	'68–69	14–39(13)	—	~0.5	—	—	—	Deag [1984]
Algeria	0	N	'83–84	16–44(3)	5	0.2–1.0	0.17–1.0	—	—	Menard et al. [1985]
Gibraltar	4	I	'36–81	20–30(2)	4.7	0.49	0.93	0.90	0.011	Fa [1984]
Salem, FRG	4	I	'77–83	81–178(1)	4.8	~0.5	0.92	—	—	Paul and Thommen [1984]
Presbytis entellus										
Kanha, India	0	N	'80–83	~30(1)	—	0.75	—	—	—	Newton [1986]
Dharwar, India	1–2	N	'61–63	17(1)	—	0.86	—	—	—	Sugiyama [1965]
Abu, India	~2	N	'71–74	14–24(2)	—	0.27	—	—	—	Hrdy [1977]
Jodhpur, India	2–3	N	'67–82	8–82(2)	3.4	0.85	0.62	—	—	Winkler et al. [1984]
Pan troglodytes										
Bossou, Guinea	0	N	'76–83	19–21(1)	—	0.23	—	—	—	Sugiyama [1984]
Gombe, Tanzania	2–3	N	'65–83	31–60(1)	14	0.19	0.73	0.59	0.08	Goodall 1983
Mahale, Tanzania	2–3	N	'65–83	~100(1)	13	0.13	0.63	—	—	Hiraiwa-Hasegawa et al. [1984]

Demographic data are collated for six species of primates studied under natural and artificially provisioned conditions. The study population is identified by its location in column A, while appropriate references are given in column K. Columns B–E describe relevant background information. B presents the intensity of provisioning on a scale from 0 to 4, with 0 = no artificial feeding; 1 = sporadic *and* light feeding; 2 = artificial foods constitute minor portion of diet; 3 = provisioning is a major dietary component; 4 = primates reliant upon provisions. Column C identifies the study population as either N (naturally occurring), I (introduced at that site), or C (captive). The duration of the study and the size of the study populations are given by columns D and E. Life-history parameters in columns F to K are discussed more fully in the text. Survival and mortality estimates are given for females when possible, but in general, male and female data are pooled.

classes, but is subject to considerable errors owing to the lability of developmental age markers. Thus in some cases the decreased age of maturity owing to provisioning may be underestimated (column F). Although long-term studies with longitudinal data on individuals give better data, such studies are rare. In no cases have primates been followed over sufficient time, relative to their long life span, to factor out the effects of environmental and demographic noise in estimates of life-history parameters. However, as a first approximation, the data suggest that provisioned populations generally increase in size, while unprovisioned ones are either relatively stable or declining as a result of man's actions.

Further support for this conclusion derives from the Koshima islet and Mt. Ryozen troops of Japanese macaques, which have histories of different artificial feeding levels (each period listed separately in Table I). At Mt. Ryozen, provisioning ceased in 1973; subsequently, the primiparous age increased, while survival and birth rates decreased [Sugiyama and Ohsawa, 1982b]. Population changes have been monitored at Koshima islet since 1952, when the macaques were semiwild, through a 1964–1971 period of provisioning, and subsequently under conditions of restricted artificial feeding. Mori [1979] notes population growth during provisioning and then a subsequent decline that can be understood in terms of changes in body weight, reproduction, and survival. Thus, the patterns of food-enhanced reproduction and survival are mirrored in both Table I and the two population manipulation experiments.

SIMPLE MODELS OF PRIMATE POPULATION DYNAMICS

The construction of a life-table remains an essential way to understand the demographic structure of any animal population [Southwood, 1978; National Research Council, 1981]. Rather than attempt this exercise for any specific primate species, we will develop a general model based upon the Leslie matrix [Leslie, 1945, 1948]. This type of model is based upon the age at first reproduction and estimates of the age-dependent survival and fertility rates of the species of interest. A variety of previously published papers and other chapters in this volume illustrate the utility of this technique when applied to specific populations [Altmann et al., 1981]. Here a range of survival and fertility values will be used to compare life-tables for hypothetical primate species. This allows the effects of changes in individual life-history parameters to be compared systematically.

Although studies on primate populations in the field have measured many parameters that relate to population growth rate, this paper will initially concentrate on how provisioning affects a few key variables. One of these is R_o, the net reproductive rate of the population, essentially, the number of female offspring that each female produces in her lifetime. The other variable is r, the intrinsic growth rate of the population. Both these parameters can be readily estimated from standard life-table analysis [Southwood, 1978]. When R_O is greater than unity, the population will grow at a rate determined by r. When R_O is less than unity, the population declines and will eventually become extinct. The mean generation time τ, is usually defined as the natural logarithm of the ratio between these two parameters [Southwood, 1978].

$$\tau = \log_e (R_O/r) \tag{1}$$

The models described in this paper are primarily concerned with the female portion of the population. Essentially we have assumed that the total population equals twice the number of females and that these females will always be mated when they come into estrus. Initially, let us assume that adult survival is constant after the first year of life and that all females have their first offspring at some age f. Although adult primate fertility ranges from four per year in the pygmy marmoset to one every 5 years in the great apes [Harvey and Clutton-Brock, 1985], let us assume that females have one offspring a year for each year after age f. Thus mature females produce female offspring every two years. While sex ratio at birth generally approaches parity [Richard, 1985], social stress in particular appears to bias this ratio in favor of the least philopatric sex [van Schaik and van Noordwijk, 1983a,b; McFarland-Symington, 1987]. The relaxation of this and other assumptions is discussed further below.

The mathematical details of the calculations of R_O, r, and τ are outlined in Appendix A. Figure 1 illustrates the effect of adult survival and age of first reproduction on mean generation time and R_O in this "simplest-case" primate life history. Two pertinent features of primate life histories emerge from this initial exercise. First, if the population is to persist, then the survival rates of adult primates are quite tightly constrained for any age at first reproduction. Secondly, it would appear that population growth rates of most primate populations depend highly upon the adult survival rate. Even changes in survival as small as 1% or 2% could make a difference between a collapsing population and one growing at a fairly healthy rate. As provisioning may influence the survival rate of adult monkeys, this suggests that the increases in numbers observed in many artificially fed primate troops may be partly due to small increases in adult survival. Unfortunately, such small changes in adult survival rates may be statistically undetectable with the data available from most comparative studies of provisioned and unprovisioned troops. Changes in adult survival do, however, produce

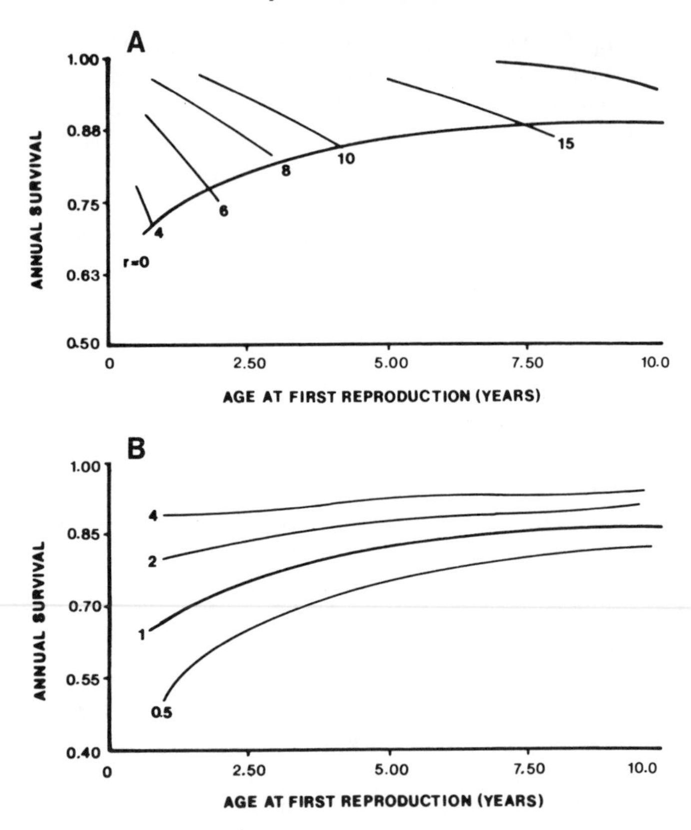

Fig. 1. These figures illustrate the influence of age at first reproduction and annual adult survival on population growth rate and generation time. The upper figure (**A**) shows contours of equal generation time, τ, associated with various combinations of adult survival and age at first reproduction. The lower figure (**B**) illustrates contours of R_O, the mean number of female offspring produced per female in the population. In both cases species are assumed to produce one offspring each year. The regions where $R_O < 1$ or $r < 0$ are indicated; populations exhibiting combinations of survival and age at first reproduction that fall in these regions will collapse.

conspicuous changes in the age-structure of primate troops. Increases in the mature-to-immature ratio are likely to be strong indicators of increased adult survival. Decreases in the age at first reproduction are also seen to lead to increases in population growth rates and shortened generation times.

If adult survival and age at first reproduction are held constant, the effects of other potential changes owing to provisioning can be examined within the context of this basic model. A reduction in interbirth intervals or infant mortality also leads to an increase in the basic reproductive rate of the population (Fig. 2). These mechanisms have one important commonality—

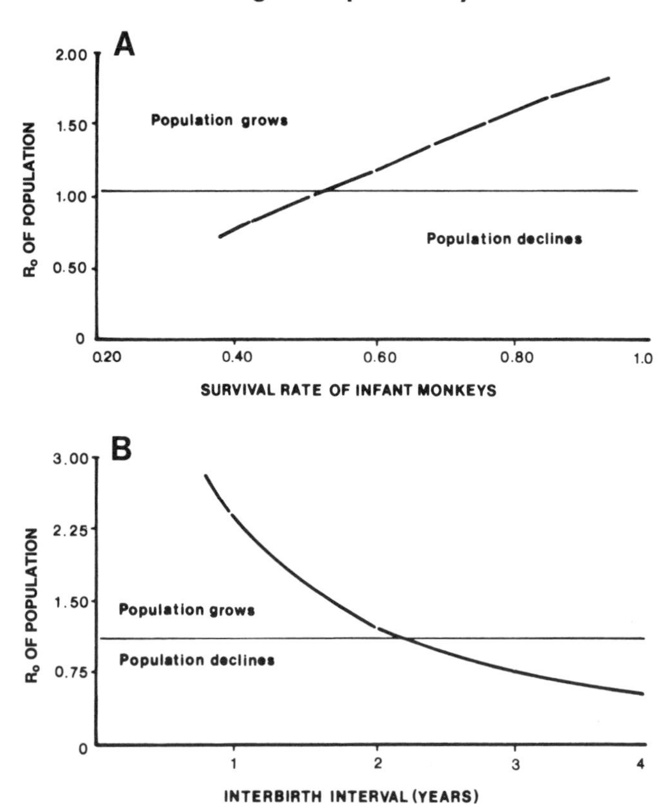

Fig. 2. The upper figure (**A**) illustrates the influence of the survival rate of infant monkeys on R_O, the basic reproductive rate of the population. Representative values for medium-sized monkeys such as vervets are used here: adult survival is held constant at 0.85, the age of first reproduction is 4 years, and one offspring is born each year. In the lower diagram (**B**) the effect of increasing interbirth interval is illustrated with values representative of a larger monkey such as a baboon. Here adult survival is assumed constant at 0.95 per annum and age at first reproduction is 7. In both graphs the line for $R_O = 1$ divides regions of parameter values where the population will grow from ones where it will decline.

they have no effect on the age-structure of the population. This contrasts with the former case, in which provisioning affects adult survival.

The analyses undertaken so far have assumed that the species concerned have a constant adult survival rate from their second year of age until they reach some fixed adult life expectancy, when they die. Although this may be a useful working approximation, survival curves for most primates indicate that senescence, or exponentially increasing mortality rates from middle-age onwards are likely to be more realistic (Fig. 3). Although it seems unlikely

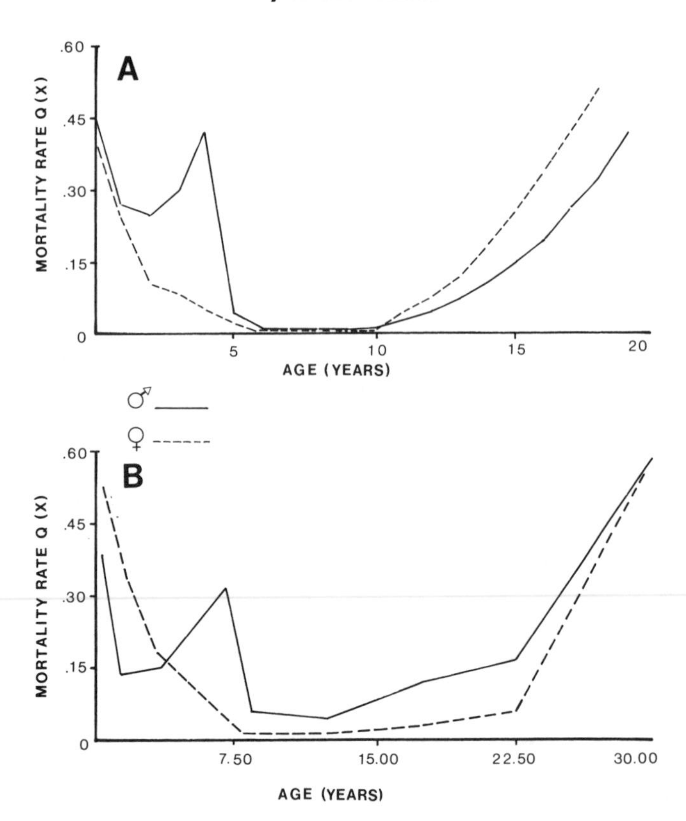

Fig. 3. This figure illustrates the age-specific rate of mortality, *q(x),* for two populations of primates that have been studied over long periods in the wild: (**A**) howler monkeys at Barro Colorado Island, Panama [after Froehlich et al., 1981]; (**B**) toque macaques at Polonnaruwa, Sri Lanka [after Dittus, 1975].

that provisioning will affect the rates of onset of senescence, it is likely to affect the number of animals that live long enough to express its effects. In models for very long-lived primates, ignoring the effects of senescence leads to inflated estimates of the population growth rates. This overestimation is primarily due to the presence of an artificial excess of old animals in the population. More detailed models for monitoring specific populations should bear this fact in mind. It is easily overcome by setting the size of the population matrix large enough and reducing the survival rates of animals in the older age classes.

Although the simple models described above are useful in ascertaining how different life-history variables influence the potential growth rates of

TABLE II. Potential Ways in Which a Variety of Factors May Affect Key Population Parameters in Primates

Density independent ↔ Density dependent	:	S	F	AM	M
Food supply	:	X	X	X	X
Infectious diseases	:	X	X		
Aggression, behavioral mechanisms determining social structure	:	X	X	X	X
Shelter	:	X			
Predators	:	X			
Weather	:	X	X		

This table lists a variety of factors that lead to changes in the basic demographic parameters of primate populations. They have been crudely classified into whether the changes they cause vary systematically with density or whether they act independently of density. Several mechanisms may operate in both ways. The final four columns give the life-history variables most likely to be affected by these mechanisms. Here S denotes survival; F, fecundity; AM, age at maturity; and M, migration.

primate populations, they ignore the fact that the magnitude of certain variables will systematically change as population density increases. This limits their utility as tools that may be used to monitor the growth of specific populations. The next section describes some ways in which these constraints may be relaxed.

DENSITY-DEPENDENT MODELS OF PRIMATE DEMOGRAPHY

Population biologists frequently refer to regulatory processes whose magnitude varies with population density as density-dependent factors. This differentiates them from density-independent factors that affect population density in an erratic manner that is not clearly dependent upon density. Table II lists some potential examples of these two classes of mechanism for primates. It is important to realize that although density-independent mechanisms may be important in determining the variation in population numbers from year to year, it is density-dependent mechanisms that ultimately regulate the size of a population, and serve to maintain it at a density that allows it to defend and utilize the available resources [Lack, 1954].

In pristine populations of primates, it seems likely that some type of resource limitation will maintain the upper bounds of abundance. In many species this resource limitation may only operate at one time of the year, as dry-season food shortages or the like [Dunbar, 1982, Terborgh, 1983]. Pathogens and predators impact populations continuously to reduce them below this upper abundance level, while density-independent events, such as bad weather, will occasionally knock the population or resources down to an unusually low level.

Resource distributions are considered important determinants of primate social systems [Wrangham, 1980], while the amount of resources and predation are likely to influence group size [Jolly, 1985; Richard, 1985]. Population regulation will thus operate at two different levels, the population and the social group. Several recent papers have discussed and reviewed a range of behavioral mechanisms that ultimately act in density-dependent fashion to regulate fertility [Dittus, 1979, 1980; Teas et al., 1981; van Schaik, 1983; van Schaik and van Noordwijk, 1983a].

As Jolly [1985] points out, our clearest indication that food limits primate population growth stems from studies of provisioned wild troops. One of the most important features of provisioned primate studies is that they provide insight into how density-dependent regulation operates. In natural primate populations the interaction between resource availability and foraging efficiency will be important in determining how different individuals in any group or population interact with each other (this volume, chapter 14). When resources are scarce, animals will spend more time searching for food and may become more aggressive when defending or dividing resources. In contrast, when resources are more plentiful, more time is available for activities such as grooming, infant care, and play (this volume, chapter 3). It seems likely that social interactions will ultimately lead to changes in fecundity or infant survival which in turn determine recruitment rates into populations. Before outlining how such density-dependent mechanisms may be included into simple models of primate populations, the effect of provisioning on some of these functions will be briefly considered.

EXAMPLES OF DENSITY-DEPENDENT MECHANISMS IN PRIMATES

Table III lists some qualitative examples of food-related mechanisms that have been suggested by various authors to explain observed changes in primate demographic parameters. These density-dependent mechanisms operate on two basic levels. At the population level, total resource availability contributes to the average condition of animals (e.g., weight, % body fat, and vitamin and nutrient levels). At the local-population level of groups, social interactions will become more important, either through differential distribution of resources according to social dominance, or through intra-troop aggression (stress, infanticide, etc.). Provisioning undoubtedly also influences primate sociality by altering rates of immigration and emigration [Ikeda, 1982; Sugiyama and Ohsawa, 1982a; Goodall, 1983], as well as the

TABLE III. Expected Effect of Provisions on Demographic Parameters When Food is Limiting

Parameter	Mechanism	Direction of change	References
Mortality			
Infants	Well-fed or heavy mothers produce nourishing milk[a]	Decrease	Altmann et al. [1978] Sugiyama and Ohsawa [1982b]
Immatures	Very sensitive to food supply at time of weaning	Decrease	Struhsaker [1976]
	Juveniles are social subordinates with restricted food access	Decrease	Dittus [1977]
	Juveniles receive aggression during competition for provisions	Increase	Wrangham [1974] Fa [1984]
Adults	Pregnant and lactating females are food-stressed	Decrease	Mori [1979]
	Heavy mothers survive longer	Decrease	Whitten [1983]
All age classes	Food-induced aggregation facilitates disease transmission	Increase	Wrangham [1974] Teas et al. [1981]
	Well-fed animals are more disease resistant	Decrease	Dunbar [1982]
Interbirth interval	Postpartum heavy females are more sexually active[a]	Decrease	Mori [1979 Whitten [1983]
	More viable, nursing infants suppress mother's fertility	Increase	Altmann et al. [1978]
	Well-fed female weans baby early	Decrease	Scucchi [1984]
	Provision-related social disruption causes stress and low fertility	Increase	Fa [1984]
	Survival of old females with low fertility improves	Increase	Fa [1984] Fairbanks and McGuire [1984]
Age at first parturition	Minimum body weight to conceive and provisioned animals are heavy[a]	Decrease	Mori [1979]
	Provisioned animals mature faster than unprovisioned ones	Decrease	Altmann et al. [1981] Sugiyama and Ohsawa [1982a,b] Loy (this volume, chapter 8)

Expected effect of provisions on demographic parameters when food is limiting. This table collates causal mechanisms gleaned from a variety of sources. In each case, the demographic parameter of interest and how artificial provisioning affected the observed population is indicated.

[a]Heavily provisioned animals have substantially increased body weights (by one-fifth to one-half in Japanese macaques: Mori [1979]; Sugiyama and Ohsawa [1982b]).

benefits of dominance behavior [Cheney et al., 1981; Sugiyama and Ohsawa, 1982b].

Although Table III provides suggestive evidence that primate mortality in all age classes declines with increased food resources, the opposite pattern might ensue if epidemic disease or aggression against immatures were common. Fertility patterns are less clear-cut as provisioning sometimes enhances reproduction by improving the physical condition of mothers who can then wean offspring faster and conceive sooner after parturition. At other times provisioning seems to depress reproduction by increasing the average interbirth interval. However, a lengthened interbirth interval does not necessarily mean that females will leave fewer offspring during their lifetimes; in most of these food-related fertility changes, well-fed females may produce fewer infants but will have more offspring survive to adulthood. An additional trend is for better-fed daughters to require less time to mature to age of first reproduction.

To a good first-order approximation it can therefore be said that mean female fertility will decline as group size or population density increases, while also tending to increase with provisioning. Although a variety of examples of potential mechanisms have been cited that may be important in regulating the size of primate populations, no attempt will be made to discuss their relative merits. This is partly because the list is still incomplete, but more specifically because the experiments necessary to determine the relative importance of these mechanisms have not been undertaken.

Instead a phenomenological model will be employed which captures the gross effects of these regulatory mechanisms at the population level, but ignores both the details of the mechanisms involved and their consequences for the individual reproductive success of different individuals in the population. A suitable function which captures the basic features of many of the mechanisms discussed above is sketched graphically in Figure 4. Here it is assumed that the total number of female offspring born to a troop containing N females, is given by the function

$$f(N) = F\,N \exp(-\,a\,N^b) \tag{2}$$

where F is the mean number of female offspring born to each sexually mature female in each breeding season.

This function was originally described by Maynard Smith and Slatkin [1973]; Bellows [1981] gives a complete list of its properties. The important point to notice for the purposes of this paper is that two basic parameters capture the changes in fertility that are produced by a wide range of

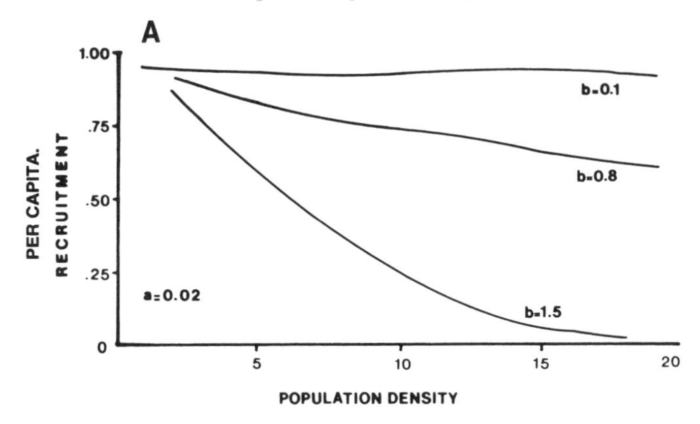

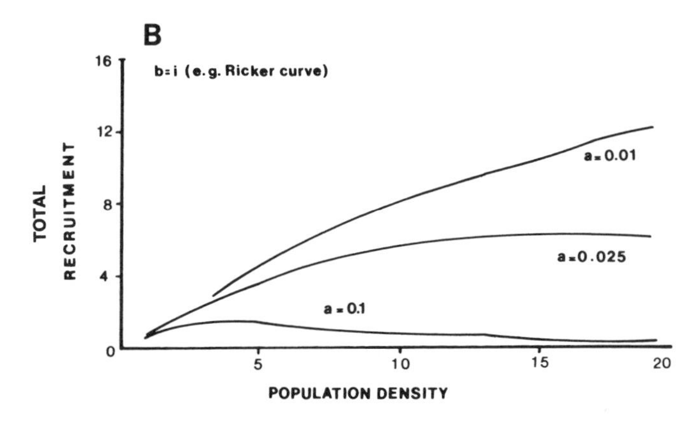

Fig. 4. These figures illustrate how the shape of the density-dependent function discussed in the main text is determined by variations in the two parameters *a* and *b*. Figure **A** shows the relationship between population density and mean female fertility for three different values of *b*. When *b* is relatively small, changes in population size have relatively little effect on mean fertility. When *b* is greater than unity, fecundity is very sensitive to changes in population density. In Figure **B** the influence of population density on total production of infants is illustrated; here the parameter *b* is held constant ($b = 1$), while *a* has been varied. Essentially *a* determines the position where this curve turns over. This parameter will essentially vary inversely with the carrying capacity of the environment, and thus small values of *a* signify a large carrying capacity. When *a* is large, density-dependent effects come into play at lower population densities.

behavioral mechanisms. One of these parameters, *b*, determines the shape of the curve. When *b* is significantly greater than unity, the production of offspring depends highly upon population density; when *b* is less than unity, fertility is less closely linked to population density (when $b = 1$ the curve is a simple Ricker function, as widely used by fisheries biologists). The former situation corresponds to instances where social interactions are highly linked to the availability of food or other limiting resources; the latter corresponds to examples where these resources are not pivotal to social behavior. The second parameter of the distribution, *a,* is inversely related to the level of available resources. In food-limited systems, this parameter should decrease commensurately with provisioning. Essentially, Equation 2 can be interpreted to mean that the social mechanisms which influence fertility increase in intensity as population densities increase.

In several primate studies, dominant females have greater reproductive success than subordinates [Drickamer, 1974; Sugiyama and Ohsawa, 1982a; Whitten, 1983; but also see Wolfe, 1984]. In other cases, reproductive success varies more between social groups than within them [Meikle et al., 1984]. Differences in the quality of territories or home ranges may lead to differences in the mean fertility of different troops. Thus, dominance behavior and environmental heterogeneity may function in a density-dependent fashion by changing patterns of reproductive success according to population size. In some years scarcity of resources limits reproduction to only highest-ranked females, while in more plentiful years, fertility will be more equitably distributed (Fig. 5a,b). Similarly for groups, at times only those occupying the most resource-rich patches, will be able to replicate, while in better years all groups in all patches will produce offspring. When population densities are low, only the best territories will be used; as population density increases, lower-quality territories are used and *net* mean fertility decreases (Fig. 5c,d). Here, it is important to notice that although a number of quite subtle behavioral mechanisms influence the relationship between population density and mean fertility, the salient features of these interactions may again be captured by a function of the form given by Equation 2.

MODELS FOR PRIMATE POPULATIONS WITH DENSITY-DEPENDENT RECRUITMENT

The models described in this section of the paper are again based upon Leslie matrices. However, instead of assuming that all the elements of these matrices remain constant at their mean values, we assume that the parameter that determines the rate at which infant females are born and become recruited to the population of adults and juveniles is dependent on population density. Thus one element of the matrix will be the function (Eq. 2) described in the previous section. We also assume that the life history can be divided into three artificial age-classes—infants, juveniles, and adults. Age at first reproduction in primates is approximately three times the mean interbirth interval (Fig. 6), although, interestingly, group living primate species tend to have slightly earlier ages at first reproduction than species that live as solitary

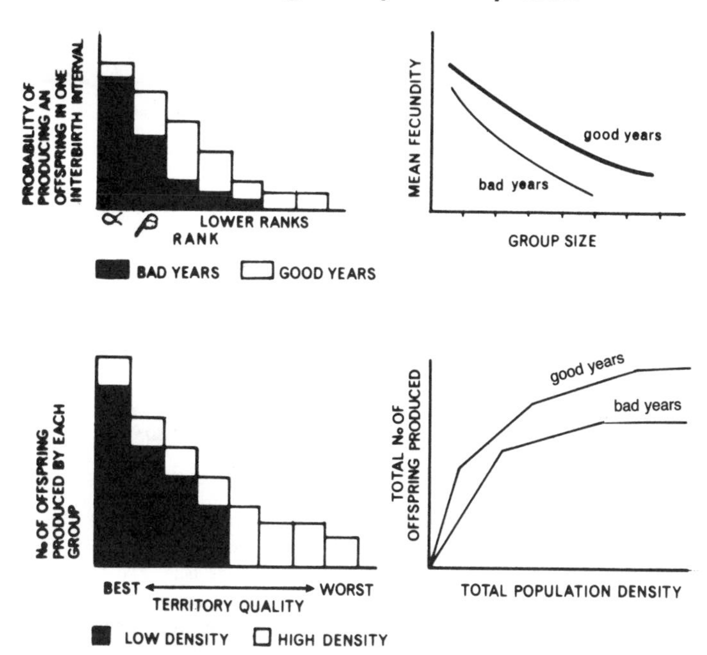

Fig. 5. This diagram illustrates some hypothetical examples of how mechanisms such as dominance hierarchies or the differential quality of territories may give rise to density-dependent recruitment curves. In the **upper figures** we have assumed that a hierarchy exists in which female dominance rank relates to the probability of producing offspring. Although reproductive success will also vary with available resources, in general when there are only a few females in the population, they will all be fairly high ranked and *mean* fecundity will be fairly high. When numbers increase, the females with lower ranks may fail to reproduce altogether and *mean* fecundity declines. Under these circumstances it is likely that the group may fission, and some members will attempt to establish their own territory. The **lower figures** illustrate how variations in territory quality also lead to density-dependent variations in *mean* fecundity. Again we assume that the territories may be ranked in some order of offspring productivity and that the best territories are occupied first. When population density is low, only the best territories will be occupied, and mean fecundity will be high. As population density increases, less productive territories are occupied and mean fecundity declines. Note that the quality of a territory is likely to vary between years and in some years a territory may allow a group to survive but not produce offspring. Although a variety of behavioral actions will determine where a female fits in the hierarchy and which territorial group she joins, the net effects of variations in population density on offspring production may still be captured by a function of the type illustrated in Figure 4.

pairs. It is thus convenient to express all rates in terms of "interbirth interval" time units. This normalization not only allows us to compare a wide range of different life histories within the same basic framework, but it also avoids the many complexities that arise when a full age-structured model is developed.

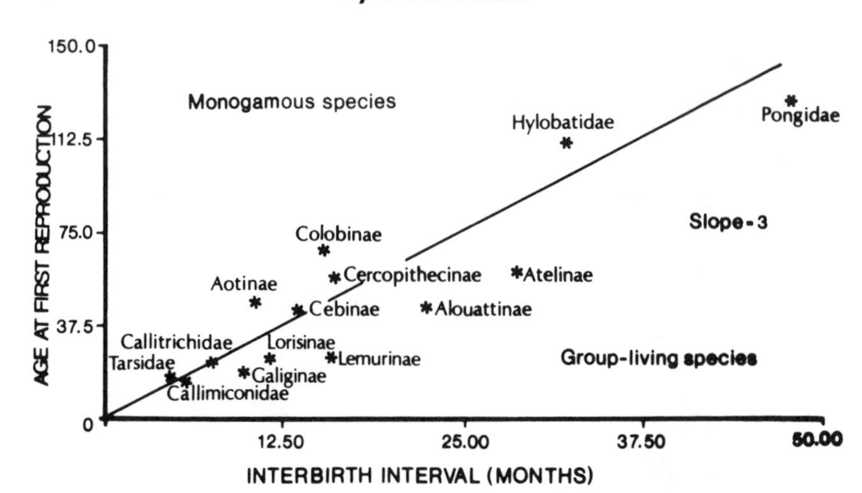

Fig. 6. This figure illustrates a plot of mean age at first reproduction against mean interbirth interval for 14 subfamilies of primates. The data are taken from Harvey and Clutton-Brock [1985]. A line passing through the origin of slope 3 has been plotted through the data. The actual major axis regression has slope 2.91, r = 0.904. Subfamilies that contain species that mainly live as monogamous pairs tend to lie above the line, species that tend to live in groups lie below it [Dobson, in prep.].

This refined model could be written as follows:

$$\begin{bmatrix} I \\ J \\ A \end{bmatrix}_{t+1} = \begin{bmatrix} 0 & 0 & sf(Nt)F \\ si & 0 & 0 \\ 0 & s & s \end{bmatrix} \begin{bmatrix} I \\ J \\ A \end{bmatrix}_{t} \tag{3}$$

where s equals the survival of an animal through one time unit, si is the density independent survival of an infant, F signifies the maximum female fecundity through one time unit, and $f(Nt)$ is a function of the type described in Equation 2. Full details of the analysis of the model are given in Appendix 3; here the more translucent results of this analysis are presented. The total number of adults and juveniles in the population at equilibrium will be

$$N^* = \frac{1}{a}\left[\ \ell n \frac{Fs^3\, i}{(1-s)}\right]^{\frac{1}{b}} \tag{4}$$

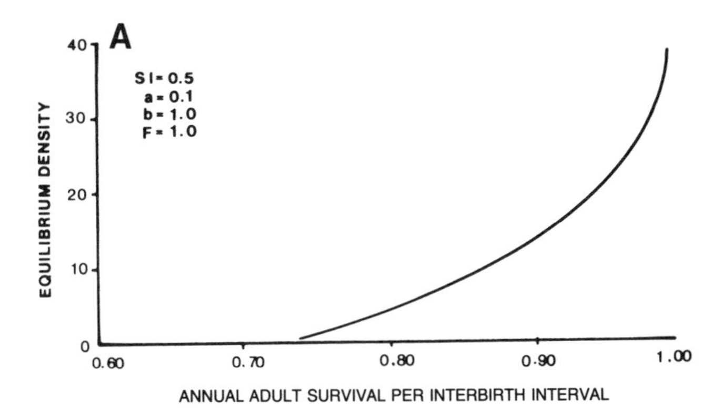

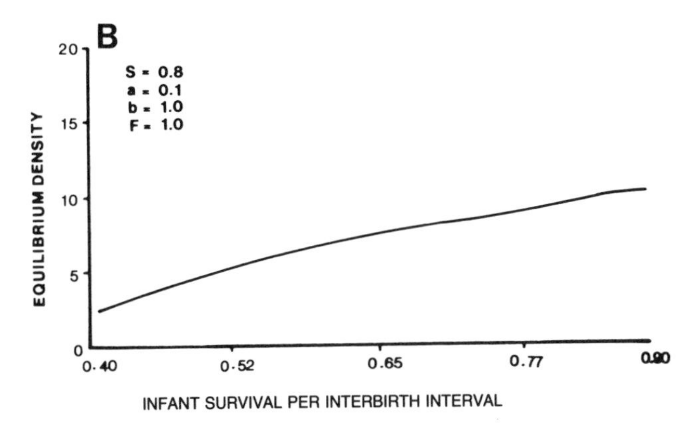

Fig. 7. This figure shows how changes in survival affect the equilibrium population density of a hypothetical primate population. (**A**) Small variations in adult survival are again seen to give large changes in population density. (**B**) Suggests that population density is less sensitive to variations in infant survival. In both cases survival rates are expressed per interbirth interval.

Total population density will thus vary inversely with both a and b. The number of individuals in each subsection of the population is given by simple proportionality; e.g., there will be sN^* adults, $(1 - s)N^*$ juveniles, and $[(1 - s)/si]N^*$ births in the population during any time interval. Changes in survival and fertility owing to provisioning will still have essentially the same effects on the population as they did in the earlier density-independent model (Fig. 7). Increases in survival lead to increases in the total number of individuals in the population and in the ratio of adult to immature monkeys. Increased fertility tends only to lead to increases in the total population density without concommitant changes in the age structure.

Provisioning may also affect the behavioral and physiological mechanisms that determine the density-dependent regulation of fertility. In general, provisioning will tend to reduce the value of *a*, by increasing the total amount of resources available to each adult female. This will always lead to increases in population density. How provisioning affects the magnitude of *b* will probably vary from species to species. In general, if increased provisioning leads to reduced aggression between group members, then *b* will decrease in value and the population may expand quite rapidly. A good example of this might be the situation in which provisioning leads to a reduction in the intensity with which the dominance hierarchy functions [Teas et al., 1981; Wolfe, 1984; Fa, 1986]. Here many more females would breed than normal and many more offspring would survive to maturity as a result of lower levels of infanticide or interfemale aggression. In contrast, if artificial provisioning leads to increased intratroop aggression, *b* will tend to increase in value and the benefits of increased fecundity and survival will be diminished.

The dynamic behavior of the population may be explored using local stability analysis (Appendix B). This exercise suggests that essentially two patterns of dynamics will be commonly observed in primate populations. Where *b* is comparatively small, the populations will tend to return asymptotically to equilibrium when perturbed. As the strength of this regulation increases, it becomes more likely that perturbations will be followed by damped oscillations that die out as the population returns to its carrying capacity; the particular dynamic displayed depends upon the relative values of *s* and *b* (Fig. 8). When adult survival during one interbirth interval is less than about 70%, the population will collapse altogether. This important result is completely independent of the form of the density-dependent function.

The models for this second section of the paper are analyzed in terms of their properties at equilibrium. Although we fully appreciate that primate populations are rarely at equilibrium in the wild [Jolly, 1985; Richard, 1985], the mathematical convenience of undertaking the analysis in this fashion allows us to ascertain properties of the population that apply more generally. Like other age-structured vertebrate populations [Fowler, 1981; May, 1985], primates are susceptible to both bad and good years which produce "gaps" and "pulses" of age cohorts whose effects on population structure are felt long after the factors causing them have died out. This is particularly true when a population has been perturbed below its "natural" carrying capacity by a variety of density-independent factors such as weather, habitat degradation, and the more direct activities of man.

DENSITY-DEPENDENT SEX RATIOS

A recent review by van Schaik and van Noordwijk [1983a] presents evidence that primate sex ratios vary with density and with provisioning [also see Rawlins and Kessler, 1986]. The authors suggest that increased social stress (i.e., conspecific aggression towards mothers and their infant daughters) tends to lead to reduced numbers of females at birth and increased mortality rates of infant females. A similar phenomenon arises in the offspring of many parasitic hymenopterous insects in which sex ratio is dependent upon the local population density [Charnov, 1982]. Hassell, Waage, and May [1983] propose that a suitable function to model changes in the proportion of female offspring $s(N)$, would be

$$s(N) = \frac{\alpha\beta}{\beta + N} \quad (5)$$

This function again has two parameters, one of which, α, acts to determine the relative position of the function, while the other, β, determines its shape (Fig. 9). As no data are available to tune these parameters for any primate system, it seems sensible to assume that the action of the behavioral mechanisms involved will be a characteristic of any species in any particular habitat. In the simplest case, provisioning may not have any influence on the general shape of the function; under these circumstances increased population density that results from increased survival or fecundity will lead directly to changes in sex ratio. If, however, provisioning reduces social stress, then it is not inconceivable that more females are produced [van Schaik and van Noordwijk 1983a,b].

As the models in this paper deal with the recruitment of individuals to the female "half" of the population, it can be assumed that in the simplest case any change in sex-ratio will simply reduce the numbers of infant females at any time. The essential features of this stress-enforced, density-dependent sex ratio will be captured by the basic function for recruitment (Eq. 2). However, in male philopatric, female-dispersing species such as the spider monkey (*Ateles paniscus*), subordinate females show increased production of daughters rather than the "normal" bias towards sons [McFarland-Symington, 1987]. This suggests that some species may be minimizing the impact of density-dependent stress on individual reproductive success by investing in offspring that have a better chance of dispersing.

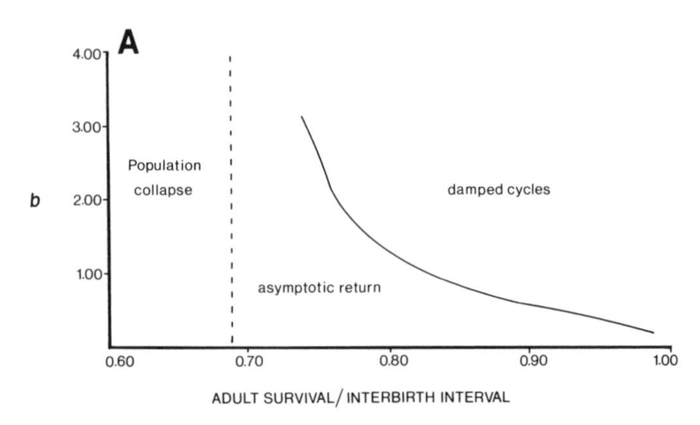
A
4.00
3.00
2.00
1.00
b
Population
collapse
damped cycles
asymptotic return
0.60
0.70
0.80
0.90
1.00
ADULT SURVIVAL/ INTERBIRTH INTERVAL

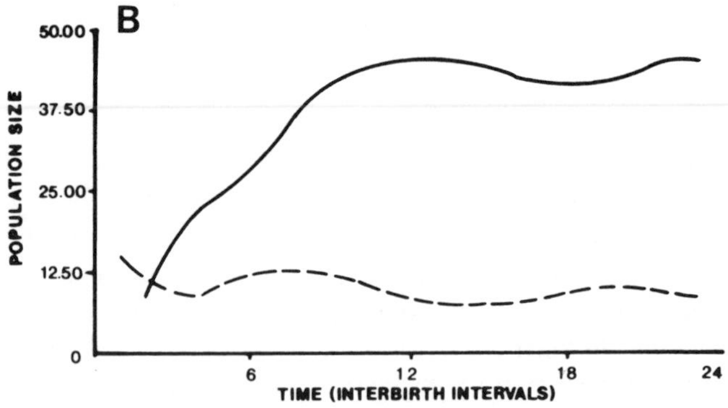
B
50.00
37.50
25.00
12.50
0
POPULATION SIZE
6
12
18
24
TIME (INTERBIRTH INTERVALS)

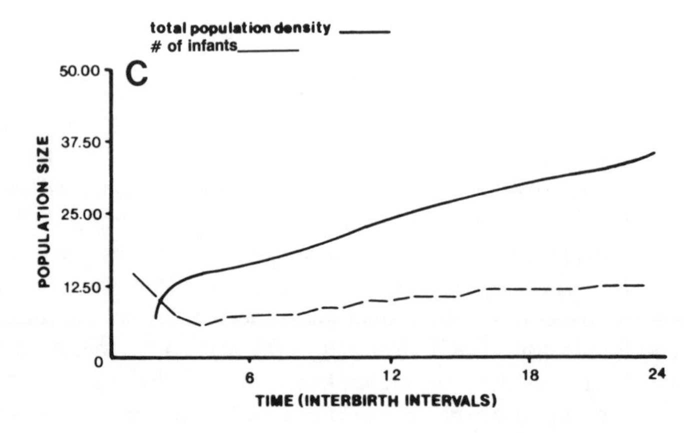
total population density
# of infants
C
50.00
37.50
25.00
12.50
0
POPULATION SIZE
6
12
18
24
TIME (INTERBIRTH INTERVALS)

Figure 8.

STOCHASTIC MODELS OF PRIMATE DEMOGRAPHY

Cohen [1969, 1971, 1972] and Keiding [1977] have presented some interesting stochastic models of group size in primates. These models rely upon fairly basic assumptions about the rates of birth, immigration, death, and emigration of primates (BIDE models), yet they generate frequency distributions of troop sizes that seem to compare favorably with those observed in actual primate populations. Interestingly, Cohen's results suggest that there should be different distributions of group sizes in provisioned and non-provisioned populations. In the former, birth rates can be expected to be relatively independent of population size since provisioning alleviates the effects of resource limitation on birth rates and recruitment. Under these conditions, the BIDE models predict that group sizes will be distributed in a truncated Poisson distribution (e.g., many small to medium-size groups in the population). In contrast, when resources are limiting, frequency distributions of groups size are more likely to be negatively binomially distributed. Here a few larger troops and a smaller proportion of small and medium-sized troops should be seen. It may prove useful to try and estimate the variance to mean ratios of group sizes for provisioned and unprovisioned populations, as these will give at least a qualitative feel for the utility of these models. Unfortunately, sufficiently large surveys of both provisioned and unprovisioned group size distributions are rare and insufficient data are available to rigorously test these ideas.

DISCUSSION AND CONCLUSIONS

Most of the above is a rather straightforward mathematical adumbration of primate population dynamics that has sacrificed considerable detail in an

Fig. 8. *(facing page)* **(A):** This diagram depicts combinations of parameter values that give rise to different patterns of population behavior. Here we have again assumed that age at first reproduction is approximately equal to three times the interbirth interval and have normalized all time-dependent parameters as rates in units of interbirth intervals. Three types of population dynamics are expected. When adult survival through one interbirth interval is less than around 70%, the population will collapse. When survival is greater than this and density-dependent regulation relatively weak (low b-values), then the population will return asymptotically to equilibrium. **(B):** Illustrates this type of behavior for $S = 0.75$, $b = 1.0$, and $a = 0.025$. As survival rates and strength of the regulatory mechanisms increase, damped oscillations are to be expected as the population returns to equilibrium. **(C):** ($s = 0.9$; $b = 2.0$). These oscillations would be more pronounced when monitored as changes of infants than when monitored as changes in total population density. In all diagrams it is assumed that si equals 0.65, the mean value in Table I. It is also important to note that the position of the line determining where the population collapses will be further to the right for monogamous species (see Fig. 6).

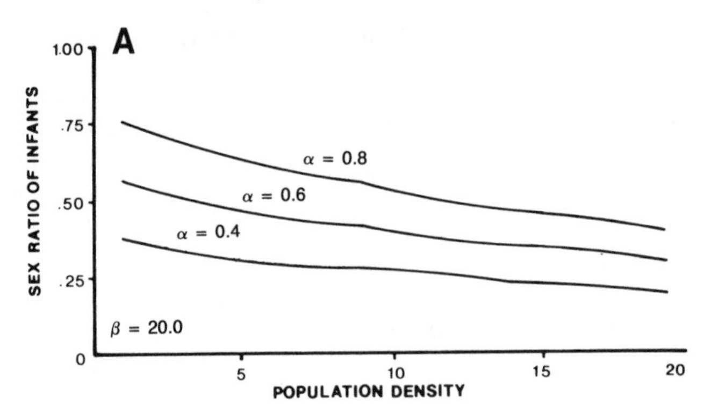

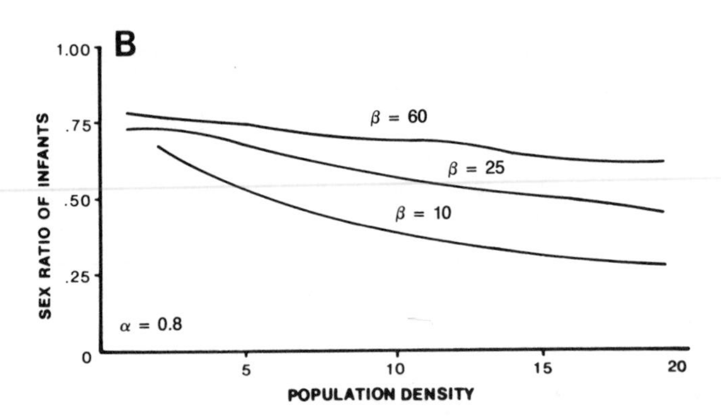

Fig. 9. This figure gives representative examples of the influence of the parameters α and β on the shape of the function that determines the influence of density on infant sex ratio. The parameter β determines the sensitivity of the sex-ratio function to changes in population density (**A**). When β is large, small changes in density produce significant changes in sex ratio; when β is small, sex ratio is less tightly coupled to population density. The other parameter, α, essentially determines the maximum value the sex ratio may take (**B**).

attempt to find a few crude generalities. Nonetheless, this exercise has suggested areas where previously collected data might be reanalyzed or where more manipulative experiments might be done. In the case of endangered primate species, some of the results may be useful as, in the absence of any specific data for these species, crude generalities are the only available information upon which to make conservation policy. In this spirit, it seems sensible to suggest that for many endangered species in areas of reduced or restricted habitat, provisioning may be the only way to sustain a viable and healthy population. A note of caution should also be sounded at this point. In general, increased survival and fecundity will be good for primate populations, as they lead to increased group sizes and hence to larger gene pools. However, in some cases, within larger groups there are increased levels of aggression or increased rates of disease transmission [Wrangham, 1974; Goodall, 1983; Teas et al., 1981].

For these reasons, provisioning should be undertaken in such a way that it mimics the natural distribution of resources. This will reduce the tendency for local populations and troops to build up to the densities where epidemiological considerations become important. Such a feeding scheme requires further detailed studies of the feeding habits of natural populations of different species. The importance of these feeding studies for making efficacious management decisions that will allow endangered populations of primates to persist in the wild cannot be emphasized.

It should also be pointed out that, if anything, the results herein err on the side of cautious optimism. In a real stochastic world it is usually better to be more pessimistic. Any attempts to manage endangered populations using the mathematical models described above should allow for a manageable margin of safety. This might most readily be achieved by hybridizing the models outlined above with the stochastic BIDE models developed by Cohen [1969, 1971, 1972]. At a time when 58% of the worlds' primate species are listed as endangered [Wolfheim, 1982; Diamond, 1985], it remains of paramount importance that conservation strategies be developed that merge the enormous field knowledge of the primatologist with the generalized simplifications of the population biologist.

SUMMARY

In this chapter we focus upon the demography of primate populations, and how provisioning may affect life-history patterns. We review demographic studies of primates and compare 43 populations belonging to six species that have been studied under varied feeding regimes. Although statistical correlation was only possible for three of the species, the data suggest that increased provisioning tends to be positively associated with birth rate and infant survival, and negatively correlated with the age of first parturition.

We then develop some simple mathematical models that attempt to capture general patterns of primate behavior and demography. We begin with the commonly used Leslie matrix model, and then add some more biologically realistic refinements, such as density-dependent infant survival, skewed sex ratios, and stochasticity. Although these models are fairly general, they emphasize several important relationships between primate life-history parameters that might be less apparent in more specific models. For example, the model suggests that adult survival must exceed about 70% per interbirth interval for a population to sustain itself.

Since provisioning tends to increase survival and fertility, it may be one temporary recourse for saving endangered species. However, the pros and cons of provisioning are not yet fully appreciated; for instance, its effects on social behavior and disease transmission are only now beginning to be understood. The models suggest that further studies into the ways that food resources determine primate survival and recruitment are essential to a fuller understanding of primate demography.

ACKNOWLEDGMENTS

We would like to thank John Fa, Tony Ives, Maria van Noordwijk, Bob May, Margaret McFarland-Symington, Richard Rawlins, John Terborgh, and Carl van Schaik for comments on a previous manuscript. A.M.L. received support from the National Science Foundation, and A.P.D. from the Rockefeller Foundation through a grant to R. M. May.

REFERENCES

Altmann J, Altmann SA, Hausfater G (1978): Primate infant's effects on mother's future reproduction. Science 201:1028–2029.

Altmann J, Altmann S, Hausfater G (1981): Physical maturation and age estimates of yellow baboons, *Papio cynocephalus,* in Amboseli National Park, Kenya. Am J Primatol 1:389–399.

Beddington JR (1974): Age distribution and the stability of simple discrete time population models. J Theor Biol 47:65–74.

Bellows TS Jr. (1981): The descriptive properties of some models for density dependence. J Anim Ecol 50:139–156.

Charnov EL (1982): "The Theory of Sex Allocation." Princeton: Princeton University Press.

Cheney DL, Lee PC, Seyfarth RM (1981): Behavioral correlates of non-random mortality among free-ranging female vervet monkeys. Behav Ecol Sociobiol 9:153–161.

Cohen JE (1969): Natural primate troops and a stochastic population model. Am Nat 103:455–478.

Cohen JE (1971): "Causal groups of monkeys and men: stochastic models of elemental social systems." Cambridge: Harvard University Press.

Cohen JE (1972): Markov population processes as models of primate social and population dynamics. Theor Popul Biol 3:119–134.

Deag JM (1984): Demography of the Barbary macaque at Ain Kahla in the Moroccan Moyen Atlas. In Fa JE (ed): "The Barbary Macaque—A Case Study in Conservation." New York: Plenum Press, pp 113–133.

Diamond JM (1985): Future of the world's primates. Nature 317:577–578.

Dittus WPJ (1975): Population dynamics of the Toque monkey, *Macaca sinica.* In Tuttle RH (ed): "Sociology and Psychology of Primates." The Hague: Mouton Publishers, pp 125–152.

Dittus WPJ (1977): The social regulation of population density and age-sex distribution in the Toque monkey. Behaviour 63:281–322.

Dittus WPJ (1979): The social regulation of primate populations. In Lindburg D. (ed): "The

Macaques: Studies in Ecology, Behavior and Evolution.'' New York: Van Nostrand Reinhold Co., pp 263–286.

Dittus WPJ (1980): The evolution of behaviours in regulating density and age-specific sex ratios in a primate population. Behaviour 69:265–301.

Drickamer LC (1974): A ten year summary of reproductive data for free-ranging *Macaca mulatta*. Folia Primatol (Basel) 21:61–80.

Dunbar RIM (1982): Demographic and life history variables of a population of gelada baboons (*Theropithecus gelada*). J Anim Ecol 49:485–506.

Fa JE (1984): Structure and dynamics of the Barbary macaque population in Gibraltar. In Fa JE (ed): ''The Barbary Macaque—A Case Study in Conservation.'' New York: Plenum Press, pp 263–306.

Fa JE (1986): Use of time and resources in provisioned troops of monkeys: Social behaviour, time and energy in the Barbary macaque (*Macaca sylvanus* L.) in Gibraltar. Contrib Primatol, Vol 23, Basel: Karger, pp 1–378.

Fairbanks LA, McGuire MT (1984): Determinants of fecundity and reproductive success in captive vervet monkeys. Am J Primatol 7:27–38.

Fedigan LM, Gouzoules H, Gouzoules S (1983): Population dynamics of Arashiyama West Japanese macaques. Int J Primatol 4:307–321.

Froehlich JW, Thorington RW Jr, Otis JS (1981): The demography of Howler monkeys (*Alouatta palliata*) on Barro Colorado Island, Panama. Int J Primatol 2:207–236.

Fowler CW (1981): Density dependence as related to life history strategy. Ecology 62:602–610.

Galat G, Galat-Luong AG (1977): Demographie et regime alimentaire d'une troop de *Cercopithecus aethiops sabeaus*. La Terre et la Vie 31:557–577.

Gartlan JS (1969): Sexual and maternal behaviour of the vervet monkey, *Cercopithecus aethiops*. J Reprod Fert Suppl 6:137–150.

Goodall J (1983): Population dynamics during a 15 year period in one community of free-living chimpanzees in the Gombe National Park, Tanzania. Z Tierpsychol 61:1–60.

Gouzoules H, Gouzoules S, Fedigan L (1982): Behavioural dominance and reproductive success in female Japanese monkeys (*Macaca fuscata*). Anim Behav 30:1138–1150.

Harvey PH, Clutton-Brock TH (1985): Life history variation in Primates. Evolution 39:559–581.

Hassell MP, Waage J, May RM (1983): Variable parasitoid sex ratios and their effect on host-parasitoid dynamics. J Anim Ecol 52:884–904.

Hiraiwa-Hasegawa M, Hasegawa T, Nishida T (1984): Demographic study of a large-sized unit-group of chimpanzees in the Mahale Mountains, Tanzania: A preliminary report. Primates 25:401–413.

Horrocks JA (1986): Life-history characteristics of a wild population of vervets (*Cercopithecus aethiops sabaeus*) in Barbados, West Indies. Int J Primatol 7:31–47.

Hrdy SB (1977): ''The Langurs of Abu.'' Cambridge, MA: Harvard University Press.

Ikeda H (1982): Population changes and ranging behaviour of wild Japanese monkeys at Mt. Kawaradake in Kyushu, Japan. Primates 23:338–347.

Itani J (1975): Twenty years with Mount Takasaki monkeys. In Bermant G, Lindburg DG (eds): ''Primate Utilization and Conservation.'' New York: John Wiley & Sons, pp 101–125.

Jolly A (1985): ''The Evolution of Primate Behavior,'' Second Edition. New York: Macmillan Publishing Company.

Keiding N (1977): Statistical comments on Cohen's application of a simple stochastic population model to natural primate troops. Am Nat 111:1211–1219.

Koford CB (1965): Population dynamics of rhesus monkeys on Cayo Santiago. In DeVore I (ed): ''Primate Behavior.'' New York: Holt, Rinehart and Winston, pp. 160–174.

Koyama N, Norikoshi K, Mano T (1975): Population dynamics of Japanese monkeys at Arashiyama. In: "5th Int Congr Primatol, Nagoya 1974." Basel: Kargar, pp 411–417.

Lack D (1954): "The Natural Regulation of Animal Numbers." Oxford: Clarendon Press.

Leslie PH (1945): On the use of matrices in certain population mathematics. Biometrika 33:183–212.

Leslie PH (1948): Some further notes on the use of matrices in population mathematics. Biometrika 35:213–245.

Malik I, Seth PK, Southwick CH (1984): Population growth of free-ranging rhesus monkeys at Tuglaqabad. Am J Primatol 7:311–321.

Maruhashi T (1982): An ecological study of troop fissions of Japanese monkeys (*Macaca fuscata yakui*) on Yakushima Island, Japan. Primates 23:317–337.

May RM (1974): "Stability and Complexity in Model Ecosystems." Princeton: Princeton University Press.

May RM (ed) (1985): "Exploitation of Marine Communities." Berlin: Springer Verlag.

Maynard Smith J, Slatkin M (1973): The stability of predator-prey systems. Ecology 54:384–391.

McFarland-Symington M (1987): Sex ratio and maternal rank in wild spider monkeys: when daughters disperse. Behav Ecol Sociobiol 20:421–425.

McGuire MT (1974): The St. Kitts vervet. Contrib Primatol, Vol 1, New York: Karger.

Meikle DB, Tilford BL, Vessey SH (1984): Dominance rank, secondary sex ratio, and reproduction of offspring in polygynous primates. Am Nat 124:173–188.

Menard N, Vallet D, Gautier-Hion A (1985): Demographie et reproduction de *Macaca sylvanus* dans differents habitats en Algerie. Folia primatol 44:65–81.

Mori A (1979): Analysis of population changes by measurement of body weight in the Koshima troop of Japanese monkeys. Primates 20:371–397.

National Research Council, Committee on Non-Human Primates (1981): "Techniques for the Study of Primate Population Ecology." Washington: National Academy Press.

Newton PN (1986) Infanticide in an undisturbed forest population of hanuman langurs, *Presbytis entellus*. Anim Behav 34:785–789.

Paul A, Thommen D (1984): Timing of birth, female reproductive success and infant sex ratio in semifree-ranging Barbary macaques (*Macaca sylvanus*). Folia Primatol (Basel) 42:2–16.

Rawlins RG, Kessler MJ, Turnquist JE (1984): Reproductive performance, population dynamics and anthropometrics of the free-ranging Cayo Santiago rhesus macques. J Med Primatol 13:247–259.

Rawlins RG, Kessler MJ (1986): Secondary sex ratio variation in the Cayo Santiago macaque population. Am J Primatol 10:9–23.

Richard AF (1985): "Primates in Nature." New York: W.H. Freeman and Company.

Scucchi S (1984): Interbirth intervals in a captive group of Japanese macaques. Folia Primatol (Basel) 42:203–208.

Smith DG (1982): A comparison of the demographic structure and growth of free-ranging and captive groups of rhesus monkeys (*Macaca mulatta*). Primates 23:24–30.

Southwick CH, Richie T, Taylor H, Teas HJ, Siddiqi MF (1980): Rhesus monkey populations in India and Nepal: Patterns of growth, decline, and natural regulation. In Cohen MN, Malpass RS, Klein HG (eds): "Biosocial Mechanisms of Population Regulation." New Haven: Yale University Press, pp 151–170.

Southwood TRE (1978): "Ecological Methods." London: Chapman & Hall.

Struhsaker TT (1976): A further decline in numbers of Amboseli vervet monkeys. Biotropica 8:211–214.

Sugiyama Y (1965): Behavioral development and social structure in two troops of hanuman langurs (*Presbytis entellus*). Primates 6:213–247.

Sugiyama Y (1984): Population dynamics of wild chimpanzees at Bossou, Guinea, between 1976 and 1983. Primates 25:391–400.

Sugiyama Y, Ohsawa H (1982a): Population dynamics of Japanese macaques at Ryozenyama: III. Female desertion of the troop. Primates 23:31–44.

Sugiyama Y, Ohsawa H (1982b): Population dynamics of Japanese monkeys with special reference to the effect of artificial feeding. Folia Primatol (Basel) 39:238–263.

Takahata Y (1980): The reproductive biology of a free-ranging troop of Japanese monkeys. Primates 21:303–329.

Tanaka T, Tokuda K, Kotera S (1970): Effects of infant loss on the interbirth interval of Japanese monkeys. Primates 11:113–117.

Teas J, Richie TL, Taylor HG, Siddiqi MF, Southwick CH (1981): Natural regulation of rhesus monkey populations in Kathmandu, Nepal. Folia Primatol (Basel) 35:117–123.

Terborgh J (1983): "Five New World Primates." Princeton: Princeton University Press.

van Schaik CP (1983): Why are diurnal primates living in groups? Behaviour 87:120–144.

van Schaik CP, van Noordwijk MA (1983a): Social stress and the sex ratio of neonates and infants among non-human primates. Neth J Zool 33:249–265.

van Schaik CP, van Noordwijk MA (1983b): Interannual variability in fruit abundance and the reproductive seasonality in Sumatran Long-tailed macaques (*Macaca fascicularis*) J Zool 206:533–549.

Whitten PL (1983): Diet and dominance among female vervet monkeys (*Cercopithecus aethiops*). Am J Primatol 5:139–159.

Wilson ME, Gordon TP, Bernstein IS (1978): Timing of births and reproductive success in rhesus monkey social groups. J Med Primatol 7:202–212.

Winkler P, Loch H, Vogel C (1984): Life history of hanuman langurs (*Presbytis entellus*). Folia primatol 43:1–23.

Wolfe LD (1984): Female rank and reproductive success among Arishiyama B Japanese macaques (*Macaca fuscata*). Am J Primatol 5:133–143.

Wolfheim JH (1982): Primates of the World. Distribution, Abundance and Conservation." Pullman, WA: University of Washington Press.

Wrangham RW (1974): Artificial feeding of chimpanzees and baboons in their natural habitat. Anim Behav 22:83–93.

Wrangham RW (1980): An ecological model of female-bonded primate groups. Behaviour 75:262–300.

APPENDIX A

The growth rate of a population with adult survival s, infant survival si, age at first reproduction f, and annual fecundity F, may be modeled using a Leslie matrix of the form

(column)		1	2	. . .	f	. . .	$n-1$	n	
(row)	1	0	0	. . .	F	. . .	F	F	
	2	si	0	. . .	0	. . .	0	0	
	3	0	s	. . .	0	. . .	0	0	(A1)
	.	.	.	. . .	.	. . .	.	.	
	.	.							
	.	.	.	. . .	.	. . .	.	.	
	n	0	0	. . .	—	. . .	s	0	

This matrix can readily be used to calculate r, the intrinsic growth rate of the population. This is given by $r = \ell n\ \lambda$, where λ is the dominant (and only) eigenvalue of the above matrix.

The basic reproductive rate of the population is calculated by using the matrix to give a life and fertility table for the species [see Southwood, 1978].

age	l_x	m_x	$l_x m_x$ $(=V_x)$
0	1	0	0
1	si	0	0
2	$s^2 i$	0	0
⋮	⋮	⋮	⋮
f	$s^f i$	F	$s^f iF$
⋮	⋮	⋮	⋮
n	$s^n i$	F	$s^n iF$

The basic reproductive rate of the population R_O is given as the sum of V_x, e.g., $R_O = \Sigma\ l_x m_x = Fi \sum_{j=f}^{n} s^j$. In the calculations performed for this paper it is assumed that $n = 30$. Models for specific populations would use more accurate estimates of s and F for each age class in the population.

APPENDIX B

The regression of age at first reproduction against interbirth interval in the main text (Fig. 6) suggests that it might be useful to rescale comparative models for primate populations so that they are normalized into time units of one interbirth interval. The population can thus be coarsely divided into three subpopulations, adults (A), juveniles (J), and infants (I). If it is also assumed that only the adults and juveniles contribute to the density-dependent mechanisms that regulate fecundity, then the dynamics of the population may be modeled by the following equation:

$$N_{t+1} = sA_t + sJ_t + s^2 iA_{t-1}\ F(N_{t-1}) \tag{B.1}$$

Here $N_t = A_t + J_t$, and other parameters are as defined in Appendix A. At equilibrium $N^* = N_{t+1} = N_t = N_{t-1}$, and the equation may be solved to determine N^*;

$$\frac{(1-s)}{s^3 i F} = f(N^*) = \exp(-aN^{*b}) \rightarrow N^* = \frac{1}{a}\left[\ell n \frac{s^3 i F}{1-s}\right]^{\frac{1}{b}} \quad \text{(B.2)}$$

The numbers of individuals in each section of the population are given by simple proportionality;

Adults	A^*	$=$	$sA^* + sJ^*$	$=$	sN^*	
Juveniles	J^*	$=$	$[(1-s)/s]A^*$	$=$	$(1-s)N^*$	(B.3)
Infants	I^*	$=$	$[(1-s)/si]A^*$	$=$	$[(1-s)/si]N^*$	

The discrete equation form of the model may be rewritten as a Leslie matrix (see Eq. 3). The stability properties of this matrix are determined by a linearized stability analysis [Beddington, 1974]. If M is the population matrix, then the eigenvalues of M_δ determine the properties of the population, where M_δ is given by

$$M_\delta = M_{(n^*)} + \mathrm{M}_{(n^*)} H \quad \text{(B.4)}$$

The matrices for the generalized life history described by Equation 3 are

$$M(n^*) = \begin{bmatrix} 0 & 0 & \frac{1-s}{si} \\ si & 0 & 0 \\ 0 & s & s \end{bmatrix}, \quad M'(n^*) = \begin{bmatrix} 0 & 0 & \Theta^* \\ 0 & 0 & 0 \\ 0 & 0 & 0 \end{bmatrix}$$

where

$$\Theta^* = F \exp[-aN^{*b}]\,[-ab(aN^*)^{b-1}]$$

and

$$H = \begin{bmatrix} \frac{(1-s)}{si}N & \frac{(1-s)}{si}N & \frac{(1-s)}{si}N \\ (1-s)\,N & (1-s)\,N & (1-s)\,N \\ sN & sN & sN \end{bmatrix}$$

The eigenvalues can be found by the characteristic equation

$$\lambda^3 + A\,\lambda^2 + B\,\lambda + C = 0 \tag{B.4}$$

For the system to be stable all the eigenvalues of M must be less than unity and lie within the unit circle. The Schur-Cohn criterion [see May, 1974] says that this will be the case when A, B, and C are less than zero, and when $1 - C^2 > |AC - B|$ and $|1 + B| > |A + C|$. In this system $A = -s$, $B = 0$, and $C = s^2 i[\Theta^* N^* + (1 - s)/si]$. As the two inequalities seem to be met for most values of s, i, F, and b, the system's behavior is dependent entirely on C. Thus when $b > 1/[\ell n(s^3 iF/(1 - s))]$ the population will tend to show damped oscillations when perturbed from equilibrium, and an asymptotic return when b is less than this (see Fig. 8).

Ecology and Behavior of Food-Enhanced Primate Groups, pages 199–228

10

Life-History Patterns of Barbary Macaques (*Macaca sylvanus*) at Affenberg Salem

Andreas Paul and Jutta Kuester

Affenberg Salem, 7777 Salem, Federal Republic of Germany

INTRODUCTION

Knowledge about life-history patterns and population dynamics is fundamental to any understanding of adaption of animals to their environment, since it is through life histories that selection operates [Dittus, 1975; Altmann and Altmann, 1979]. During the past two decades our knowledge about population dynamics of nonhuman primates has grown rapidly [see, for example, Richard, 1985]. Nevertheless, the highly complex and interdependent relationships among ecological, demographic, and social factors, as discussed by Altmann and Altmann [1979] and Dunbar [1979, 1985], are still poorly understood. This is largely because long-term studies on primate population biology and behavior are still rare and restricted to few species living under specific environmental conditions [see Richard, 1985]. Quality of the habitat has a significant influence on life-history patterns, and it is well known that food enhancement in general promotes fertility, longevity, and survival. However, food enhancement may also have deleterious effects on life-history variables by promoting competition between individuals [Sugiyama and Ohsawa, 1982; Wrangham, 1974]. In this paper we describe life-history patterns of Barbary macaques (*Macaca sylvanus*) living in a seminatural environment, and discuss implications of food-enhancement on demographic processes and the social organization of the population.

MATERIALS AND METHODS

Since 1976 the Barbary macaque population of "Affenberg Salem" has been living in a 14.5-ha outdoor enclosure in southwest Germany, near Lake Constance (Plate I). Climate in this region is temperate, but temperatures can reach +30°C during summer, and fall to −20°C during winter. The monkeys live outdoors throughout the year. Because of the hilly ground, the actual surface area of the enclosure is increased to about 18 ha. The area is

Plate I. Barbary macaques in the Salem enclosure.

surrounded by a 2.75-m-high fence with an electrified wire bordering the top. The park's vegetation consists mainly of a beech/spruce mixed forest (*Fagus sylvatica* and *Picea abies*). The park is free of predators, but dogs near the enclosure and low-flying birds of prey (buzzards, hawks, and kites) regularly elicit alarm calls. Similarly, movements of a roe deer (*Capreolus capreolus*) living in the park cause alarm barks.

Between March and October the park is open to visitors. Movement of visitors is restricted to a pathway. Most parts of the park are out-of-bounds for visitors. Visitors are allowed to feed the monkeys with unsweetened popcorn offered by the staff of the park. Contact frequency between monkeys and visitors varies greatly among individuals, different age/sex classes, as well as among groups. In general, contact is restricted to a very short period of the daily routine (unpubl. data).

The flora and fauna within the park offer a variety of natural foods, which the monkeys routinely utilize [see De Turckheim and Merz, 1984; Kaumanns, 1978]. Despite this, monkeys are fed daily. Wheat or oat grains and fruit or vegetables are spread widely on the ground. Commercial monkey chow and water are available ad lib in several feeding boxes and water basins, respectively (Plate II). Popcorn offered by visitors is an attractive

Plate II. Adult male (left, with infant) and adult female, feeding.

food, but not as attractive as peanuts or bananas. Its caloric value is negligible [De Turckheim and Merz, 1984; Kaumanns, 1978]. Highly attractive food normally is neither offered by the park staff nor by visitors.

Human interventions with the population during the period 1976–1983 were restricted to a minimum. In 1984 and 1985 large numbers of monkeys had to be removed from the enclosure because habitat destruction and escapes owing to severe intergroup aggressions increased. Escaped monkeys usually stayed near the park and returned within a few hours, but in some cases they traveled over large distances for several days and had to be recaptured. Outside the context of intergroup aggression the fence served as an effective barrier.

Routine medical treatment was restricted to a yearly vaccination of all animals against rabies, which is endemic in the area, and to preventive antihelminthic treatment performed 3–4 times a year. The influence of these measures on animal survival cannot be estimated. Veterinary treatment of sick or injured animals was seldom necessary and seemed to have little influence on demographic processes of the population. No outbreak of an epidemic or other severe infectious diseases that may have endangered the population ever occurred.

Intense, individually focused observations on demographic events were started in 1978. Data from the period before this date were thus only partially included in the analysis. Pregnant females were checked daily so that the majority of newborns were discovered within 24 hr of birth. Ages of all monkeys born in France (see below) and in Salem were known. Ages of older individuals, born in Morocco, are based on estimates. In the age classification, individuals less than 1 year old are considered infants. Juveniles are all individuals 1–3 years old. Four-year-old females, whether they gave birth or not, and 4–6-year-old males are classified as subadults. Males aged 7 years and older and females aged 5 years and older are considered adults.

Dominance rank of females was ascertained by the outcome of dyadic agonistic encounters and displacements. In one group (group B) the structure and dynamics of female dominance relations were recorded continuously throughout the whole study period by focal and ad lib observations. Dominance relations of females in two other groups (A2 and C) were ascertained by random tests and discontinuous observations. While it was not possible to determine the exact rank position of each female in these latter groups over the whole study period, data were sufficient to rank females as either high ranking or low ranking.

RESULTS

Population Size and Structure

Population growth and density. The original stock of the Salem Barbary macaque population consisted of 164 monkeys that were translocated from two large outdoor enclosures in France to the Salem enclosure in 1976. Owing to high birth rates and low mortality rates the population steadily increased to more than 500 monkeys at the end of 1983 (see Fig. 1). The observed exponential rate of increase during this time period on the average was $r = 0.17$ (SD = .010), corresponding to a doubling time of 4.1 years [calculated after Caughley, 1977], or an annual population growth of 18.3%. In 1984 and 1985, four social groups (see below) were removed from the park, resulting in an actual decrease of 35%. The remaining population still increased with $r = .19$ in 1984 and $r = .17$ in 1985.

Age/sex composition. The age/sex-composition of the entire population after the translocation (January 1977) and 7 years later (i.e., before large numbers of monkeys were removed) is shown in Table I. With a percentage of 56.1 immatures in 1977 and 52.5 in 1984 the population was a relatively young one. About one-third of the original 164 animals were feral born; the others were born in the French enclosures. The overall sex ratio was well

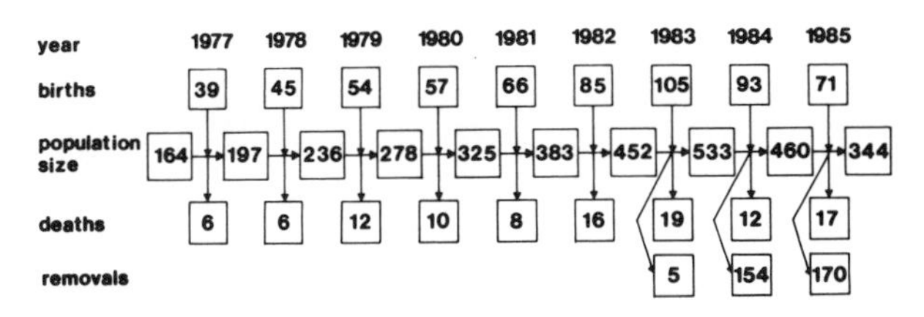

Fig. 1. Flow chart of demographic events in the Barbary macaque population 1977–1985.

TABLE I. Age/Sex Composition of the Salem Barbary Macaque Population 1977 and 1984 (Census Date: January 1)

	1977					1984				
	Males		Females			Males		Females		
Age class	n	%	n	%	F : M	n	%	n	%	F : M
Adults	15	18.5	39	47.0	2.6	64	25.1	113	40.6	1.8
Subadults	11	13.6	7	8.4	0.6	46	18.0	30	10.8	0.7
Juveniles	44	54.3	26	31.3	0.6	97	38.0	87	31.3	0.9
Infants	11	13.6	11	13.3	1.0	48	18.8	48	17.3	1.0
Total	81		83		1.0	255		278		1.1

balanced. Among adults (see Table I) and sexually mature members of the population, there was a slight preponderance of females (1 male per 1.8 females in 1977, and 1 male per 1.3 females in 1984). All these data are within the range of wild populations living in Morocco and Algeria (Table III).

Origin, size, and age/sex-composition of social groups. In 1976 the population consisted of one large, well-established group of 144 monkeys (70 males, 74 females), which was translocated from "La Montagne des Singes," Kintzheim, France (group I in Merz, [1976], later called group A1, see Fig. 2), and a small party of 20 (11 males, 9 females) juvenile and subadult individuals from "La Foret des Singes," Rocamadour, France. During the early stages of the park (in 1976 and early 1977) the "Rocamadour monkeys" were frequently harassed and chased by members of group A1 [Kaumanns 1978]. Some of the older "Rocamadour" males remained "outcasts" until the end of 1977.

During the early months of autumn 1977 group A1 underwent a process of fissioning. About the same time, the "Rocamadour monkeys" were integrated into the two emerging groups. All "Rocamadour" males immigrated into the splitting group B, while all but two of the "Rocamadour" females

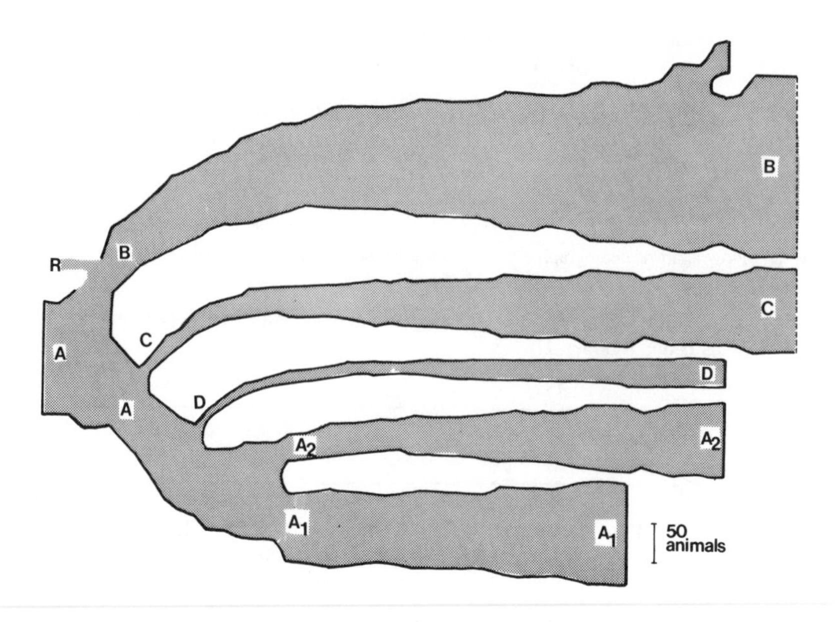

Fig. 2. History of social groups at Salem.

joined group A1, which contained the most central and high-ranking males of the former group I (Merz, pers. comm.).

The spatial and social separation of group B from A1 was complete by November 1977. At this time, group A1 contained 113 individuals; group B had 84 members. The sex ratio of the sexually mature members in the two groups was 1.9 females per male in group A1, and 1.5 females per male in group B (complete information on age/sex-composition of all social groups summarized in Tables II and III, respectively).

In the following years, group A1 underwent several other fissions (see below) and oscillated at around 100 individuals. In January 1984, the group was removed from the park.

Shortly after the formation of the two groups, during the winter months 1977/78, three adult males emigrated one after another from group B. They joined in January 1978 and interacted primarily with females and their young at the periphery of group A1. The social and spatial separation of this new formation, group C, was complete in spring 1978. At the onset of the 1978 birth season the group had 18 members with a sex ratio of sexually mature members of 2.3 females per male. By the end of the study, in December 1985, group C had grown to 110 individuals.

A third fission occurred in the winter months 1978/79, when one adult

TABLE II. Composition of Social Groups After Fissioning (M: Males; F: Females)

Group	A1		A2		B		C		D		E	
Census date	1978		1980		1978		1978		1979		1985	
Age class[a]	M	F	M	F	M	F	M	F	M	F	M	F
Adults	10	26	2	7	8	19	3	7	1	2	9	9
Subadults	6	4	3	4	7	3	0	1	0	0	5	1
Juveniles	27	19	1	6	21	11	0	1	0	5	3	8
Infants	11	10	2	5	8	7	2	4	1	1	5	4
Total	54	58	8	22	44	40	5	13	2	8	22	22

[a]Group C: April 1; all others, January 1.

TABLE III. Demographic Characteristics of Barbary Macaque Groups in Different Habitats (n = Counted Groups)

	Group size		SR adults		SR matures		% juveniles	
	Mean	Range	Mean	Range	Mean	Range	Mean	Range
Moyen Atlas n=11	25.3	12–39	1.8	0.6–2.7	1.2	0.5–2.5	46.0	33.3–55.6
Rif n=27	18.3	7–59	1.2	0.3–3.0	1.1	0.3–3.0	48.4	33.3–71.4
Grande Kabylie n=3	34.0	16–46	1.2	1.0–1.6	0.9	0.7–1.4	37.7	25.0–45.6
Salem n=6	84.3	10–243	2.3	1.0–7.0	1.3	0.6–2.7	52.5	38.9–70.0

Sources: Moyen Atlas: Deag [1984]; Taub [1984]; Rif: Fa [1982]; Mehlman [1986]; Whiten and Rumsey [1973]. Grande Kabylie: Menard et al. [1985]; Salem data based on 40 censuses.

male together with two adult females (presumably a mother and her adult daughter) and their immature offspring separated from group A1. This group, called group D, initially contained only 10 animals. It remained a one-male group for about 1 year, until another adult male transferred from group B to group D (later other males, too, joined group D). Group D increased to 35 individuals in 1984, but remained the smallest group in the park. Having had a relatively undisturbed life during the first years, the group was subjected to increasing social pressure from other groups beginning in spring of 1982. During the early spring months especially, the group was frequently chased and harassed by other groups. Many individuals were wounded, and three newborns were kidnapped by members of other groups (two of them, kidnapped by adult males, died within a few days, a third was adopted and raised by an adult female who already had a baby). In spring 1985 group D was removed from the enclosure.

During the winter 1979/80 another fission of group A1 took place. The new group, A2, initially had 30 members (sex ratio of the sexually matures: 2.2 females per male). The group increased to 93 individuals in January 1985, and was thereafter removed from the enclosure.

The last group, group E, originated from a fission of group B that was completed in winter 1984/85. Group B at this time had more than 200 members. Group E consisted of 44 individuals with a sex ratio of 0.7 sexually mature females per mature male. Like group D, but in this case shortly after the formation, group E came under social pressure from former groupmates. As with groups A2 and D, group E was removed from the enclosure in April 1985.

Summarizing data on group size and composition (Table III), we can clearly see that the Salem groups differ most strikingly from wild groups in size, a feature commonly associated with food enhancement [e.g., Koyama et al., 1975]. Nonetheless, group size was highly variable, and size of most fissioned groups initially fell well within the range of wild groups. Sex ratios of the adult and sexually mature group members were slightly higher in the Salem population than in wild populations, but were clearly within the range of wild groups. This holds true also for the percentage of immatures, but there exists a more substantial difference. In Salem the percentage of immatures was above 50%, indicating *increasing* population size, while in all wild populations there were less than 50% immatures, suggesting *decreasing* population size [Southwick et al., 1980].

Natality

Reproductive seasonality. Distribution of births in the Salem Barbary macaque population (data 1978–1985) was highly seasonal (Kolmogorov-Smirnov goodness-of-fit test, $D = 0.46$, $P << 0.001$, $n = 568$) (Fig. 3). Birth seasons started mid-March (first births March 10–21) and ended late June to mid-August (last births June 17 to August 23). More than 90% of all births occurred in the months of April to June. Mean birth date was May 24 ($SD = 24.8$ days); median birth date, May 9. The mating season in Salem lasted from the middle of August to the end of March, with a peak of mating activity in November. The vast majority of females conceived until the end of December [Kuester and Paul, 1984].

Date of birth depended on age and reproductive history of the mother. Young primiparous females (4 years old) gave birth on the average 3 weeks later than older primiparous females. Multiparous females with an infant from the preceding birth season gave birth later than multiparous females who had a sterile year or infant loss in the preceding season (Table IV).

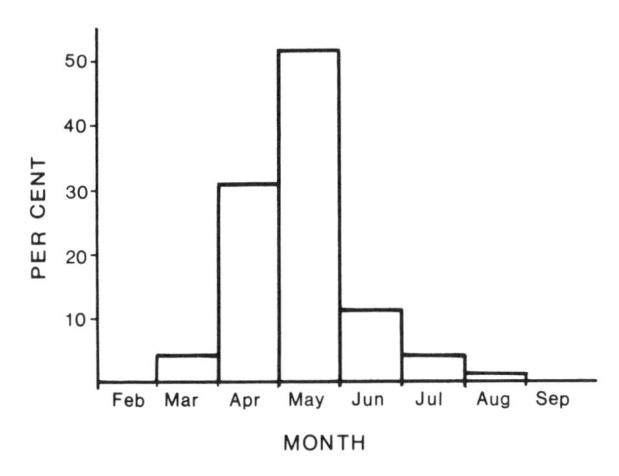

Fig. 3. Distribution of births (1977–1985).

TABLE IV. Timing of Birth and Age/Reproductive History of Mothers

Reproductive history	Age (yr)	Mean birth date	SD (days)	Median birth date	n
Primiparous	4	4 June	13.9	5 June	40
Primiparous	5+	20 May	16.5	19 May	67
Multiparous[a]	5–9	3 June	18.9	2 June	172
Multiparous[a]	10–14	21 May	24.4	17 May	111
Multiparous[a]	15+	16 May	28.8	12 May	69
Multiparous[b]	5+	16 May	18.0	16 May	29
Multiparous[c]	5+	3 May	24.9	1 May	61

[a]With surviving infant.
[b]After infant loss.
[c]After sterile year.

When the infant from the last season survived, older multiparous females tended to give birth earlier than younger ones [see also Paul and Thommen, 1984]. Rank of the mother apparently had no influence on timing of birth [Paul and Thommen, 1984; and unpubl. data].

Age-specific fertility rates. Fertility or birth rates refer to the number of stillborn (n = 12) and liveborn (n = 603) infants per sexually mature female (aged 4 years and older). Two births (0.33%) that consisted of twins were considered as singletons. Two sterile females were excluded from analysis.

Barbary macaque females at Salem delivered their first baby at the age of 4–7 years, mean age at first birth was 4.71 years (SD = 0.602, n = 129) (Note that our age classification differs somewhat from that of Fa [1984].

TABLE V. Age-Specific Birth Rates of Barbary Macaque Females at Salem (Data 1978–1985)

Age (yr)	Reproductive years (n)	Births (n)	Birth rate (bx)
3	151	0	.000
4	128	48	.357
5	107	81	.757
6	93	73	.785
7	69	64	.928
8	47	45	.957
9	40	37	.925
10	36	34	.944
11	31	29	.935
12	26	25	.962
13	25	20	.800
14	23	20	.870
15	28	23	.821
16	23	23	1.000
17	14	13	.929
18	16	12	.750
19	13	9	.692
20+	44	17	.386

Because we do not know exact birth dates of monkeys born before 1978, a female born in July 1980 who delivered her first baby in June 1984 is categorized here as a 4-year-old female. In Fa [1984], this female would be categorized as a 3-year-old female.) Females continued to reproduce until the age of about 25 years. The postreproductive lifespan may be as long as 7 years (one case, 5 years in two other cases). Birth rates were relatively low in young females (aged 4–6 years, Table V). This was due to the fact that only a minority of females had the physical capability to conceive at the age of 3.5 years, and to a high frequency of sterile years after the first infant [Paul and Thommen, 1984]. High birth rates were found among middle-aged females (7–12 years). Older females (13–20 years) had slightly lower birth rates, and after the age of 20 years, birth rates declined sharply.

Demography and birth rates. Yearly variation between birth rates in the population was low ($\bar{x}$ = 0.758, SD = 0.030; chi-square goodness-of-fit test, $\chi^2 = 0.961$, df = 8, $P > 0.99$). Similarly, variation among social groups was low ($\bar{x}$ = 0.758, SD = 0.045; $\chi^2 = 1.361$, df = 4; $0.8 < P < 0.9$).

There was a slight, but statistically nonsignificant, negative correlation between population size (censused during the preceding mating season, January 1) and birth rate in the population as a whole ($r_s = -0.417$, n =

TABLE VI. Annual Birth Rates in the Salem Colony

Group	1977	1978	1979	1980	1981	1982	1983	1984	1985	Total
A1	.765	.731	.800	.750	.773	.833	.727			.767
A2				.909	.769	.706	.818	.778		.789
B		.773	.731	.767	.694	.660	.679	.703	.783	.718
C		.889	.875	.778	.692	.842	.818	.731	.880	.809
D			1.000	.400	.714	.750	.700	.727		.705
Total	.765	.772	.791	.760	.725	.739	.734	.727	.812	.752

TABLE VII. Spearman Rank Correlations between Birth Rates and Demographic Parameters (HF-Females = Highly Fecund Females, see text.)

	Group				
Demographic parameter	A1	A2	B	C	D
Population Size	0.000	−.500	−.238	−.357	−.143
Group Size	−.402	−.500	−.238	−.238	−.143
Breeding Sex Ratio	+.257	−.300	+.095	+.500	−.171
Prop. HF-Females	+.321	+.100	+.690*	+.190	+.243

*$P<0.05$.

9, $P > 0.1$). This trend was even less marked when groups were analyzed separately (Table VII). Notably, group D, which was the only group under obvious social pressure from other groups (except group E, which cannot be considered here), showed a negligible correlation.

Group size did not affect birth rates in any obvious manner (Table VII). Since group size was proportional to the number of adult females/group, neither of these parameters apparently influenced birth rates. Similarly, there was no correlation between birth rates and the number of sexually mature females (age 3.5+ years) per adult males (age 7.5+ years) during the preceding mating season ("breeding sex ratio," Table VII).

Because fecundity in Barbary macaques is age dependent (see above), it seems reasonable to assume a relationship between birth rates and the age structure of the female population [Dunbar, 1979]. There was, indeed, a significant positive correlation between the proportion of highly fecund females ("HF"-females, aged 7–12 years, see Table VII) and birth rates in the population as a whole ($r_s = +0.667$, $n = 9$, $P < 0.05$). At the group level, this correlation was significant only in the largest group (group B, Table VII). The lack of significance in other groups seems plausible, because a large sample size is necessary when the variation is low. Differences in birth rates among groups were apparently related to differences in the mean proportion of highly fecund females ($r_s = +0.900$, $n = 5$, $P = 0.05$).

TABLE VIII. Female Rank and Birth Rate in Three Social Groups at Salem (RY = Reproductive Years)

Group	Rank	RY (n)	Births (n)	Birth rate	χ^2
A2	High	45	36	.800	0.000
	Low	45	35	.778	
B	High	163	121	.742	0.451
	Low	179	126	.704	
C	High	66	58	.879	4.860*
	Low	65	46	.708	
Total	High	274	215	.785	0.415
	Low	289	207	.716	

*$P<0.05$.

Female rank and fertility. In all three groups, where dominance rank of females could be determined, high-ranking females had higher birth rates than low-ranking females (Table VIII). However, differences were small, and only in one group (group C) did the difference reach statistical significance. The degree of difference did not depend on group size (it may be a difference to be Omega among 5 females or to be Omega among 20 females [see Dunbar, 1979]). Rank-related differences were not biased by differences in age structure of high- and low-ranking females [see data in Paul and Kuester, 1987].

Part of the rank-related variation in birth rates was clearly attributable to differences in maturation or growth rates. Daughters of high-ranking mothers delivered their first infant more frequently at the age of 4 years (14 out of 24), while daughters of low-ranking mothers began to reproduce more frequently at the age of 5 years or later (21 out of 28; $\chi^2 = 4.98$, df = 1, $P < 0.05$).

Secondary sex ratio. With 91.8 males per 100 females the sex ratio at birth in the population was slightly female-biased ($P = 0.312$, binomial test, two-tailed). A female-bias occurred in 6 out of 9 years, and in four out of five social groups (Table IX). Differences in the proportion of males (PM) between social groups were low ($\bar{x} = 0.48$, SD = 0.019, range = 0.45–0.50). There was no correlation between group size and the proportion of males born into these groups ($r_s = 0.071$, n = 9). Similarly, there was apparently no correlation between population density and overall secondary sex ratios ($r_s = 0.038$, n = 9). However, such a correlation may be masked by the effect of group fissions that took place between 1977 and 1979.

The PMs of newly formed groups in the year after formation were

TABLE IX. Proportion of Males Born at Salem 1977–1985

Group	1977	1978	1979	1980	1981	1982	1983	1984	1985	Tot.
A1	.51	.63	.26	.43	.41	.55	.58			.49
A2				.40	.60	.42	.44	.43		.45
B		.28	.42	.61	.56	.45	.45	.44	.51	.47
C		.25	.57	.29	.78	.69	.50	.53	.36	.50
D			.00	.50	.80	.17	.83	.43		.48
Total	.51	.42	.35*	.48	.58	.49	.51	.46	.46	.48

*$P<0.05$, binomial test.

TABLE X. Proportion of Males (PM) Born to Mothers of Different Age Classes (Data 1978–1985)

Age (yr)	PM	Births (n)
4–5	.492	126
6–7	.482	137
8–9	.451	82
10–11	.531	64
12–13	.467	45
14–15	.465	43
16–17	.444	36
18–19	.476	21
20+	.353	17

significantly lower than PMs in well-established groups (Mann-Whitney U-test, n1 = 4, n2 = 30, U = 113, $P < 0.005$, two-tailed). If we exclude PMs of these newly formed groups from analysis, PMs tended to decrease with increasing population density ($r_s = -0.417$, n = 9, $P > 0.1$). Thus, the results suggest that high population density and unstable social relationships caused by the process of group fissioning may lead to a preponderance of females in the secondary sex ratio.

Both female age and rank apparently had an influence on sex of progeny. Secondary sex ratios decreased with increasing age of mothers ($r_s = -0.633$, n = 9, $P < 0.05$, Table X). In all groups where female rank could be determined, high-ranking females delivered higher proportions of males than low-ranking females (Table XI) (for a more detailed analysis see Paul and Kuester [1987]).

Mortality

Crude death rates. Crude death rates refer to the yearly number of deaths as the percentage of the average group/population size (Table XII). Between 1977 and 1985 a total of 115 deaths (including 12 stillbirths) in the population was recorded. Nine deaths of animals belonging to groups that

TABLE XI. Female Rank and Secondary Sex Ratio in Three Social Groups at Salem (Births of 4-Year-Olds Excluded; PM = Proportion of Males)

Group	Births (n)	Rank	
		PM high	PM low
A2	53	.50	.44
B	211	.52	.45
C	91	.55	.46
Total	355	.53	.46

TABLE XII. Crude Death Rates in the Salem Barbary Macaque Population 1977–1985

Group	1977	1978	1979	1980	1981	1982	1983	1984	1985	Mean (SD)
A1	4.67	1.74	5.33	7.10	0.00	3.81	5.83			4.07 (2.459)
A2				2.86	0.00	6.87	0.00	3.61		2.65 (2.831)
B		3.33	4.88	0.83	3.57	3.13	2.69	2.27	4.65	3.17 (1.299)
C		4.55	3.17	4.30	3.39	1.45	3.59	1.14	6.93	3.57 (1.827)
D			0.00	0.00	5.13	8.16	13.79	9.09		6.03 (5.434)
Total	3.32	2.77	4.67	3.32	2.26	3.83	3.86	2.86	5.38	3.59 (.976)

were removed from the enclosure during the year of deaths were excluded from the sample. Variation in annual mortality rates in general (range = 2.26–5.38), as well as variation in mean mortality rates between groups was low (range = 2.56–6.03, see Table XII). There was no general relationship between population density and mortality rate in the population as a whole, as well as in most social groups (Table XIII). However, mortality rate in group D, which was frequently chased and harassed by members of other groups, was significantly correlated with increasing population density (mortality rate in this group might have been even higher if the group had not been sometimes protected against attacks from other groups). Thus, subunits of a population can suffer from crowding effects, while no damaging effect may be detectable in the entire population.

Infant mortality. Analysis of infant mortality was based on 569 infants born between 1977 and 1985. Infants born into groups that were removed from the enclosure before completion of the first year of life were excluded from the sample.

TABLE XIII. Spearman Rank Correlations between Crude Death Rates and Average Population Size

Group	r_s
A1	+.143
A2	−.025
B	−.214
C	−.071
D	+.986*
Population	+.379

*$P<0.01$.

Annual infant mortality rates ranged from 4.5% to 13.0% (mean = 9.6%, SD = 2.99). Altogether infant mortality was not correlated with population density (r_s = +0.233, n = 9, not significant [NS]), but in group D mortality rate increased with increasing population density (r_s = +0.955, n = 7, $P < 0.01$). Aside from this, survival rates of liveborn infants in all groups were high ($\bar{x}$ = 90.8%, SD = 4.76, range = 82.8–95.5%, n = 5), and overall survival rate of infants born in group D (82.8%) did not differ significantly from survival rates in the other groups (χ^2 = 2.304, df = 1, NS, data from other groups pooled).

Age- and sex-specific mortality and survival rates of infants are presented in Table XIV. In five cases in which the exact date of death was not known we prorated sample size *fx* over those months of life during which death was most likely to have occurred. In general, mortality rate of male infants was somewhat although not significantly higher than that of female infants (χ^2 = 0.255, df = 1, NS). A significant difference, however, was found in neonatal deaths (i.e., during the first month of life), where mortality risk for male infants was considerably higher than for female infants (χ^2 = 6.509, df = 1, $P < 0.02$). Later on, mortality rates were slightly, but not significantly, higher for female infants (χ^2 = 1.858, df = 1, NS).

In all three groups, where female rank could be determined, infant mortality was slightly, but not significantly, higher in offspring of low-ranking females than in offspring of high-ranking females (Table XV). Altogether, mortality rate was 7.0% in offspring of high-ranking females, and 10.6% in offspring of low-ranking females (χ^2 = 1.33, df = 1, NS). The difference in rank-related mortality rates was not dependent on group size (see also rank-related birth rates).

Life table. The life-table data presented here are based on age-specific mortality rates (*qx:* proportion of animals alive at age *x* that die before age *x+1*), because of the limitations of calculating life tables from standing age distributions [see Caughley, 1977; Dunbar, 1979]. Key-data of age *x* was

TABLE XIV. Survivorship of Male and Female Infants Born at Salem 1977–1985

Age × month	Sample *fx* ♂♂	Sample *fx* ♀♀	Mortality rate *qx* ♂♂	Mortality rate *qx* ♀♀	Survival rate *px* ♂♂	Survival rate *px* ♀♀	Survival *lx* ♂♂	Survival *lx* ♀♀
0[a]	271	298	.018	.017	.982	.983	1.000	1.000
0,1	266	293	.056	.014	.944	.986	.982	.983
1,2	251	289	0	.007	1.000	.993	.927	.969
2,3	251	287	0	0	1.000	1.000	.927	.962
3,4	251	287	.001	.002	.999	.998	.927	.962
4,5	250.8	286.5	.001	.007	.999	.993	.926	.961
5,6	250.6	284.5	.005	.014	.995	.986	.925	.955
6,7	249.4	280.5	.017	.012	.983	.988	.921	.940
7,8	245.2	277	.001	0	.999	1.000	.905	.929
8,9	245	277	0	.007	1.000	.993	.904	.929
9,10	245	275	0	.005	1.000	.995	.904	.923
10,11	245	273.5	.004	.002	.996	.998	.904	.918
11,12	244	273	0	0	1.000	1.000	.900	.916

[a]Stillbirths.

TABLE XV. Female Rank and Infant Mortality

Group	Rank	Births (n)	Deaths (n)	Mortality rate	χ^2
A2	High	36	1	.028	.922
	Low	35	4	.114	
B	High	121	9	.074	.326
	Low	113	13	.103	
C	High	58	5	.086	.003
	Low	46	5	.109	

May 9 (i.e., the median of the birth season) because exact birth data of older animals were not known (only during the first year of life did we use exact birth dates).

Life-table data (Table XVI) show low mortality and high survival rates in both sexes (Fig. 4). More than 80% of all males and females reached adulthood, and nearly 45% of the males and 60% of the females were still alive at the age of 20 years. During the juvenile stage mortality rate was slightly higher for females than for males (1.8% vs. 1.3%), but this difference was not statistically significant ($\chi^2 = 0.174$, df $= 1$, $P > 0.50$). Aside from comparably high mortality rates in both sexes during infancy and late adulthood, mortality rates were relatively high for older subadult (6-year-old) and young adult (7–11-year-old) males.

TABLE XVI. Life Table for Barbary Macaques at Salem (1977–1985)

Age x	Sample *fx*		Mortality rate *qx*		Survival rate *px*		Survival *lx*	
years	♂♂	♀♀	♂♂	♀♀	♂♂	♀♀	♂♂	♀♀
0	271	298	.018	.017	.982	.983	1.000	1.000
0,1	266	293	.083	.068	.917	.932	.982	.983
1,2	199	228	.020	.035	.980	.965	.900	.916
2,3	181	183	.006	0	.994	1.000	.882	.884
3,4	163	151	.012	.013	.988	.987	.877	.884
4,5	138	126	0	.008	1.000	.992	.867	.873
5,6	111	113	.009	0	.991	1.000	.867	.866
6,7	85	91	.047	.033	.953	.967	.859	.866
7,8	70	68	.043	.044	.957	.956	.818	.837
8,9	55	51	.073	0	.927	1.000	.783	.800
9,10	46	45	.043	.022	.957	.978	.726	.800
10,11	38	37	.053	0	.947	1.000	.695	.783
11,12	28	31	.036	0	.964	1.000	.658	.783
12,13	23	29	0	0	1.000	1.000	.634	.783
13,14	20	29	.050	0	.950	1.000	.634	.783
14,15	15	30	0	0	1.000	1.000	.603	.783
15,16	10	28	0	.036	1.000	.964	.603	.783
16,17	9	25	.111	.080	.889	.920	.603	.754
17,18	6	23	.167	.087	.833	.913	.536	.694
18,19	4	18	0	.056	1.000	.944	.446	.634
19,20	1	14	0	0	1.000	1.000	.446	.598
20+	3	44	0	.068	1.000	.932	.446	.598

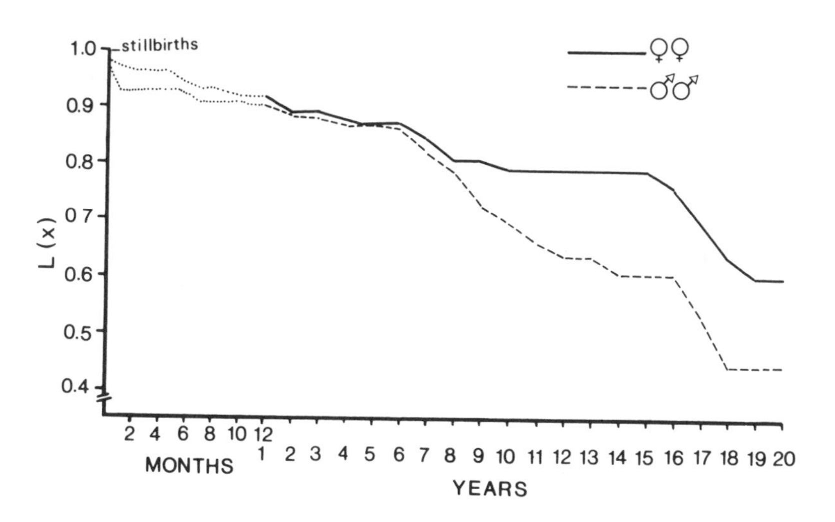

Fig. 4. Survivorship curves for male and female Barbary macaques at Salem.

TABLE XVII. Annual Migration Rates of Barbary Macaque Males at Salem 1978–1985

Year	No. males[a]	No. migrations	Migration rate
1978	79	5	.063
1979	97.5	6	.062
1980	111	19	.171
1981	122.5	5	.041
1982	145	7	.048
1983	175.5	21	.120
1984	145.5	6	.041
1985	127	2	.016

[a]Average number of males 2 years and older.

Although there was no obvious relation between population density and overall mortality, many deaths were behaviorally induced, and these cases were responsible for much of the age/sex differences in mortality rates. At least 12 (probably more) deaths of neonates (8 males, 3 females, 1 unsexed) were due to starvation caused by "allomothering to death." Kidnappers were mostly males; only three of the infants were kidnapped by females, one of them by his grandmother. Male and female Barbary macaques regularly handle infants, and in general do this in a very gentle and careful manner. Thus, in all cases of "allomothering to death" it was apparently the inability of mothers to get their infants back that caused the death of infants.

Behaviorally induced mortality was also responsible for the comparably high death rates in the older subadult and young adult males. At least seven of these males (50%) died through heavy injuries, indicating that intermale competition substantially influences mortality rates in these age classes. In this respect, it is worth noting that males of these age classes are regularly peripheralized during the mating season [see also Taub, 1980; Fa, 1984].

Migrations

Male transfer. Between 1978 and 1985—i.e., the period in which two or more fully established social groups were present in the park—71 intergroup transfers of males (and only one female transfer) occurred. Annual migration rates (Table XVII) varied from year to year; on the average 7% of all males 2 years and older (SD = 0.051) emigrated from their previous social group each year.

Age-specific migration rates are shown in Table XVIII. Intergroup transfer was absent in infant and yearling males, and rare in 2-year-old males, but

TABLE XVIII. Age-Specific Migration Rates of Barbary Macaque Males at Salem 1978–1985

Age	No. monkey-years of observation	No. of migrations	Migration rate
0	255.5	0	.000
1	197	0	.000
2	180.5	4	.022
3	162	22	.136
4	138	15	.109
5	111	10	.090
6	83	3	.036
7	68.5	7	.102
8	53	1	.019
9	45	3	.067
10+	154	6	.039

common around sexual maturity, which is reached at the age of 4.5 years (ejaculatory copulations were rarely observed in 3.5-year-old males but frequently in 4.5-year-old males). Three- and four-year-old males often migrated in the company of peers, while adult males generally migrated alone (except in the cases of group fissions; see below). Several males transferred more than once between groups, but no male was known to return to his natal group [see also Pusey and Packer, 1987].

The apparent effect (and one of the suggested ultimate causes [see, for example, Pusey and Packer, 1987]) of sex-biased dispersal, typical of most Old World monkeys and Salem Barbary macaques also, is the separation of sexually mature male and female relatives. However, the emigration of single adult, or small cohorts of juvenile and subadult males, seems not to be the only way leading to the separation of male and female relatives. Group fissions in Barbary macaques may serve the same function. Evidence for this hypothesis comes from the recent fission of group B, where genealogical relationships of all members of the fissioned group were known. The fissioned group E contained three female genealogies, but none of the adult and subadult male members of these genealogies (n = 4) transferred into the new group. One of the three juvenile male members of the three genealogies returned to group B a few weeks after the fission. On the other hand, four natal males from group B (besides nine other mature males born in other groups) joined group E (note that this form of intergroup transfer is not included in the migration rates presented above). One of the residing B genealogies lost all her mature males by this fission.

Although we have no complete information about the genealogical relationships of all adult monkeys of the population, the composition of all

TABLE XIX. Percentages of Natal Sexually Mature Males in Social Groups at Salem (Census Date: May 1, 1985)

Group	A2	B	C	D	E
Adults	0.0(0/10)	50.0(12/24)	0.0(0/10)	0.0(0/3)	0.0(0/9)
Subadults	69.2(9/13)	95.0(19/20)	46.2(6/13)	50.0(2/4)	20.0(1/5)
Total	39.1(9/23)	70.5(31/44)	26.1(6/23)	28.6(2/7)	7.1(1/14)
Group size	92	205	94	35	40
n female genealogies	8	17	7	1	3

other fissioned groups together with our knowledge about the history of their members strongly suggest that this pattern of sex separation also took place in the other group fissions.

Paul and Kuester [1985] suggested that most Barbary macaque males leave their natal group before reaching adulthood. This notion is further supported by the data on the percentage of sexually mature natal males in all groups during the birth season 1985 (Table XIX). Only in group B, which had at this time about 200 members, was there a high proportion of adult natal males. Similarly, the proportion of subadult natal males was considerably higher than in the other groups. Moreover, all subadult natal males in the small groups D and E were 4-year-olds; i.e., they had not yet mated. Thus, male emigration rates may be strongly influenced by group size, or, probably much more important, by the number or proportion of closely related females (see Table XIX).

Female transfer. Only one female transferred between groups as a result of natural causes. Because female transfer is a seldom-observed event in most Old World monkey species, this case, and the integration of two other females, belonging to groups whose other members had been removed, should be mentioned briefly.

The female that transferred owing to natural causes was 7 years old. She had a female infant, two juvenile sons, and an 8-year-old sister. Her mother died 1 year before. In December 1985—i.e., in the mating season—she transferred from her natal group B to group C. Having sometimes mated with young males of group C at the periphery of the group during her first estrus, she completely transferred into group C during her second estrus, which was also the conceptional estrus. After the mating season she remained in group C. While her infant was seen frequently in contact with her, interactions with her other relatives were observed only occasionally. Resident females of group C sometimes attacked her, but intensity and frequency of these aggressions remained low. While she was a middle-ranking female in her old group, she did not gain more than the Omega-status among the adult females in the new group. Similarly, a 6-year-old daughter of the Alpha-female of the removed group D that also immigrated into group C during the birth season 1985 became the Omega-female in the new group (until the immigration of the B-group female). This female was integrated at the day she gave birth, indicating that the black infant promoted the integration process. She also received little aggression from resident females.

A third female that belonged to the removed group, E, was rapidly reintegrated into her former group, B. She immediately gained her old status in the female hierarchy in group B.

DISCUSSION

The Salem Barbary macaque colony is characterized as a rapidly growing population with one of the highest growth rates ever observed in a macaque population for a longer time period [compare Malik et al., 1984; Richard, 1985]. Rhesus monkey populations of Cayo Santiago and La Parguera, which live under similar conditions as the Salem macaques, had growth rates of 13–16% [Drickamer, 1974; Koford, 1966; Rawlins et al., 1984]. Southwick et al. [1980] suggested that under favorable conditions macaque populations can increase at rates of 10–16% per year. The recently reported rate of increase of 21.4% over a 3-year period in free-ranging rhesus monkeys at Tughlaqabad, India, was considered as a record growth rate by the authors [Malik et al., 1984].

Between 1976 and 1985, on the average, 375 monkeys lived in the Salem enclosure, that is, about 20 monkeys per ha (considering an actual surface area of about 18 ha). This is a 40-fold higher density than in "high-density" regions in the central zone of the Moyen Atlas, where population density varies between 40 and 70 monkeys/km^2 (i.e., about 0.5 monkeys/ha [see Deag, 1974, 1984; Fa et al., 1984; Taub, 1977]). Home ranges of wild groups seem to be at least tenfold larger than the Salem enclosure [Deag, 1974; Taub, 1978; Mehlman, 1986].

Thus, from a standpoint of pure counting, the Salem Barbary macaque population is clearly a crowded one when compared with wild populations. However, the term "crowding" is not only a descriptive one. It implies several destructive effects concerning social, behavioral, and physiological parameters [Christian, 1961]. Before discussing if and to what extent biology and behavior of the Salem macaques was affected by crowding effects we will consider some other aspects of population dynamics.

Social and reproductive units in the wild as well as in Salem are multimale/multifemale groups. One-male groups, like group D during the first year, are also seen occasionally in the wild [Fa, 1982; Mehlman, 1986]. Age/sex-composition of the Salem groups did not differ largely from that of wild groups (see Table III). However, group size of the Salem population was considerably higher than in the wild, where group size rarely exceeds 30 individuals.

Data on reproductive seasonality resemble those of the Moyen Atlas region where most births seem to occur in the months of April to June [Deag, 1984]. Similar data come from the Barbary macaque populations in Gibraltar and Kintzheim, respectively [De Turckheim and Merz, 1984; Fa, 1984; MacRoberts and MacRoberts, 1966]. Median birth dates in Gibraltar lay in June and July, i.e., 1–2 months later than in Salem [Fa, 1984]. In Kintzheim, a peak of births was observed in April, i.e., one month earlier than in Salem [De Turckheim and Merz, 1984].

The pattern of age-specific fertility, i.e., an increase of fertility during the first years of the reproductive life and a decrease in aging females, corresponds to most other data on fertility rates of nonhuman primates available now [e.g., Dittus, 1975; Drickamer, 1974; Dyke et al., 1986; Fa, 1984; Koyama et al., 1975; Sade et al., 1976; Sugiyama and Ohsawa, 1982; Strum and Western, 1982]. Age of onset of breeding also corresponds well to data from the Gibraltar population [Burton and Sawchuk, 1982; Fa, 1984]. There is some evidence that Barbary macaques in Algeria deliver their first infant at a later age than in the food-enhanced colonies of Salem and Gibraltar [Menard et al., 1985]. Such an effect of food supply on rate of maturation was also found in other macaques and baboons [Altmann et al., 1977; Sugiyama and Ohsawa, 1982]. However, Mehlman (personal communication) also found high birth rates of 4-year-old females in wild Barbary macaques, suggesting that a quick maturation cannot only be achieved under conditions of food-enhancement, but—at least occasionally—also under natural conditions.

A similar age of last reproduction as we found (about 25 years) was reported for captive or provisioned rhesus and Japanese macaques [e.g., Koyama et al., 1975; Van Wagenen, 1972]. In contrast, Fa [1984] reported that Barbary macaque females in Gibraltar stopped reproducing at the age of 20 years. Moreover, he classified females older than 13 years already as "very old" ones, because of their low birth rates. However, birth rates were considerably lower in all age classes (except the youngest) in Gibraltar than in Salem [compare Fa, 1984: Table 11.13]. This point is further stressed below.

When regarding the "menopause debate" [see Small, 1984], it is noteworthy to point out the length of the postreproductive life span in apparently most of the Salem females [compare Dittus, 1975]. This is certainly influenced by food enhancement and absence of predators, since all these postreproductive females showed signs of senility and weakness.

It is frequently assumed that demographic factors such as population density, group size and composition, and even intergroup relations affect female fecundity patterns [e.g., Dunbar, 1979, 1985; Dunbar and Sharman, 1983; Van Schaik, 1983; Van Schaik and Van Noordwijk, 1983; Wolfe et al., 1986]. The underlying idea of many of these suggestions is the concept of female competition, as elucidated by Silk [1983], Wasser [1983], and others. Female competition theory predicts that socially mediated reproductive suppression is the favored mechanism through which females compete to improve the relative quality of their offspring [Wasser, 1983]. Moreover, in species showing male-biased dispersal, as most Old World monkeys, females are expected to limit selectively the numbers of females born and raised in the group [Silk, 1983]. The competition intensity clearly depends on socioenvironmental and demographic factors like population density, resource availability, and group size—which is inversely correlated with the average degree of relatedness between individuals. Food enhancement does not necessarily prevent female competition but could, on the other hand, promote it [Sugiyama and Ohsawa, 1982]. The results of this study did not confirm the predictions of female competition theory. Birth rates depended neither on population density [see also Strum and Western, 1982], group size [see also Dittus, 1975], nor on high numbers of adult males per female, which may protect lower-ranking females from harassment by higher-ranking females [Dunbar and Sharman, 1983]. The only demographic factor apparently influencing variations in birth rates in the Salem population was the age structure of the female population.

These findings are even more accentuated by the fact that birth rates in the low-density Gibraltar colony were significantly lower than in the high-density Salem colony: 0.752 (Salem) vs. 0.549 (Gibraltar, $\chi^2 = 55.04$, $P < 0.001$, Gibraltar data calculated after Fa [1984]). This contrast does not refer to possible differences in the age structure of both populations, because nearly all age classes in Gibraltar showed lower fertility rates than in Salem (see above). Fa [1984] referred the low fertility rates in Gibraltar to deleterious effects of the manner of provisioning (this volume, chapter 3), but it also seems reasonable that population size and structure [i.e., the high degree of relatedness, see Burton and Sawchuk, 1984] may have some negative impact on fertility rates in Gibraltar.

Female competition theory predicts that intense competition should promote social differentiation between females, since low-ranking females should suffer more from stress induced by harassment of high-ranking females. They are, therefore, expected to deliver and raise fewer, and especially fewer female, offspring than do high-ranking females. There is little evidence from data on natality and mortality to support this prediction. Indeed, in all groups low-ranking females delivered and raised fewer offspring than did high-ranking females. Differences, however, in general were low, and the only significant difference between high-ranking and low-ranking females was the age at first birth. This suggests that different fertility rates were more attributable to differences in maturation rates and thereby food competition than to reproductive suppression. Consistently, harassment of matings by females was rarely observed (unpubl. data). Moreover, the vast majority of high-ranking females, as well as low-ranking females, conceived during their first estrus in the mating season; and there was no indication that low-ranking females suffered from induced abortions [Paul and Kuester, 1987].

There was no support from the data for the prediction that high-ranking females should deliver higher proportions of female offspring than low-ranking ones. In fact, low-ranking females produced even more female offspring per reproductive year than high-ranking females [Paul and Kuester, 1987]. These results (and the observed relation between female age and secondary sex ratio) confirm the prediction of Trivers and Willard [1973] that mothers in good condition should produce predominantly sons, while mothers in poor condition should produce predominantly daughters (for a more detailed discussion on the relation between secondary sex ratio and socioenvironmental factors see Paul and Kuester [1987]).

Female competition theory is partly based on observations that female infants and juveniles suffer more from wounding and high mortality rates than their male peers [Dittus, 1979; Silk et al., 1981; Struhsaker, 1973]. Infant and juvenile female Barbary macaques apparently receive more aggression from unrelated adult females than their male peers [Paul, 1984]. However, this difference seems largely due to a higher contact frequency between adult and young females. Most aggression was of low intensity and did not appear to harm the young females [see also Eaton et al., 1985]. Consistently, mortality of infants and juveniles was not female biased. Among neonates, however, which are most sensitive to harmful handling by others, mortality rate was significantly higher in males. Handling of infants is a common feature in Barbary macaque social behavior and sometimes results in the death of the infant [see also Fa, 1984]. However, neither the sex of the damaged infants, nor identity of the kidnappers confirmed the prediction of female competition theory that females should damage unrelated, female infants [Silk, 1980; Wasser and Barash, 1981]. Finally, it is worth mentioning the apparent tolerance of resident females to immigrating females.

All this does not mean that competition in the Salem Barbary macaque is absent. Females clearly compete over access to resources, and a strict female hierarchy as well as rank-related differences in reproductive success, as found in all social groups, are an expression of competition. The differences found in this study resemble those found by Sugiyama and Ohsawa

[1982] under natural conditions. This suggests that it is not the fact of provisioning per se but rather the manner of provisioning (e.g., attractiveness and dispersal of provisioned food) that may promote competition. In Salem, food items (including provisioned foods) are widely dispersed, while pictures of Japanese macaques at feeding sites [e.g., Richard, 1985: p. 251] give a good impression of the crowding behavior induced by provisioning.

Direct competition over access to food seems not to be the actual cause of intergroup aggressions in Salem. The small groups, D and E, were often chased outside a purely feeding context. Probably, male competition was to some extent the underlying cause of such aggressions, since in general males were the most aggressive individuals during these intergroup encounters. A clear indication of the occurrence of male competition were the high mortality rates of older subadult and young adult males, caused mainly by heavy injuries inflicted by other males. During this age period males reach full size and their dominance relations become to some extent unstable and disputed, resulting in heavy fights during the mating season—note that male deaths were not concentrated in the mating season, but injuries were (unpubl. data). Dominance relations of younger males are age-dependent and they are easily displaced by older males, resulting in lower overt competition and lower mortality rates. High migration rates in these younger males did not increase mortality rates, as suggested by Koford [1966].

Until very recently, it was suggested that intergroup mobility in Barbary macaque males is greatly reduced or more-or-less absent [Taub, 1984]. Such a pattern would not only be very unusual for Old World monkeys—in all baboons and macaques studied long enough, males regularly leave their natal group and breed in other groups [see, for example, review by Pusey and Packer, 1986]; but it would also have important consequences on inbreeding coefficients and social relationships in Barbary macaque groups. Taub [1978] believed that the assumed high degree of relatedness between members of Barbary macaque groups would reduce intermale (and consequently also interfemale) competition and enhance close male infant association (which in fact occur in Barbary macaques). Recent studies on semi-free-ranging and wild Barbary macaques [Kuester and Paul, 1986; Mehlman, 1986; Paul, 1984; Paul and Kuester, 1985], however, have cast doubt on Taub's inbreeding hypothesis and its consequences. Pusey and Packer [1987] have stressed the fact that accurate estimates of dispersal rates can only be gained from long-term data on known individuals. This view is supported by the remarkable differences in annual migration rates of Barbary macaque males at Salem. Nevertheless, while it may be true that migration rates of male Barbary macaques are somewhat lower than in rhesus or Japanese macaques living under similar conditions [e.g., Drickamer and Vessey, 1973; Sugiyama, 1976], results presented here show that most males leave their natal group before adulthood [see also Paul and Kuester, 1985].

The question remains if migration rates at Salem are an artifact of provisioning. It may be argued that high-density, repeated group fissions, and frequent contact between groups promotes intergroup mobility, because it is known from many species that males prefer to join adjacent groups [see Pusey and Packer, 1987]. Transfer into neighboring groups may minimize the risk of predation and promote the choice of, and the integration into, the new group via a better knowledge about its composition and social structure [Cheney and Seyfarth, 1983]. This argument, however, does not seem to be sufficient, since as in Salem, Barbary macaques in the wild have neighboring groups, and intergroup encounters are quite frequent [Deag, 1973]. Moreover, several other factors influence intraspecific variation in migration rates [Pusey and Packer, 1987]. If it is true that Barbary macaques avoid mating with close relatives [Paul and Kuester, 1985], migration rates in the wild should be even higher than in Salem. Most wild groups contain no more than five adult (probably often closely related) females. Consequently, most males maturing in wild groups are closely related to all female group members. Therefore, they may have no potential mating possibilities in their natal group or at least would have much more in other groups. Natal males in the large Salem groups have many unrelated sexual partners. Thus, if Barbary macaque males in the wild are believed to behave in order to enhance their fitness (we suppose here is a consensus), we can expect similar or even higher migration rates than in the Salem population. This view is supported not only by the proportion of natal males in different-sized Salem groups, but also by Sugiyama's [1976] observation that male Japanese macaques in abnormally large groups stay longer in their natal groups than males born in smaller groups.

In conclusion, food enhancement, restriction of the area, and the absence of predators clearly affect life-history patterns and, thereby, social organization and behavior of Barbary macaques at Salem. Direct effects were high fertility, early onset of breeding, and high survival rates. These factors caused a rapid population growth, high density, and large group sizes, which in turn may have caused repeated group fissions. Restriction of the area and increasing density led to several severe intergroup aggressions. The increasing mortality rate in group D showed that subunits of a population can suffer from density effects. On the other hand, fertility rates were not affected by group size, density, or intergroup competition. The manner of provisioning allowed subordinate females to exploit their reproductive potential nearly as well as dominant females. High fertility and survival rates clearly enhance the possibility of interacting with close relatives [Altmann and Altmann, 1979]. Noteworthy, males at Salem nevertheless did not prefer to interact with closely related infants as Taub [1978] predicted [see Kuester and Paul, 1986]. This, as well as other aspects of behavior already discussed, allowed us to assume that social organization of the Salem Barbary macaques does not differ substantially from the social organization of wild populations.

SUMMARY

Data were collected on the structure and dynamics of a population of Barbary macaques (*Macaca sylvanus*) living under seminatural conditions in a large outdoor enclosure at Salem/FRG. Demographic events in the population like births, deaths, intergroup migrations, and group fissions were recorded continuously over a period of 9 years. Social and reproductive units were multimale/multifemale groups, but one-male groups occasionally represented a transitional stage in group development. Age/sex-composition of social groups resembled those of wild groups, while size of most Salem groups was considerably higher than in the wild. The population was characterized by high fertility and survival rates, and early onset of reproduction, resulting in a rapid population growth, high density, and several group fissions. Intergroup transfer of males was a common phenomenon, while female transfer was rarely observed. It is concluded that food enhancement affects life-history patterns of the Salem Barbary macaques, but that basic features of their social organization do not differ substantially from the social organization of wild populations.

ACKNOWLEDGMENTS

We are deeply indebted to Walter Angst, Ellen Merz, Gilbert de Turckheim, and Christian Vogel, without whose support and encouragement this work would not have been possible. We are grateful to Carola Borries, Werner Kaumanns, Dieter Thommen, and the staff of the Affenberg Salem, who contributed to census data. The financial support of the Deutsche Forschungsgemeinschaft (grant An 131/1-5) is gratefully acknowledged.

REFERENCES

Altmann SA, Altmann J (1979): Demographic constraints on behavior and social organization. In Bernstein IS, Smith EO (eds): "Primate Ecology and Human Origins." New York: Garland Publ., Inc., pp 47–63.

Altmann J, Altmann SA, Hausfater G, McCuskey SS (1977): Life history of yellow baboons: Infant mortality, physical development, and reproductive parameters. Primates 18:315–330.

Burton FD, Sawchuk LA (1982): Birth intervals in *M. sylvanus* of Gibraltar. Primates 23:140–144.

Burton FD, Sawchuk LA (1984): The genetic implications of effective population size for the Barbary macaque in Gibraltar. In Fa JE (ed): "The Barbary Macaque: A Case Study in Conservation." New York: Plenum Press, pp 307–315.

Caughley G (1977): "Analysis of Vertebrate Populations." London: Wiley.

Cheney DL, Seyfarth RM (1983): Nonrandom dispersal in free-ranging vervet monkeys: Social and genetic consequences. Am Nat 122:392–412.

Christian JJ (1961): Phenomena associated with population density. Proc Natl Acad Sci USA 47:428–449.

Deag JM (1973): Intergroup encounters in the wild Barbary macaque *Macaca sylvanus* L. In Michael RP, Crook JH (eds): "Comparative Ecology and Behaviour of Primates." London: Academic Press, pp 315–373.

Deag JM (1974): "A Study of the Social Behaviour and Ecology of the Wild Barbary Macaque *Macaca sylvanus* L." PhD thesis. University of Bristol.

Deag JM (1984): Demography of the Barbary macaque at Ain Kahla in the Moroccan Moyen Atlas. In Fa JE (ed): "The Barbary Macaque: A Case Study in Conservation." New York: Plenum Press, pp 113–133.

De Turckheim G, Merz E (1984): Breeding Barbary macaques in outdoor open enclosures. In Fa JE (ed): "The Barbary Macaque: A Case Study in Conservation." New York: Plenum Press, pp 241–261.

Dittus WPJ (1975): Population dynamics of the toque monkey, *Macaca sinica*. In Tuttle RH (ed): "Socioecology and Psychology of Primates." The Hague: Mouton, pp 125–152.

Dittus WPJ (1979): The evolution of behaviors regulating density and age-specific sex ratios in a primate population. Behaviour 69:265–302.

Drickamer LC (1974): A ten-year summary of reproductive data for free-ranging *Macaca mulatta*. Folia Primatol (Basel) 21:61–80.

Drickamer LC, Vessey SH (1973): Group changing in free-ranging male rhesus monkeys. Primates 14:359–368.

Dunbar RIM (1979): Population demography, social organization, and mating strategies. In Bernstein IS, Smith EO (eds): "Primate Ecology and Human Origins." New York: Garland Publ. Inc., pp 65–88.

Dunbar RIM (1985): Population consequences of social structure. In Sibly RM, Smith RH (eds): "Behavioural Ecology: Ecological Consequences of Adaptive Behaviour." Oxford: Blackwell Scientific Publications, pp 507–519.

Dunbar RIM, Sharman (1983): Female competition for access to males affects birth rate in baboons. Behav Ecol Sociobiol 13:157–159.

Dyke B, Gage TB, Mamelka PM, Goy RW, Stone WH (1986): A demographic analysis of the Wisconsin Regional Primate Center rhesus colony, 1962–1982. Am J Primatol 10:257–269.

Eaton GG, Johnson DF, Glick BB, Worlein JM (1985): Development in Japanese macaques (*Macaca fuscata*): Sexually dimorphic behavior during the first year of life. Primates 26:238–248.

Fa JE (1982): A survey of population and habitat of the Barbary macaque *Macaca sylvanus* L. in North Morocco. Biol Conserv 24:45–66.

Fa JE (1984): Structure and dynamics of the Barbary macaque population in Gibraltar. In Fa JE (ed): "The Barbary Macaque: A Case Study in Conservation." New York: Plenum Press, pp 263–306.

Fa JE, Taub DM, Menard N, Stewart PJ (1984): The distribution and current status of the Barbary macaque in North Africa. In Fa JE (ed): "The Barbary Macaque: A Case Study in Conservation." New York: Plenum Press, pp 79–111.

Kaumanns W (1978): Berberaffen (*Macaca sylvana*) im Freigehege Salem. Z Kölner Zoo 21:57–66.

Koford CB (1966): Population changes in rhesus monkeys: Cayo Santiago (1960–1964). Tulane Studies Zool 13:1–7.

Koyama N, Norikoshi K, Mano T (1975): Population dynamics of Japanese monkeys at Arashiyama. In Kondo S, Kawai M, Ehara A (eds): "Contemporary Primatology: 5th Int. Congr. Primat., Nagoya 1974." Basel: Karger, pp 411–417.

Kuester J, Paul A (1984): Female reproductive characteristics in semifree-ranging Barbary macaques (*Macaca sylvanus* L. 1758). Folia Primatol (Basel) 43:69–83.

Kuester J, Paul A (1986): Male-infant relationships in semifree-ranging Barbary macaques (*Macaca sylvanus*) of Affenberg Salem/FRG: Testing the "male care" hypothesis. Am J Primatol 10:315–327.

MacRoberts MH, MacRoberts BR (1966): The annual reproductive cycle of the Barbary ape (*Macaca sylvana*) in Gibraltar. Am J Phys Anthropol 25:299–304.

Malik I, Seth PK, Southwick CH (1984): Population growth of free-ranging rhesus monkeys at Tughlaqabad. Am J Primatol 7:311–321.

Mehlman P (1986): Male intergroup mobility in a wild population of the Barbary macaque (*Macaca sylvanus*), Ghomaran Rif Mountains, Morocco. Am J Primatol 10:67–81.

Menard N, Vallet D, Gautier-Hion A (1985): Demographie et reproduction de *Macaca sylvanus* dans different habitats en Algerie. Folia Primatol (Basel) 44:65–81.

Merz E (1976): Beziehungen zwischen Gruppen von Berberaffen (*Macaca sylvana*) auf La Montagne des Singes. Z Kölner Zoo 19:59–67.

Paul A (1984): "Zur Sozialstruktur und Sozialisation semifreilebender Berberaffen (*Macaca sylvanus* L. 1758)." Dissertation, University of Kiel.

Paul A, Kuester J (1985): Intergroup transfer and incest avoidance in semifree-ranging Barbary macaques (*Macaca sylvanus*) at Salem (FRG). Am J Primatol 8:317–322.

Paul A, Kuester J (1987): Sex ratio adjustment in a seasonally breeding primate species: Evidence from the Barbary macaque population at Affenberg Salem. Ethology 74:117–132.

Paul A, Thommen D (1984): Timing of birth, female reproductive success and infant sex ratio in semifree-ranging Barbary macaques (*Macaca sylvanus*). Folia Primatol (Basel) 42:2–16.

Pusey AE, Packer C (1987): Dispersal and philopatry. In Smuts BB, Cheney DL, Seyfarth RM, Wrangham RW, Strothsaker T (eds): "Primate Societies." Chicago: University of Chicago Press, pp 250–266.

Rawlins RG, Kessler MJ, Turnquist JE (1984): Reproductive performance, population dynamics and anthropometrics of the free-ranging Cayo Santiago rhesus macaques. J Med Primatol 13:247–259.

Richard AF (1985): "Primates in Nature." New York: Freeman.

Sade DS, Cushing K, Cushing C, Dunaif J, Figueroa A, Kaplan JR, Lauer C, Rhodes D, Schneider J (1976): Population dynamics in relation to social structure on Cayo Santiago. Yearbook Phys Anthropol 20:253–262.

Silk JB (1980): Kidnapping and female competition among captive bonnet monkeys. Primates 21:100–110.

Silk JB (1983): Local resource competition and facultative adjustment of sex ratios in relation to competitive abilities. Am Nat 121:56–66.

Silk JB, Clark-Weathley CB, Rodman PS, Samuels A (1981): Differential reproductive success and facultative adjustment of sex ratios among captive female bonnet macaques (*Macaca radiata*). Anim Behav 29:1106–1120.

Southwick CH, Richie T, Taylor H, Teas HJ, Siddiqi MF (1980): Rhesus monkey populations in India and Nepal: Patterns of growth, decline, and natural regulation. In Cohen MN, Malpass RS, Klein HG (eds): ''Biosocial Mechanisms of Population Regulation.'' New Haven: Yale University Press, pp 151–170.

Small MF (1984): Aging and reproductive success in female Macaca mulatta. In Small MF (ed): ''Female Primates: Studies by Woman Primatologists.'' New York: Alan R. Liss, Inc., pp 249–259.

Struhsaker TT (1973): A recensus of vervet monkeys in the Masai-Amboseli Game Reserve, Kenya, Ecology 54:930–932.

Strum SC, Western JD (1982): Variations in fecundity with age and environment in olive baboons (*Papio anubis*). Am J Primatol 3:61–76.

Sugiyama Y (1976): Life history of male Japanese monkeys. In Rosenblatt JS, Hinde RA, Shaw E, Beer C (eds): ''Advances in the Study of Behaviour,'' Vol. 7, New York: Academic Press, pp 255–284.

Sugiyama Y, Ohsawa H (1982): Population dynamics of Japanese monkeys with special reference to the effect of artificial feeding. Folia Primatol (Basel) 39:238–263.

Taub DM (1977): Geographic distribution and habitat diversity of the Barbary macaque. *Macaca sylvanus* L. Folia Primatol (Basel): 27:108–133.

Taub DM (1978): ''Aspects of the Biology of the Wild Barbary Macaque (Primates, Cercopithecinae, *Macaca sylvanus* L. 1758): Biogeography, the Mating System and Male-Infant Associations.'' PhD dissertation, University of California, Davis.

Taub, DM (1980): Female choice and mating strategies among wild Barbary macaques (*Macaca sylvanus* L.). In Lindburg DG (ed): ''The Macaques: Studies in Ecology, Behavior and Evolution. New York: Van Nostrand Reinhold, pp 287–344.

Taub DM (1984): Male caretaking behavior among wild Barbary macaques (*Macaca sylvanus*). In Taub DM (ed): ''Primate Paternalism.'' New York: Van Nostrand Reinhold, pp 20–55.

Trivers RL, Willard DE (1973): Natural selection of parental ability to vary the sex ratio of offspring. Science 179:90–92.

Van Schaik CP (1983): Why are diurnal primates living in groups? Behaviour 87:120–144.

Van Schaik CP, Van Noordwijk MA (1983): Social stress and the sex ratio of neonates and infants among non-human primates. Neth J Zool 33: 249–265.

Van Wagenen G (1972): Vital statistics from a breeding colony: Reproduction and pregnancy in Macaca mulatta. J Med Primatol 1:3–28.

Wasser SK (1983): Reproductive competition and cooperation among female young yellow baboons. In Wasser SK (ed): ''Social Behavior of Female Vertebrates.'' New York: Academic Press, pp 349–390.

Wasser SK, Barash DP (1981): The selfish ''allomother'': A comment on Scollay and DeBold (1980): Ethol Sociobiol 2:91–93.

Whiten A, Rumsey TJ (1973): ''Agonistic buffering'' in the wild Barbary macaque. *Macaca sylvana* L.. Primates 14: 421–425.

Wolfe LD, Schilling P, Jones TD (1986): Reproductive rates in two breeding colonies of rhesus monkeys. Am J Primatol 10:441.

Wrangham RW (1974): Artificial feeding of chimpanzees and baboons in their natural habitat. Anim Behav 22:83–93.

SECTION III: BEHAVIOR AND SOCIAL ORGANIZATION

Ecology and Behavior of Food-Enhanced Primate Groups, pages 231–246

11

Impact of Feeding Practices on Growth and Behavior of Stump-Tailed Macaques (*Macaca arctoides*)

Arnold S. Chamove and James R. Anderson

Department of Psychology, University of Stirling, Stirling FK9 4LA, United Kingdom (A.S.C.), and Laboratoire de Psychophysiologie, Universite Louis Pasteur, 67000 Strasbourg, France (J.R.A.)

INTRODUCTION

Social relations between sympatric groups of primates, as well as between primates and nonprimates, often reflect competition between the animals over access to limited food supplies [Richard, 1985]. Within-group feeding competition occurs too, and it seems likely that pressures arising from such competition have played an important role in the evolution of primate social structures [Jolly, 1985].

Fluctuations in food supply can have considerable demographic consequences. For example, Altmann et al. [1985] reported that over a 15-year period during which there was high mortality of an important species of food tree, there was a 95% loss of a population of baboons (*Papio cynocephalus*). Dittus [1980] described a 13.5% decrease in a population of toque macaques (*Macaca sinica*) coinciding with a drought-related decrease in food supply. In these two cases mortality was particularly high among juveniles. Immature and adult female chacma baboons (*Papio ursinus*) also suffered high mortality as a consequence of a 5-month period of extreme food shortage [Hamilton, 1985]. In contrast to such population decreases following food shortages, one group of toque macaques with access to extra food (at a garbage dump) grew at an annual rate of 12.5% [Dittus, 1980]. Increased population growth during a period of supplementary feeding has also been reported in Japanese macaques (*Macaca fuscata*) [Mori, 1979; Sugiyama and Ohsawa, 1982].

Behavioral mechanisms influencing the distribution of food items among members of a group seem to be an important aspect of group processes, especially where attractive, spatially restricted food is concerned [e.g., Feistner and Chamove, 1986]. Interindividual spacing patterns play a role here. Mori [1977] and Furuichi [1983] reported that adult Japanese macaques foraging for (abundant) natural foods tend to maintain interindividual distances of several meters, which reduces the likelihood of agonistic interactions. Aggression is more common when individuals converge on concentrated food sources, and it was reported to increase markedly when the monkeys crowded into a 14 × 9.5-m area to eat wheat given by humans [Mori, 1977]. In small cages, aggression owing to a single food source may increase by a factor of 4; the increase in the most dominant animals is even greater—aggression being about seven times greater than during baseline periods [Chamove and Bowman, 1978]. During such competition plasma cortisol values are almost doubled, suggesting heightened stress along with the aggression. Competition over food can also exacerbate abnormal behaviors such as stereotyped movements and self-aggression [Anderson and Chamove, 1981; Chamove and Anderson, 1981; Chamove et al., 1984].

That dominant members of a group have priority of access to spatially restricted food is a fairly robust finding. In rhesus macaques (*Macaca mulatta*) and Barbary macaques (*Macaca sylvanus*) dominants are able to control (actively or passively) the area around the food [e.g., Southwick, 1967; Fa, 1986]. In one study on rhesus monkeys' responses to piled and dispersed fruit and vegetables, a subgroup of dominant individuals generally ate earlier and for longer regardless of the experimental conditions, whereas in other subgroups access to food and agonistic behavior varied according to the type and distribution of the food [Belzung and Anderson, 1986]. However, dominance-related effects were less clear in these animals when fear-producing stimuli (model snakes) were presented along with food (Brennan and Anderson, in prep.). Iwamoto [1974] demonstrated a direct relationship between the social rank of adult females and the amount of wheat obtained during provisioning in free-ranging Japanese macaques: dominants ate most. As will be seen in the present chapter, a tendency to monopolize prized food items is also a characteristic of dominant members in captive stump-tailed macaque (*Macaca arctoides*) groups.

A series of studies by the present authors has examined the role of dominance relations in the distribution of food in captive stump-tails and the extent to which modifying their captive environment could lead to both a greater approximation to natural foraging patterns and a reduction in abnormal behaviors. Methods to equalize distribution were also examined because in captivity monopolizing of food by dominants can become a serious management problem. To set this work in perspective, information regarding normal feeding and foraging in *Macaca arctoides* is presented below.

FEEDING BEHAVIOR IN STUMP-TAILED MACAQUES

Despite their popularity as a laboratory primate, stump-tailed macaques have not yet been the subject of any detailed field study. Therefore, not much is known about their basic foraging and feeding techniques, let alone the influence of social relations on food-related behavior. Bertrand's [1969] brief study of a group released near a village in Thailand still provides the best descriptive account of feeding in this species. Additional elementary information is given by Roonwal and Mohnot [1977] and by Fooden et al. [1985] with regard to stump-tails in Yunnan province, China. From these reports it emerges that stump-tailed macaques are omnivorous. Vegetal matter consumed includes fruits, leaves, leaf buds, stems, tubers, seeds, grasses, and bamboo shoots. Bertrand [1969] did not see wild stump-tails feeding on animal matter, but Fooden et al. [1985] list birds, eggs, and larvae as forming part of the diet of Chinese stump-tails. Cultivated crops may also be eaten [see also McCann, 1933].

In Thailand, over 50% of the day was spent foraging and feeding. This compares with between 27% and 52% "feeding" times in Himalayan rhesus, depending on locality [Wada, 1982], 35% of time in long-tailed macaques (*Macaca fascicularis*) [Aldrich-Blake, 1980], and 37% to 49% in *M. fuscata* [Iwamoto, 1982]. Comparable figures for some other primate genera are 33% in arboreal red colobus monkeys (*Colobus badius*) [Marsh, 1981], between 16% and 60% in howler monkeys (*Alouatta palliata*), depending on the habitat [Clutton-Brock, 1977; Milton, 1980], 50% in green monkeys (*Cercopithecus sabaeus*) [Harrison, 1983], 42% in mangabeys (*Cercocebus albigena*) [Waser, 1975], 60% in *Galago senegalensis,* and 19% in *Galago crassicaudatus*. Time spent traveling accounted for most of this difference [Crompton, 1983].

As for feeding techniques, Bertrand [1969] noted that when feeding, stump-tails usually transfer items to the mouth by hand, but like other macaques they probably also remove leaves and fruit directly with the mouth, using the hands to bend and pull in branches. Bertrand also described a "factory pattern" of feeding on grasses, where the monkeys shuffle along the ground on their hindlimbs while the two hands work alternately at plucking food items from the substrate and transferring them to the mouth. While food is being collected some of it may be stored in the cheek pouches, later to be pushed into the mouth and masticated. This is a useful adaptation for a crop-raiding species with a "retrieve-and-retreat" feeding pattern, and it also appears to be related to intragroup feeding competition [Murray, 1975].

Further useful information on stump-tail feeding adaptations comes from a group released to range freely over a small island in Mexico [Estrada and

Estrada, 1976, 1977]. Reports concerning this group are mainly interesting for the diversity of items eaten by the group (which was also provisioned). During the first 3 months after being released, the monkeys made use of 18 major plants out of 65 identified on the island, with fruit, leaves, flowers, stems, seeds, tree bark, and roots being eaten. They also systematically searched for animal prey. Freshwater snails were obtained by traveling along the lake shore, sometimes wading in the water, and lifting or moving aside rocks and pebbles. Stones and layers of earth were also removed when the monkeys hunted for terrestrial spiders and earthworms, which were also dug out of the earth. For one type of earthworm, the monkeys' technique consisted of sweeping with the hands to spread out the layer of topsoil, thereby exposing the worms just below the surface. Vertebrates were also caught and eaten, including birds, mice, frogs, and lizards. Individuals not responsible for capturing the prey could sometimes peaceably obtain parts of it from the possessor, but prized food items such as meat led to aggressive episodes, involving both the possessor and individuals waiting on the periphery [Estrada et al., 1978].

In summary, stump-tailed macaques appear to follow the omnivorous macaque general feeding pattern. They are opportunistic feeders and are susceptible to interindividual aggression over prized, clumped food sources. The present brief literature review also reveals some basic foraging techniques that could be simulated in attempts to improve captive environments.

VARYING FEEDING CONDITIONS IN CAPTIVITY

Recently there have been a number of studies devoted to the effects of different provisioning methods on monkeys housed in captivity. Topics of interest have included mother-infant relations [Rosenblum and Sunderland, 1982], aggression [e.g., Wasserman and Cruikshank, 1983], and methods for "improving" the behavior of the animals [e.g., Tripp, 1985; our studies described below]. For the most part improvement has been taken to mean a) a reduction in behaviors which the caretakers feel are undesirable in that animals may be injured (e.g., social or self-directed aggression) or which may otherwise indicate emotional stress (excessive immobility, stereotyped motor acts), and b) an increase in alternative, "desirable" behaviors, e.g., overall activity, more equal use of the cage area.

Other forces have influenced this type of research. Animal rights' proponents encourage the development of improved captive conditions. However, the institutions concerned are sometimes resistant to innovations that increase the cost, workload, or variability of established regimes. One way of making progress in this area is to identify relatively cheap and simple methods of improving the captive environment. Where possible, these can be based on a sound knowledge of the natural social organization and gross environmental requirements of the species, such as temperature ranges [McGrew, 1982]. As will be seen below, feeding procedures are also important.

HEALTH AND GROWTH

One clear indication of the impact of feeding procedures on captive populations comes from a review of three years of morbidity and mortality reports for 112,600 primates [Chamove et al., 1979]. Put simply, current feeding procedures are often inadequate in protecting primates from digestive-related illness. Fully half (8%) of the ill population (16%) in any yearly quarter were diagnosed to have a digestive problem. In *Macaca arctoides* this was just below the average at 13%, possibly reflecting this species' relatively uncomplicated nutritional requirements, and was highest in orangutans, *Pongo pygmaeus,* 44%. One-fourth of the deaths per quarter (3%) were digestive-related in the 33 species considered, and this was as high as 10% in the cotton-top tamarin, *Saguinus oedipus.* This is not surprising when one considers the diets sometimes offered to these animals. In the wild, tamarins spend about 40% of the time feeding on fruits, 40% on high-protein insects, and 15% on plant exudates, i.e., gums and resins [Garber, 1984]. In captivity they are almost never offered plant gums, are fed fruits of a type, of a size, and located quite differently from that in the wild. Protein may also be fed in an unpalatable form [Pereira and Resende, 1986]. Consequently, tamarins often consume a high proportion of high-carbohydrate fruit and relatively little necessary protein. Of the 12 diagnostic categories considered by Chamove et al. [1979], the digestive category was consistently the highest for both illness and death over the 3 years.

In evaluating findings such as those above, it seems very likely that the nutritional needs of certain species have not been met. In addition to choice of foods, however, the amount consumed, the rate of consumption, and social factors certainly play a role; and these are often unfortunately far removed from those recorded in the wild. Commonly, variety is reduced in favor of consistent ''high''-quality food, and large or even ad lib amounts are favored over restricted intake. The folly of similar practices is being recognized in the human diet, but, somewhat paradoxically, subjecting captive animals to rigorous feeding regimes more likely to approximate feeding in the wild, including presenting food with marked variations in quality and quantity, would certainly be unpopular in some circles and possibly even illegal. Social factors also seem a likely candidate for some digestive-related diseases in captive primates.

The results of excessive intake of food can be seen from the numbers of obese animals encountered in laboratories or in heavily provisioned, urban groups of macaques. Obesity in macaques is primarily an adult problem [Kemnitz, 1984], but some effects of extra food availability can be seen at a surprisingly early age. When stump-tailed macaque infants are able to leave their mothers and gain uninterrupted access to unrestricted amounts of cow's milk, they consume about half the amount of this milk as infants fed totally by hand. This suggests that they do not obtain as much milk as they would like from their mothers. Further, the average laboratory-reared monkey which is not reared by its mother and with unlimited access to milk consumes about twice the amount of a mother-reared infant [Chamove, 1981; Chamove and Anderson, 1982; Scheffler and Kerr, 1975]. The slightly higher carbohydrate and slightly lower protein content of the commercially available milk formulas cannot account for the intake differences, although milk production in mothers is of course limited by nutritional factors [Altmann, 1983]. The high milk intake of hand-fed stump-tails might be expected to lead to heavier juveniles and adults compared with mother-fed infants. Surprisingly, where data have been collected this does not seem to be the case during adolescence [Faucheux et al., 1978; Chamove and Anderson, 1982]. Between 200 and 300 days of age the weight pattern changes from one where mother-reared infants are two standard deviations lighter than hand-reared infants to one where they are over two standard deviations heavier. Diets are different and growth is different.

Alterations in the quality of food taken in infancy may also have effects later in the life of macaques. When infant stump-tailed monkeys were given supplementary feedings of SMA, a proprietary milk formula for human infants, while still receiving most of their nourishment from their mothers, this led to a greater preference for SMA when offered at a later age [Chamove and Anderson, 1982]. The amount of such supplementary feeding did not seem to greatly influence the degree of preference; but it did influence the amount of SMA consumed over a 1-month period after weaning from the mother. In conclusion, while obesity arising from standard captive feeding practices is a recognized problem, the relative influence of developmental, nutritional, and social as well as nonsocial environmental factors remain to be elucidated [see Kemnitz, 1984].

BEHAVIOR

For several years the present authors have been involved in research into improving captive environments for primates, concentrating on stumptailed macaques. These large, highly social monkeys have a clear advantage for research over our second main type of research subject—the small, arboreal callithrichids; stump-tails appear to need a more complex environment, showing behavioral deterioration when housed in small enclosures. Therefore any behavioral improvements resulting from enrichment procedures can be easily measured.

Two primary aims have guided this work: 1) to choose enrichment techniques that encourage the expression of more naturalistic behavior patterns, and 2) to identify techniques that would be widely acceptable to people keeping primates in captivity. The work of Markowitz [1982] exemplifies interesting enrichment procedures for captive animals, but they require specialist technical skills and can be quite costly to construct and implement. Our approach has used food as a motivator for the monkeys—not via food deprivation, but rather by using rarely presented foods that were highly desirable and that could be presented in small amounts. Because food is such an important aspect of the environment, the results of these studies should be of general interest.

During our initial studies mixed grain was offered to the animals, either scattered on the bare floor of the pen or buried in sawdust, woodchips, or woodwool [Chamove and Anderson, 1979; Chamove et al., Anderson, 1982; Anderson and Chamove, 1984]. Plate I illustrates the outdoor pens when clear of floor covering and with grain recently scattered. Feces can be seen, although the floor areas were cleaned only 1 hr previously. The same area with woodchips on the floor is shown in Plate II. More group members are attracted to the floor in this condition, and feces are rapidly covered and dessicated by the litter. Plate III shows animals with the same weight of woodwool. As can be seen, there is a clear preference for the half with the floor covering. Because of the springy nature of the deeper woodwool, younger individuals incorporate this substrate more into their activity, especially play, than is the case with the denser coverings.

Scattering grain in the floor coverings mentioned above produced quite dramatic changes in behavior. In pens housing adult stump-tailed monkeys, aggression was almost twice as frequent when the floor was bare than when it was covered; in pens housing juveniles, aggression was ten times more frequent with no floor covering. The litter-induced reduction in agonistic episodes was most marked in intermediate- and subordinate-ranked individuals, a finding in agreement with the greater variability of response to changing feeding circumstances in nondominant macaques reported by Belzung and Anderson [1986] and by Rosenblum and Smiley [1984].

Other effects of the floor covering included a tendency for affiliative behaviors to increase, again most noticeably in lower-ranking subjects, and for all subjects to spend more time on the ground. Another effect that may be of significance for other laboratory groups of primates was an overall reduction in abnormal behaviors, by about half [Chamove et al., 1984].

Not surprisingly, all the animals fed more rapidly when the floor of the pen

Plate I. Outdoor pens with no floor covering and grain scattered on the floor.

was bare and the grain thus easily visible on the floor (see Plate I); this was especially so for dominant individuals. The difference in feeding rates between monkeys occupying differing hierarchical positions was reduced in the presence of woodchips, to the extent that it became detectable only in the first minute after grain was introduced.

Subsequently, different-sized grains were used, varying in size from just over 8 mm in diameter (maize) down to just under 2 mm (millet) [details in Chamove and Anderson, 1979]. However, the total weight of grain offered was kept constant over conditions. As grain size was reduced, feeding rate increased by a factor of about three, this being measured by the number of hand-to-mouth contacts per unit time. In bare floor conditions the stump-tails' feeding behavior resembled the "factory pattern" described by Bertrand [1969]. At the same time, the reduction in personal space on the bare floor led to frequent disputes. With the floor coverings, on the other hand, the animals' attention seemed to be diverted from their neighbors and instead channeled more to the task of searching for and extracting the food items from the substrate, leading to greater tolerance of intrusions into personal space. Notably, in the deep-litter condition the monkeys used sweeping motions with one hand to remove the top layer of the litter and

Plate II. Outdoor pens with woodchips covering the floor area and grain added.

expose the food items, which were picked up with the other hand. This is reminiscent of the pattern described in semi free-ranging stump-tails by Estrada and Estrada [1976, 1977].

Another series of experiments looked at the generality of the above findings to other species of primates. For example, it was expected that more arboreal primates would be less affected by the presence or absence of a floor covering. Eight different species were tested, using groups at Edinburgh Zoo [Chamove et al., 1982]. The amount of time spent on the ground when the floor was bare or covered with woodchips is depicted in Figure 1. Use of the floor increased clearly for *all* species [see also McKenzie et al., 1986], with only slight evidence of a greater effect for the more terrestrial species. The reduction in aggression found in the earlier work was also extended to other species, albeit less dramatically (see Fig. 2).

Whenever the idea of floor coverings is mooted, the possibility arises of disease and parasite transmission and reinfection. To assess the safety of deep-litter techniques and to compare it with the usual practice of bare floors, we collected samples of the floor covering at weekly intervals for 8 weeks. Two assessment procedures were used: a) One measured naturally occurring bacterial levels of three common bacteria; b) The second involved innocu-

Plate III. Outdoor pens with woodwool covering half the floor area and grain added.

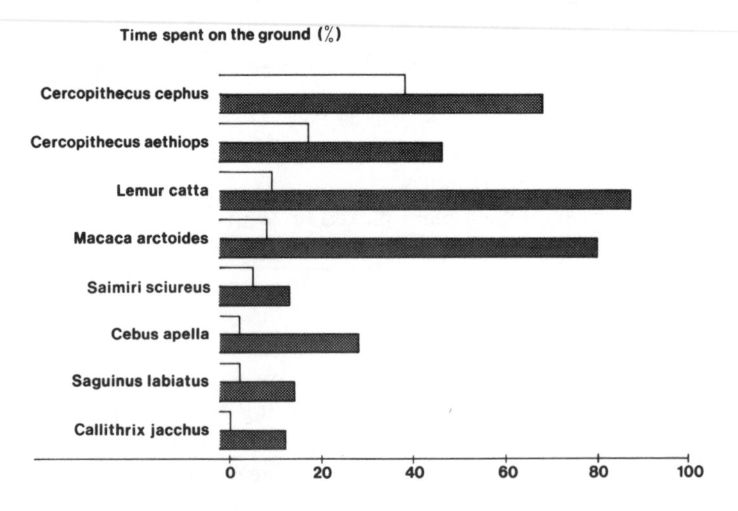

Fig. 1. Percentage of time spent on the floor when uncovered (open) and when covered with woodchips (hatched).

lating floor-covering samples with *Salmonella typhimurium* and measuring how well this organism survived. The results showed that, in common with similar tests using poultry litter, as the floor covering is used more over the weeks, it becomes more inhibitory to bacteria. Total bacteria count actually

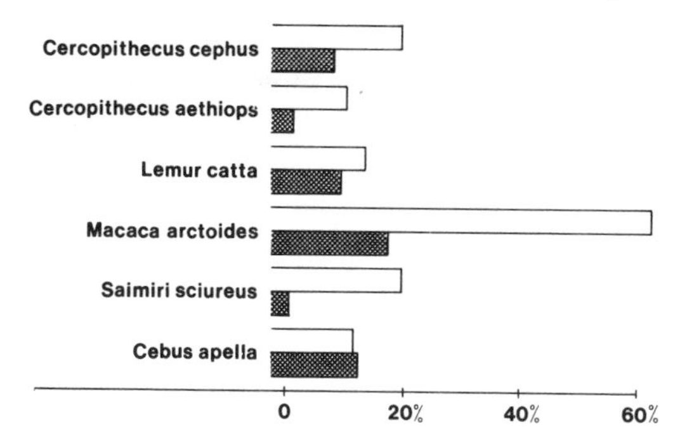

Fig. 2. Percentage of time spent in agonistic behavior when floor was bare (open) and when covered with litter (hatched).

decreased over the 8 weeks, and salmonella survival was reduced as the floor covering became more soiled [details in Chamove et al., 1982].

A further series of manipulations involved varying the spatial distribution of food. These were based on procedures developed for the maintenance of the animals and efforts to allow subordinate individuals to obtain a larger proportion of food when fruit and vegetables were presented. For example, when fruit and leafy vegetables were fed to the macaques, the food was often placed outside the mesh of the enclosures, often on the roof, so that the animals had to climb and pull it in through the mesh [Chamove, 1981]. This led to longer feeding times and better distribution of the food.

The main experiment involved the following conditions: 1) To assess the influence of the desirability of the food, three types were offered—namely, banana, apple, and carrot—in decreasing order of preference. 2) To assess the effect of mode of distribution, food was either massed in two piles or distributed evenly over the floor area. 3) To assess the influence of interanimal visibility, the food was either distributed in a single area with clear visibility throughout or in an area which was divided by four opaque partitions, these having only small openings for the monkeys to pass through. 4) The influence of the visibility of the food was assessed either by distributing it on the bare floor area or burying it under woodchips. 5) Food was presented either fresh or solidly frozen. With regard to the last condition, feeding lasted about 2 min when fresh food was used, compared with about 24 min when the same food was frozen. An adult stump-tailed monkey eats a fresh apple in about 1.8 min and a banana in half this time; the frozen fruit takes about six times as long to eat.

In general, the greatest amount of aggression was shown when the food was presented distributed over the floor. Aggression was reduced when food was massed into two piles, further reduced when buried in the woodchips, and reduced to its lowest level when the monkeys could not see many other animals because of the opaque partitions. The finding of increased aggression with spatially distributed food is interesting, and goes against some of the traditional findings of reduced aggression with less spatially concentrated food items. This suggests that there is a minimum area, still to be determined, beyond which scattering food will reduce aggression but below which the same practice may increase conflict. This is due to attempts by dominants to control more-or-less all the food, as was the case in this experiment when the food was fresh, distributed, and visible. Also, aggression levels were relatively higher when the more preferred fruits were offered than when carrot was used [details in Chamove et al., 1982]. Comparable results have recently been reported by Caba-Vinagre et al. [1986] for a group of stump-tails in seminatural conditions, where there were 93 agonistic episodes per hour during provisioning of the 24 monkeys, compared with only 2 per hour during free-foraging time.

A final set of comparisons assessed the role of food in the maintenance of foraging behavior. One might expect that some reinforcement by finding food would be essential in maintaining foraging, or that frustration if food was not found would lead to the appearance of other behaviors. The following conditions were compared: 1) bare floor, 2) woodchips on the floor but with no food, 3) woodchips with grain mixed into it, and 4) woodchips containing grain plus an unlimited amount of the same grain freely available in food hoppers.

The results are illustrated in Figure 3. The presence of clean grain-free litter altered the monkeys' behavior. Not only did the animals spend more time on the ground in the presence of litter alone, but they played more there and searched through the litter, apparently looking for things to eat. Both aggression and abnormal behaviors were reduced in the presence of litter [see Anderson and Chamove, 1984]. Were this a frustrating or competitive situation, abnormal behaviors would be expected to increase [Chamove et al., 1984; Nash, 1982], but they did not.

A similar picture emerged when free grain was available: The monkeys still foraged through the woodchips, searching for and eating any grain they found even though handfuls of the same food were readily available in food hoppers. As before, aggression and abnormal behaviors were less frequent than in the bare condition, with play on the floor and intense foraging activity being prominent. It appears that in captive conditions like these, animals will seek out situations where they can search for food, even when free food is easily available or even when no food is present. Interestingly, other experimental paradigms have shown persistent searching for food in preference to freely available but otherwise identical food in rodents and birds

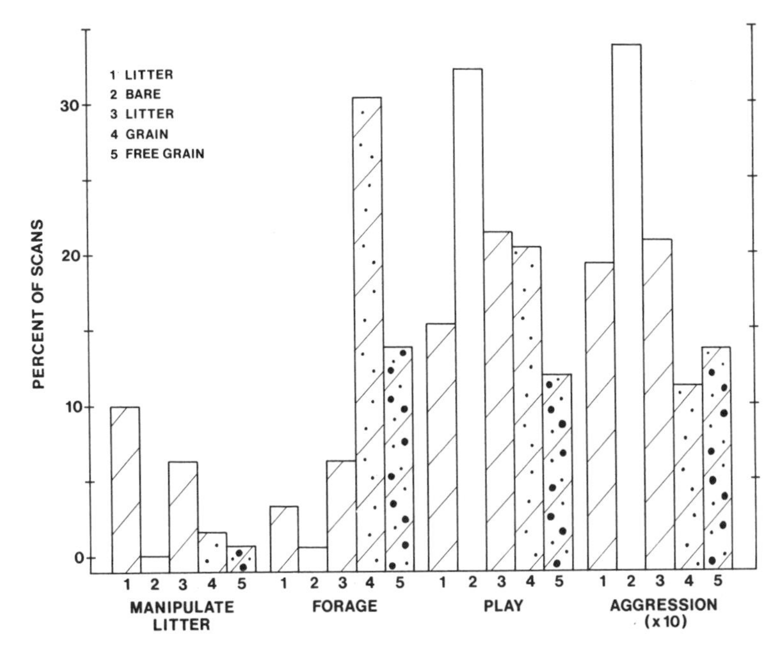

Fig. 3. Percentage of scans in which an animal was observed in a particular activity. Clear histograms indicate no floor covering, small dots indicate grain, large dots grain in food hoppers.

[Inglis and Ferguson, 1986]. In other words, there is a residual motivation to search for food, and this motivation is resistant to extinction or to satiation by feeding. Therefore, simple feeding techniques of the type described in this chapter also have the advantage of permitting the animals to engage in information-gathering exercises that are otherwise largely obviated by traditional captive-feeding regimes.

CONCLUSIONS

The findings described in the present chapter show that variations in how food is presented to nonhuman primates can have marked effects on behavior. Importantly for the management of captive primates, some simple food-related techniques can be used to enhance life in captivity: reducing abnormal behaviors, encouraging the expression of more natural foraging patterns, and increasing the complexity of the environment and thus reducing boredom. Recent studies have confirmed the beneficial effects of floor coverings on activity of apes and monkeys in captive settings [Tripp, 1985; Westergaard and Fragaszy, 1985], and it is gratifying to see that the use of deep-litter substrates containing edibles is now being more widely recommended [e.g., Canadian Council on Animal Care, 1984]. There is room for further research in this area. For example, variety in both fresh and dry commercial foods is considered desirable [Chamove, 1981], but the effects have yet to be measured. The potential of simple alterations to the appearance of food could be assessed [see Barbiers, 1985], as could the effects of improved food presentation methods on some physiological indices of stress. Such research is economical and has the advantage of leading to better captive environments and to a better understanding of behavior.

REFERENCES

Aldrich-Blake FPG (1980): Long-tailed macaques. In Chivers DJ (ed): "Malayan Forest Primates: Ten Years' Study in Tropical Rain Forest." New York: Plenum, pp 147–165.

Altmann J (1983): Costs of reproduction in baboons (*Papio cynocephalus*). In Aspey WP, Lustick SI (eds): "The Cost of Survival in Vertebrates." Ohio State University Press, pp 67–88.

Altmann J, Hausfater G, Altmann SA (1985): Demography of Amboseli baboons, 1963–1983. Am J Primatol 8:113–125.

Anderson JR, Chamove AS (1981): Self-aggressive behaviour in monkeys. Curr Psychol Rev 1:139–158.

Anderson JR, Chamove AS (1984): Allowing captive primates to forage. In "Standards in Laboratory Animal Management. Part 2." Potters Bar: Universities Federation for Animal Welfare, pp 253–256.

Barbiers RB (1985): Orangutans' color preference for food items. Zoo Biol 4:287–290.

Belzung C, Anderson JR (1986): Social rank and responses to feeding competition in rhesus monkeys. Behav Proc 12:307–316.

Bertrand M (1969): "The Behavioral Repertoire of the Stumptail Macaque." Biblio Primatol No. 11. Basel: Karger.

Caba-Vinagre M, Castro-Barradas J, Lopez-Dominguez R, Menza-Diaz S (1986): Social relationships in a troop of free-ranging stumptail macaques (*Macaca arctoides*). Primate Rep 14:69–70.

Canadian Council on Animal Care (1984): Non-human primates. In "Guide to the Care and Use of Experimental Animals, Vol. 2." Ottowa: CCAC, pp 163–173.

Chamove AS (1981): Establishment of a breeding colony of stumptailed macaques. Lab Anim 15:251–259.

Chamove AS, Anderson JR (1979): Woodchip litter in macaque groups. J Inst Anim Tech 30:69–74.

Chamove AS, Anderson JR (1981): Self-aggression, stereotypy and self-injurious behaviour in man and monkeys. Curr Psychol Rev 1:245–256.

Chamove AS, Anderson JR (1982): Hand-rearing infant stumptailed macaques. Zoo Biol 1:323–331.

Chamove AS, Anderson JR, Morgan-Jones SC, Jones SP (1982): Deep woodchip litter: Hygiene, feeding, and behavioral enhancement in eight primate species. Int J Stud Anim Prob 3:308–318.

Chamove AS, Anderson JR, Nash VJ (1984): Social and environmental influences on self-aggression in monkeys. Primates 25:319–325.

Chamove AS, Bowman RE (1978): Rhesus plasma cortisol response at four dominance positions. Aggressive Behav 4:43–55.

Chamove AS, Cameron G, Nash VJ (1979): Primate disease and breeding rates. Lab Anim 13:313–316.

Clutton-Brock TH (ed) (1977): "Primate Ecology: Studies of Feeding and Ranging Behaviour in Lemurs, Monkeys, and Apes," London: Academic Press.

Crompton RH (1983): Age differences in locomotion of two subtropical Galaginae. Primates 24:241–259.

Dittus WPJ (1980): The social regulation of primate populations: A synthesis. In Lindburg DG (ed): "The Macaques: Studies in Ecology, Behavior and Evolution." New York: Van Nostrand, pp 263–286.

Estrada A, Estrada R (1976): Establishment of a free-ranging colony of stumptail macaques (*Macaca arctoides*): Relations to the ecology 1. Primates 17:337–355.

Estrada A, Estrada R (1977): Patterns of predation in a free-ranging troop of stumptail macaques (*Macaca arctoides*): Relations to the ecology II. Primates 18:633–646.

Estrada A, Sandoval JM, Manzolillo D (1978): Further data on predation by free-ranging stumptail macaques (*Macaca arctoides*). Primates 19:401–407.

Fa JE (1986): "Use of Time and Resources by Provisioned Troops of Monkeys: Social Behaviour, Time and Energy in the Barbary Macaque (*Macaca sylvanus* L.) in Gibraltar." Contrib Primatol 23. Basle: Karger.

Faucheux B, Bertrand M, Bourlière F (1978): Some effects of living conditions upon the patterns of growth in the stumptailed macaque (*Macaca arctoides*). Folia Primatol (Basel) 30:220–236.

Feistner ATC, Chamove AS (1986): High motivation towards food increases food-sharing in cotton-top tamarins. Dev Psychobiol 19:439–452.

Fooden J, Quan G-q, Wang Z-r, Wang Y-x (1985): The stumptail macaques of China. Am J Primatol 8:11–30.

Furuichi T (1983): Interindividual distance and influence of dominance on feeding in a natural Japanese macaque troop. Primates 24:445–455.

Garber PA (1984): Use of habitat and positional behavior in a neotropical primate, *Saguinus oedipus*. In Rodman PS, Cant JGH (eds): "Adaptations for Foraging in Nonhuman Primates." New York: Columbia University Press, pp 112–133.

Hamilton WJ III (1985): Demographic consequences of a food and water shortage to desert chacma baboons, *Papio ursinus*. Int J Primatol 6:451–462.

Harrison MJS (1983): Time budget of the green monkey, *Cercopithecus sabaeus:* Some optimal strategies. Int J Primatol 6:351–376.

Inglis IR, Ferguson NJK (1986): Starlings search for food rather than eat freely-available food. Anim Behav 34:615–617.

Iwamoto T (1974): A bioeconomic study on a provisionized troop of Japanese monkeys (*Macaca fuscata fuscata*) at Koshima Islet, Miyazaki. Primates 15:241–262.

Iwamoto T (1982): Food and nutritional condition of free ranging Japanese monkeys on Koshima Islet during winter. Primates 23:153–170.

Jolly A (1985): "The Evolution of Primate Behavior," 2nd Edition. New York: Macmillan.

Kemnitz JW (1984): Obesity in macaques: Spontaneous and induced. Adv Vet Sci Comp Med 28:81–114.

Markowitz H (1982): "Behavioral Enrichment in the Zoo." New York: Van Nostrand Reinhold.

Marsh CW (1981): Time budget of Tana River red colobus. Folia Primatol (Basel) 35:30–51.

McCann C (1933): Notes on some Indian macaques. J Bomb Nat Hist Soc 33:796–810.

McGrew WC (1982): Social and cognitive capabilities of nonhuman primates: Lessons from the wild to captivity. Int J Stud Anim Prob 2:138–149.

McKenzie SM, Chamove AS, Feistner ATC (1986): Floorcoverings and hanging screens alter arboreal monkey behavior. Zoo Biol 5:27–39.

Milton K (1980): "The Foraging Strategy of Howler Monkeys. A Study in Primate Economics." New York: Columbia University Press.

Mori A (1977): Intra-troop spacing mechanisms among Japanese monkeys. Primates 14:113–159.

Mori A (1979): Analysis of population changes by measurement of body weight in the Koshima troop of Japanese monkeys. Primates 20:371–397.

Murray P (1975): The role of cheek pouches in cercopithecine monkey adaptive strategy. In Tuttle RH (ed): "Primate Functional Morphology and Evolution." The Hague: Mouton, pp 151–194.

Nash VJ (1982): "Dominance and Personality in Stumptailed Macaques." Ph.D. thesis, University of Stirling.

Pereira LH, Resende DM (1986): Gelatin as a vehicle for food and vitamin administration to marmosets. Lab Anim Sci 36:189–190.

Richard AF (1985): "Primates in Nature." New York: Freeman.

Roonwal ML, Mohnot SM (1977): "Primates of South Asia." Cambridge: Harvard University Press.

Rosenblum LA, Smiley J (1984): Therapeutic effects of an imposed foraging task in disturbed monkeys. J Child Psychol Psychiatry 25:485–497.

Rosenblum LA, Sunderland G (1982): Feeding ecology and mother-infant relations. In Hoffman LW, Gandelman R, Schiffman HR (eds): "Parenting: Its Causes and Consequences." Hillsdale, NJ: Lawrence Erlbaum Associates, pp 75–109.

Scheffler G, Kerr GR (1975): Growth and development of infant *M. arctoides* fed a standard diet. J Med Primatol 4:32–44.

Southwick CH (1967): An experimental study of intragroup agonistic behavior in rhesus monkeys (*Macaca mulatta*). Behaviour 28:182–209.

Sugiyama Y, Ohsawa H (1982): Population dynamics of Japanese monkeys with special reference to the effect of artificial feeding. Folia Primatol (Basel) 39:238–263.

Tripp JK (1985): Increasing activity in captive orangutans: Provision of manipulable and edible materials. Zoo Biol 4:225–234.

Wada K (1982): Ecological adaptation in rhesus monkeys at the Kumaon Himalaya. J Bomb Nat Hist Soc 80:469–498.

Waser P (1975): Monthly variations in feeding and activity patterns of the mangabey, *Cercocebus albegina* (Lydekker). East Afr Wildl J 13:249–263.

Wasserman FE, Cruikshank WW (1983): The relationship between time of feeding and aggression in a group of captive Hamadryas baboons. Primates 24:432–435.

Westergaard GC, Fragaszy DM (1985): Effects of manipulatable objects on the activity of captive capuchin monkeys (*Cebus apella*). Zoo Biol 4:317–327.

Ecology and Behavior of Food-Enhanced Primate Groups, pages 247–268

12

Relationship Between Foraging and Affiliative Social Referencing in Primates

Michael W. Andrews and Leonard A. Rosenblum

Department of Psychiatry, State University of New York, Brooklyn, New York 11203

INTRODUCTION

Social and foraging aspects of life history have provided key foci in studies of animal behavior in the last two decades. These studies have ranged from those in which only naturally occurring food items were available to the animals [e.g., Post et al., 1980; Rodman, 1977; Struhsaker, 1974] to those in which most, or even all, of the food available to the animals was provisioned [e.g., Loy, 1970; Menzel and Juno, 1982; Southwick, 1967]. An important goal in the study of animals that forage in the company of one or more of their own species, i.e., social foragers, has been the attempt to understand ways in which social and foraging variables interact in shaping individual behavior. A majority of such studies appear to derive from one of two, broad, but not mutually exclusive perspectives: the social-competition perspective and the time and energy budget perspective. The social-competition point of view is based on the premise that similar organisms searching for similar resources are potential competitors. The somewhat different premise of the time- and energy-budget point of view is that an organism has a finite amount of time and energy that it may budget to such tasks as foraging and establishing and maintaining social relationships.

COMPETITION

Several approaches, based on the viewpoint that association with conspecifics while foraging provides a competitive social milieu, have contributed to our understanding of the behavior of social foragers. One approach, which derives quite directly from the social-competition perspective, examines the influence of such parameters as food availability and food distribution upon agonism and/or tension within the group. For example, it has been observed with rhesus macaques (*Macaca mulatta:* Southwick, 1967], chimpanzees and baboons [*Pan troglodytes* and *Papio anubis,* respectively: Wrangham,

1974] that, compared to widely distributed food, restriction of provisioned food to a small area results in an increase in agonistic interactions. Increased difficulty in access to food may have similar effects. Plimpton et al. [1981], in an experimental study on bonnet macaques (*Macaca radiata*), found that when food treats (raisins) were not readily available, i.e., they were buried, there was more aggression within the group than when the treats were readily available. Overt aggression, however, is not the only reflection of increased tension when food access is limited. For example, de Waal [1984] found that limited availability of food resulted in group tension that was not necessarily expressed as overt aggression. In fact, the male rhesus macaques he studied, like male chimpanzees, actively engaged in behaviors such as grooming in an apparent effort to reduce social tension. These few examples suggest that ecological conditions in which the same food items are highly attractive to a number of individuals, e.g., highly clumped or scarce food, result in increases in group tension that may be expressed either as overt agonism or as activities that serve to reduce the tension.

A second approach to studying the behavior of social foragers from the social-competition viewpoint is essentially the converse of the first, i.e., the examination of the impact social factors have on one or more aspects of foraging. As will be considered below, many studies have investigated the influence of dominance position on foraging, but other social factors have been considered as well. For example, Jaeger and his associates have shown that red-backed salamanders (*Plethodon cinereus*), a territorial species, will sacrifice foraging efficiency until a marked territory has been established [Jaeger et al., 1981]; both foraging time and diet are influenced by competitive threat in this species [Jaeger et al., 1983]. Similarly, for the great tit (*Parus major*), a territorial bird, Kacelnik et al. [1981] demonstrated that under experimental conditions these birds will sacrifice food intake for territorial vigilance. In a more general sense, Slatkin and Hausfater [1976] found that the feeding-bout length for a solitary male yellow baboon (*Papio cynocephalus*) was 67% longer than the average for group-living males. They argued that food patch size is effectively reduced for group-living animals as a result of social factors such as supplantation and co-consumption. Not only can foraging patterns be influenced by competition within a group, but as Struhsaker [1974] has demonstrated, extragroup social relations can also influence an important aspect of foraging, i.e. ranging pattern. For a red colobus (*Colobus badius*) group he found that ranging-pattern diversity correlated with proximity to or aggressive encounters with another group.

These and related findings illustrate that social factors have an important influence on foraging. This is no more clearly demonstrated than in the study of social-dominance status and foraging. Although the definition of dominance is a rather controversial issue [e.g., Bernstein, 1981], used in the general sense of priority of access to incentives or success in agonistic encounters, there is evidence from a wide variety of species indicating a foraging advantage may accrue to the more dominant individual in a group. Consider the following representative examples from quite disparate taxonomic forms: lower-ranking spotted hyenas (*Crocuta crocuta*) in the Namib Desert may be entirely excluded from smaller carcasses [Tilson and Hamilton, 1984]; high-ranking juncos (*Junco hyemalis*) obtain more food than low-ranking individuals in the flock [Baker et al., 1981]; sticklebacks (*Gasterosteus aculeatus*) that are less-successful competitors exhibit an increasing tendency to select small prey in a competitive situation [Milinski, 1982].

For primates there are also important linkages between status and foraging patterns. In their study of yellow baboons, Post et al. [1980] have shown that high-ranking animals are less likely to suffer aggressive interruption of feeding bouts than are lower-ranking individuals. With regard to access to specific, presumably highly desirable food types, Hamilton and Busse [1982] found that the meat-eating potential of baboon (*Papio ursinus*) troop members is constrained by the social-dominance hierarchy, with dominant males exhibiting a pronounced advantage. Thus, it is clear that for those animals living and foraging in social groups, the feeding ecology and social structure are constantly interacting to influence individual patterns. Social factors affect the response to a given foraging situation, and each situation alters the course of social relations.

A third approach to examining the relationship between social and foraging variables from the social-competition point of view is concerned with identifying the functional significance for the individual of foraging within a group. Inasmuch as group living is generally presumed to have the automatic detriment of increased competition for resources [Alexander, 1974], this approach confronts the apparent paradox that group-living and social-foraging patterns are quite common. The goal of those studies that adopt the functional approach, therefore, is to provide an ultimate evolutionary explanation as to why individuals forage in a group. These studies generally focus on social versus nonsocial foraging. Attempts to relate specific aspects of social organization, such as group size or the nature of the reproductive unit, to foraging (and other dimensions of an animal's ecology), e.g., Crook and Gartlan [1966], have been largely unsuccessful.

Bertram [1978] has reviewed many of the potential advantages of group foraging. Although the advantages, e.g., increased detection of predators, are not exclusively in terms of obtaining food, a large number of possible advantages do relate to foraging efficiency. The advantages noted here by no means are intended to be exhaustive, nor is any particular one necessarily relevant for the individuals of a selected group; they are simply intended to be indicative of the range of possibilities which may be applicable to the members of a group. As compared with solitary individuals, individuals searching in groups may have an improved chance of locating food, catching prey once it has been located, catching large prey, and/or competing with "extragroup" individuals for food. In light of these and related advantages for many social foragers, the disadvantage of intragroup competition for located resources may be more than compensated by the advantage of having more resources available as a group forager than as a solitary forager. In fact, there is evidence [e.g., Krebs, 1974] that in some instances foraging success increases with group size, at least to a point.

For whatever ultimate reasons animals do live and forage in groups, and despite the competition that may exist among group members over limited resources, it is abundantly evident that attraction between individuals and towards the group as a whole, with the concomitant affiliative social interactions, is characteristic of a great many species. It is reasonable to presume that the diverse array of social activities that are observed play an important role in maintaining group stability and cohesion in a variety of foraging and nonforaging situations, and among other "goals," facilitate the social maturation of young group members. Engaging in social interactions, however, requires time and energy.

TIME AND ENERGY BUDGETING

The second broad perspective on the relationship between social and foraging variables follows from the premise that an organism has a finite amount of time and energy to devote to its activities. As social and foraging activities appear to be predominant activities in the lives of social animals, e.g., the vast majority of primate species, it follows that increases in time and energy devoted to one class of activities may result in a decrease in time and energy available for the other. For example, Galdikas and Teleki [1981] noted that chimpanzees devoted 55% of their time to feeding and traveling, whereas the less-social orangutans (*Pongo pygmaeus*) devoted 79% of their time to such "subsistence" activities. These authors suggested that an inverse relationship may exist between social and nutritional prerequisites for survival.

Researchers commonly report that under conditions of food shortage or conditions that make food difficult to obtain, social activities, such as play, greeting, and grooming, may be reduced significantly. Thus, baboons stranded on an island that did not appear to offer adequate nutrition exhibited reduced cohesive activities such as social greetings and vocalizations and devoted their activity almost exclusively to the search for food [Hall, 1963]. Similarly, Loy [1970] recorded reduced grooming and play (as well as fights) in social-living rhesus monkeys whose provisioned food was suddenly and dramatically reduced. Such social curtailment often was observed during normal provisioning on those days on which food was not supplied. In a rather dramatic demonstration of the interdependence of foraging requirements and social behavior, Baldwin and Baldwin [1972] reported that they did not see any social play in over 260 hr of observations of squirrel monkeys (*Saimiri oerstedi*) under natural conditions of food scarcity. When squirrel monkey (*Saimiri sciureus*) behavior was further examined under controlled conditions in which food was made difficult to obtain, play dropped to as little as 1.2% of baseline [Baldwin and Baldwin, 1976]. Similar results have been reported for lemurs (*Propithecus verreauxi*). Richard [1974] found that play ceased during the dry season and suggested that this finding was related to available energy. These findings lend strong support to the view that when more time is devoted to feeding, and/or when less total food is obtained from feeding activities, there is likely to be a reduction in affiliative social interactions. This view, of course, must be balanced by an appreciation of the fact that brief episodes of enhanced social interaction in the form of agonism may result in situations which promote direct conflict over specific items.

CHALLENGES TO THE ADEQUACY OF CURRENT VIEWS

Clearly, the perspectives discussed above have resulted in considerable contributions to our understanding of the behavior of social foragers. These points of view, in fact, have played an important role in the design of our own research projects. Our findings on the bonnet macaque and on the titi monkey (*Callicebus moloch*) have suggested, however, that current hypotheses regarding factors shaping the behavior of social foragers may need to be modified and/or expanded. These studies have helped emphasize that, quite apart from competition and agonism amongst social partners, affiliative or attachment bonds between individuals, although based on attraction, may nonetheless alter foraging patterns and efficiency. That the demands of the foraging environment may also affect these relationships only complicates further the intricate interplay of social and physical aspects of the environment in affecting individual behavior. Our studies have focused upon the mother-infant dyad in the bonnet macaque and upon the adult male-female pair in *Callicebus*.

MOTHER-INFANT DYADS IN THE BONNET MACAQUE

The relationship of mother to infant in the bonnet macaque may be briefly characterized as one of tolerant support. When living in stable groups, mothers appear to be highly permissive and accepting, with the majority of changes in mother-infant contact due to the comings and goings of the developing infants. Bonnet mothers do not exhibit much overt restraint, rejection, or punishment of the offspring. Such a positive relationship offered a promising baseline against which to analyze the possible effects of imposing a demanding foraging task on the experimental dyads.

In a study using bonnet macaques, Rosenblum and Sunderland [1982] compared a group of five mother-infant dyads living in a high-foraging-demand (HFD) condition with a group of five dyads living in a low-foraging-demand (LFD) condition. Each group of five dyads was housed in a double pen which had 11.2 m^2 of floor space and was 2.1 m high. Animals in the HFD condition were not food deprived, but each food item they obtained was more costly than that obtained by the LFD group because HFD animals had to search in more spatial locations, on average, than LFD animals. That this manipulation was successful was indicated by the observation that foraging scores were five times higher for the group in the HFD condition than for the group in the LFD condition. Moreover, unlike the animals of the LFD group that wasted more than half of the food obtained, HFD animals consumed all the food they found.

Consistent with the social-competition perspective that has been outlined, it was found that within the HFD group there was more hierarchical behavior (i.e., behaviors relating to dominance and subordination) than in the LFD group; in general, there was more tension in the HFD group, partly reflected in high levels of visual checking of other group members. It is likely that increased agonism, rather than reduced affiliative social interactions, was more prominent because the animals were required to search for food within the restricted spatial setting of the laboratory, i.e., a total of 2.2 m^2 per dyad. When attention is turned to the mother-infant dyads within these groups, however, the results do not fit so easily within either of the perspectives outlined previously.

There is compelling evidence that the infant-mother bond was adversely affected by the HFD condition in comparison with the LFD condition. When infants of the two groups were subsequently separated from their mothers, infants from the HFD group showed significantly more depression than did infants from the LFD group [Plimpton and Rosenblum, 1983]. The causes of this insecurity, however, were not readily apparent. In fact, Rosenblum and Sunderland [1982] reported that play was comparable in both groups, with infants from the HFD group appearing to function more independently of their mothers than infants from the LFD group. The finding that HFD infants appeared to be more independent of their mothers, yet less secure in their maternal attachment, than LFD infants is inconsistent with the well-established position that one of the most important criteria of a secure mother-infant attachment is the ability of the infant to use the mother as a base from which to explore the environment [e.g., Ainsworth and Wittig, 1969].

From the time- and energy-budgeting perspective, however, it follows that the fivefold difference in foraging activity between the HFD and LFD mothers suggests a way to reconcile the findings of this study with the standard position regarding security of attachment and independence. If mothers in the HFD group were spending more time away from their infants (and perhaps rejecting them more) because of foraging demands on their time and energy, then the apparent "independence" of the infants might actually be a reflection of a form of imposed separation. It was found, however, that although mother-infant contact scores were slightly lower in the HFD group during the observations, there was no significant difference in mother-infant contact scores between the two groups. Furthermore, as is typical of bonnets, rejection of infants was infrequent and comparable in both groups.

The general view that bonnet macaque mothers do not readily alter time spent with their infants is supported by earlier findings on mother-infant dyads. Despite the normally active role of infants in regulating contact-time, mothers will compensate for a lack of infant response (when infants are anesthetized) by maintaining age-appropriate levels of contact with them [Rosenblum and Youngstein, 1974]. The stability and resilience of bonnet dyads were also seen in a comparison of bonnet and pigtail dyads over a series of 24-hr periods [Rosenblum et al., 1964]. In this study it was shown that whereas mother-infant separation increased with the approach of feeding time in the pigtails, as might be expected from the time- and energy-budgeting perspective, there was no such effect for the bonnets. Moreover, in another study involving periods of imposed food deprivation on bonnet mother-infant dyads [Rosenblum et al., 1969], there was no significant change in mother-infant contact during periods of deprivation.

Although the evidence does not support the view that bonnet mothers become systematically more rejecting of their infants when confronted with a chronically high foraging demand, the fact remains that the mother-infant relationship appears to be influenced by the foraging demand, as evidenced in the separation data. Despite the general acceptance of the essentially symbiotic, rather than competitive nature of the mother-infant dyadic relationship, we propose, in addition, that a simple budgeting perspective is not sufficient for dealing with the effects of foraging demands. Foraging and mothering are tasks with which a mother must deal simultaneously. What is suggested is that a bonnet mother faced with a difficult foraging task may not quantitatively reduce her maternal activities, but the qualitative nature of her maternal responsivity may be altered. In other words, a bonnet mother engaged in a demanding foraging task may be less sensitive to the moment-to-moment status of her infant, and therefore her responses may correlate less closely with the infant's condition. Viewed in terms of mother-infant relations in humans, the bonnet mother is less contingently responsive to her infant's needs and communications [Lewis and Goldberg, 1969].

The findings on the HFD and LFD groups have been extended by the addition of a variable-foraging-demand (VFD) group [Rosenblum and Paully, 1984]. In the VFD group the animals were given alternate 2-week periods of high- and low-demand conditions. The results of the comparision of the groups reinforce the view that environmental changes that alter the foraging conditions may adversely affect the qualitative nature of the mother-infant relationship without quantitatively decreasing the degree of positive association between mother and infant.

In comparison with the HFD group, foraging in the VFD group during observation periods was generally quite low and did not differ between high- and low-demand periods. On the other hand, the VFD group exhibited the highest levels of hierarchical behavior of the groups, and these high levels were sustained across the 14 weeks of the imposed condition; in contrast, the levels of hierarchical behavior in the HFD group tended to diminish over time, despite initial high levels. In addition, the VFD subjects exhibited significantly more mother-infant contact and mother-leave-contact than either of the other groups. Furthermore, infants of the VFD group exhibited significantly less play than those of either of the other groups. It is quite likely that for the VFD group the higher incidence of mother-leave-contact than in the other two groups was due to the extremely high levels of mother-infant contact that provided more potential occasions for the mother to break contact. In fact, the degree of mother-infant contact exhibited a trend opposite to that normally observed; i.e., it increased with infant age. The prevalence of mother-infant contact likely reflects an insecurity of attachment between infant and mother.

It follows from these results that the influence of the environmental conditions on the infant-mother attachment is not simply the reflection of a foraging task requiring the mother to spend more time away from her infant. VFD mothers, in fact, spent more time with their infants than did either the LFD or HFD mothers. Furthermore, time spent foraging in the VFD group during observation periods was quite low and does not seem to account directly for the effects noted. It does appear, however, that much of what little time VFD infants spent out of contact was the result of separation initiated by the mother. Not surprisingly then, when off the mother VFD infants played least, were most emotionally disturbed, and were off for the briefest time. Unpredictable variations in the foraging conditions may have resulted in mothers attending more strongly to environmental factors and consequently being less attentive to the immediate condition of their infants.

This division of maternal attention could have an adverse effect on the security of the infant's attachment to its mother and retard the pace of infant development.

We are currently engaged in a further extension of the earlier findings. In this study bonnet cohorts consisting of three mother-infant dyads each under one of two conditions, LFD or VFD, are being compared. Each cohort is housed in a 5.2-m^3 indoor pen and is visually isolated from other animals. A feeding device (1.3 m × 0.5 m × 0.7 m) that accommodates eight food pans is located in each pen. The feeding structure is completely enclosed when the pen door is closed, except for eight holes (5 × 5 cm) on either side that permit the animals to reach into the food pans. The latter are replaced daily, except on the weekends when adequate food is supplied on Friday to last through the weekend. In the LFD condition, at least twice the amount of food consumed daily by the group is distributed among the pans. In the VFD condition, 2-week LFD periods as described above are alternated with 2-week HFD periods in which the food placed in the pans is covered by approximately 12 cm of clean woodchip bedding and the amount of food is reduced so that only 10–20% of the total amount provided remains in the pans the following day (as is typical of high-demand conditions, all food removed from the pans is consumed). For a total of 14 weeks, beginning and ending with 2 weeks of LFD for both groups, each animal is observed on 4 days each week for a period of 5 min (10 min per dyad). At the conclusion of the 14 weeks, each cohort is given eight 1-hr sessions in a novel environment (16.3-m^3, with climbing cables and other unfamiliar objects) in which each individual is the focal animal for two 5-min periods (20 min per dyad). All observations are conducted through one-way glass.

Thus far, one cohort in each of the two conditions has completed the study. For these two cohorts, the mean ages of the infants at the beginning of the study were 10.7 weeks and 13.0 weeks in the VFD and LFD cohorts, respectively. As a result of this age difference, there was an overall group difference in the proportion of time infants and mothers were separated by more than 0.3 m, the values being 38.5% and 51.0% for VFD and LFD, respectively. As might be expected, the difference was greatest during the first 2 weeks of observations, but was no longer evident after 14 weeks (when a 2-week difference in age would be of less developmental significance than it was at the younger age at the beginning of observations). Overall, therefore, it appears that the foraging conditions, LFD vs. VFD, had no significant effect on mother-infant separation in the home pen. Similarly, no significant group differences in infant social play or in maternal rejections were found during the 40 min per week of observations in the home pen under the experimental conditions.

In contrast to the apparent absence of foraging condition effect on behavior

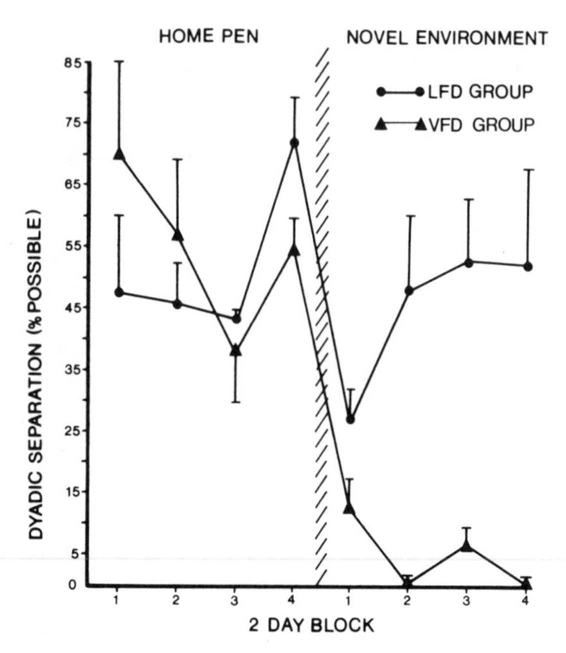

Fig. 1. Mean dyadic separation (>0.3 m) in the low-foraging-demand (LFD) and variable-foraging-demand (VFD) groups in the home pen and the novel environment.

in the home pen, when behavior in the novel environment was compared with an equal number of sessions (eight sessions) at the end of the 14 weeks in the home pen, a clear group difference emerged (Fig. 1). For the LFD group, no significant difference between the home pen and the novel environment with respect to mother-infant separation was found; in contrast, the difference between home pen and novel environment for the VFD group was significant [$F(1,2) = 35.59$, $P < .05$]. In fact, the mean dyad score for separation in the novel environment was 5.5% of time samples for the VFD group, a value that was about a fifth of the lowest value obtained for the group in the home pen during the first 2 weeks of observations. This strongly suggests that the environmental foraging conditions did differentially influence infant-mother attachments in the two groups, but this latent effect did not become evident until the relationships were challenged by introduction to a novel environment. It follows that changes in a relationship may be so subtle that they remain undetected until the relationship is placed under appropriate stress.

In Figure 2 the number of times infants departed from contact with the mother per 10 min of observation is plotted for the home pen and novel environment. Although an analysis of variance failed to reveal any signifi-

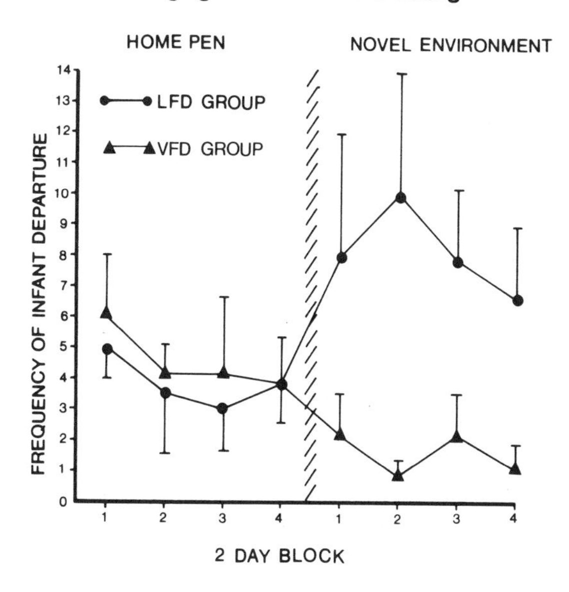

Fig. 2. Mean frequency of departures by an infant from contact with its mother per 10 min for LFD and VFD groups in the home pen and the novel environment.

cant difference between groups for the home pen vs. novel environment conditions, the figure suggests that infants of the LFD group were much less reluctant to separate from their mothers in the novel environment than were the VFD infants. These findings help to confirm that, as in previous studies with bonnet dyads, foraging conditions that are likely to distract the mother from attention to her infant, although not necessarily apparent in the relationship during extended experience with the conditions, are likely to result in an infant that is less securely attached to its mother.

Further insight into the mother-infant relationship in the two groups is provided by Figure 3. Mothers' departures from contact with their infants are graphed here on the same scale as infants' departures in Figure 2. As would be expected for bonnet macaques, the overall frequency of departures was considerably lower for mothers than for infants. Furthermore, the graph reveals that mothers' departures were differentially affected by condition (home pen vs. novel environment) in the two groups. Although there were no group differences evident in the home pen, mothers' departures in the LFD group tended to rise in the novel environment compared to the home pen, whereas there was, in fact, a drop in mean departure rate between the home pen and novel environment for the VFD group. The differential effect of conditions on the two groups was confirmed by a significant group ×

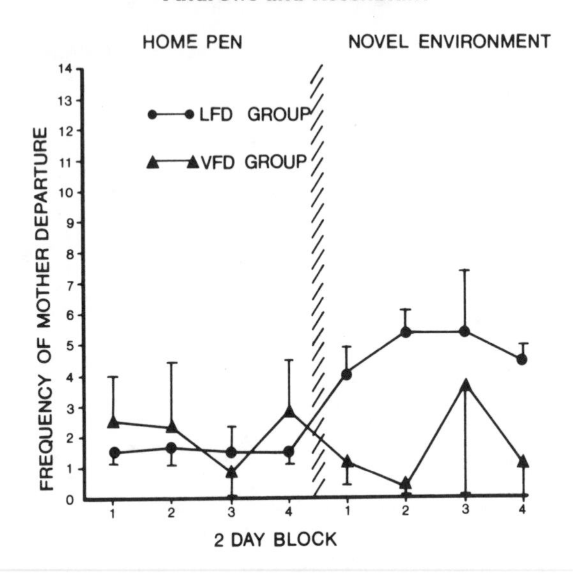

Fig. 3. Mean frequency of departures by a mother from contact with her infant per 10 min for LFD and VFD groups in the home pen and the novel environment.

condition interaction for mothers' departures [$F(1,4) = 18.3$, $P < .05$]. Exhibiting a trend exactly opposite that of VFD mothers, LFD mothers initiated separation more frequently in the novel environment than in the home pen, as did their infants. However, since separations tended to be much more brief in the novel environment than in the home pen, the overall probability of finding LFD mothers and infants separated during a time sample remained relatively unchanged between the conditions, contrasting sharply with the drop in separation observed for VFD mothers and infants in the novel environment. These differences, which became evident only when the relationships were challenged, may reflect the greater security of attachment between LFD mothers and infants than between VFD mothers and infants. It appears that activities that may chronically distract the mother's attention from her infant are likely to adversely affect the security of the infant-mother bond and the development of the infant.

TITI MONKEY MALE-FEMALE PAIRS

As with the strong positive bond that exists between a bonnet macaque infant and its mother, a strong bond between opposite-sexed adults is typical of the single-dyad, monogamous social structure found in *Callicebus moloch*. Mason and his associates have highlighted the strong bond characteristic of the monogamous relationship in *Callicebus* by contrasting the behavior of captive male-female pairs of this species with captive male-female pairs of the squirrel monkey (*Saimiri sciureus,* Colombian and Peruvian mix). In contrast to *Callicebus, Saimiri* typically live in large groups with little evidence of bonding between opposite sexed individuals.

Observations of pairs of the two species in their captive living areas revealed that in comparison with *Saimiri* pairs, *Callicebus* dyads were spatially closer and more often in contact, they groomed each other more, their activity was more coordinated [Mason, 1974], and they spent more time looking at their cagemates [Phillips and Mason, 1976]. In addition, experimental testing has demonstrated a stronger and more specific attraction to cagemate [Mason, 1975] and higher levels of arousal (change in heart rate) resulting from separation in *Callicebus* than in *Saimiri* [Cubiciotti and Mason, 1975]. The results of these and other studies clearly indicate that bonds within male-female pairs of *Callicebus* play an important role in their daily activities, whereas male-female bonds do not appear to be very important in the daily activities of *Saimiri.* Given the strong difference between *Saimiri* and *Callicebus* in male-female bonding, an important question to consider was: Do species differences in social referencing influence foraging efficiency by pairs of the two species?

In part to determine what influence differences in social attachment might have on the efficiency of response to a foraging task, captive male-female pairs of *Callicebus* and *Saimiri* were released into a large outdoor enclosure (36 × 18 × 2.1 m high) containing ten feeding stations interconnected by a runway system elevated 0.85 m above ground level [Andrews, 1984, 1986]. Each feeding station was designed to represent an artificial tree and consisted of a vertical post and six horizontal perches; a plastic cup was located at the end of four of the perches. Thus, with ten stations and four cups per station, a total of 40 food loci were available on each trial. Previous testing had established that the food available (one marshmallow piece per cup) was highly attractive to both species and that all pairs of each species could easily consume the entire quantity of food available. Prior to release into the large enclosure, animals were partially habituated to the area by release into a small, temporary cage (1.8 m high × 0.75 m wide × 3.7 m long) placed within the outdoor enclosure; a feeding station similar to those in the large enclosure was located within the small cage. All animals were given experience in the small cage until they demonstrated a readiness to approach and obtain food from the feeding station; this required two sessions for most *Saimiri,* whereas most *Callicebus* required at least twice this number of sessions.

Eight pairs of each species were tested, with an observer for each member of the pair making focal observations continuously through a session. For a

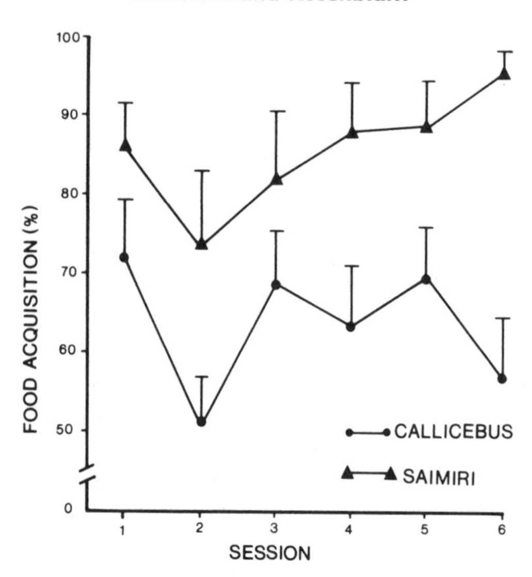

Fig. 4. Mean percentage of available food items acquired for consumption by *Callicebus* and *Saimiri* pairs.

given pair, a session lasted 1 hr and each pair received a total of six sessions, each separated from the previous one by a minimum of 2 weeks. To facilitate behavioral observations, the test area was divided into 72 square quadrats that were 3.6 m on a side. Each time an individual entered a quadrat or a feeding site in which its pairmate was located, an "approach" was scored for that focal animal. In addition to social approaches, for each focal animal a record was made of all feeding sites that were visited and any food items that were eaten. A record was also made of all quadrats entered by an individual during a session.

Using quadrat entries as a measure of the use of space, it was found that *Saimiri* pairs did use a slightly greater, but not significantly greater percentage of the available foraging area than did *Callicebus* pairs (83.6% vs. 76.9%). Despite the absence of a significant species difference in the use of space, *Saimiri* pairs obtained a significantly greater percentage of the available food than did *Callicebus* pairs, as reflected in Figure 4 [85.8% and 63.5% for *Saimiri* and *Callicebus,* respectively; $F(1,14) = 15.5$, $P < .01$].

In light of the absence of clear species differences in food motivation or in the use of the foraging space, other sources of the species difference in foraging success must be considered. The data suggest that at least a partial explanation may lie in the species difference in attraction between opposite sexed individuals. Conceivably, *Callicebus* attraction to the pairmate com-

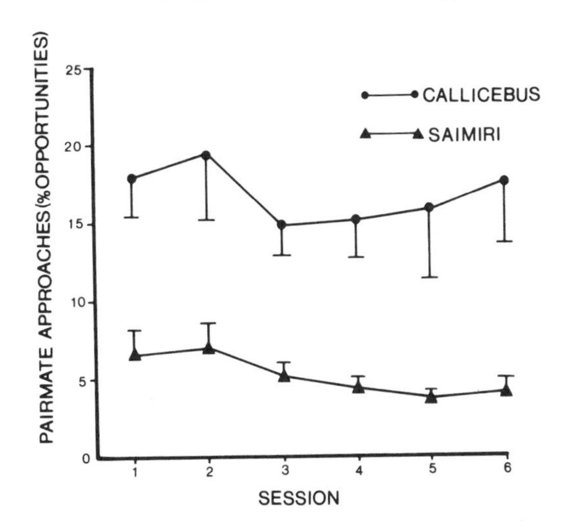

Fig. 5. Entries into foraging areas occupied by the partner as the mean percentage of opportunities to do so, for members of *Callicebus* and *Saimiri* pairs.

peted with attraction to the feeding locations to a much greater extent than it did for *Saimiri,* and thereby constrained free movement by *Callicebus* to the feeding locations.

As a measure of social attraction while foraging, entries of an individual into a quadrat or feeding station in which its pairmate was located were determined as a percentage of the total number of quadrat and feeding station entries made during the session (i.e., the number of ''opportunities'' to approach the partner). Figure 5 clearly indicates that the attraction of *Callicebus* to their pairmates during a foraging session was much greater than that between *Saimiri* pairmates. *Callicebus* individuals were approximately three times as likely to approach the pairmate as were *Saimiri* individuals (16.8% vs. 5.2% of opportunities), and this difference between the species in partner attraction during foraging was highly significant [$F(1,30) = 30.99$, $P < .001$].

Further evidence for the possible influence of social attraction on foraging success was found when first visits to each feeding station were selectively considered. First visits to feeding stations may be presumed to reflect times when an individual's activity was most strongly focused on foraging. In fact, both species obtained over 90% of the food they consumed on first visits to feeding stations. These data indicate that *Callicebus,* compared with *Saimiri,* were much more likely to visit a feeding station with their pairmate. Figure 6 shows the mean proportion of feeding stations visited in a session in which

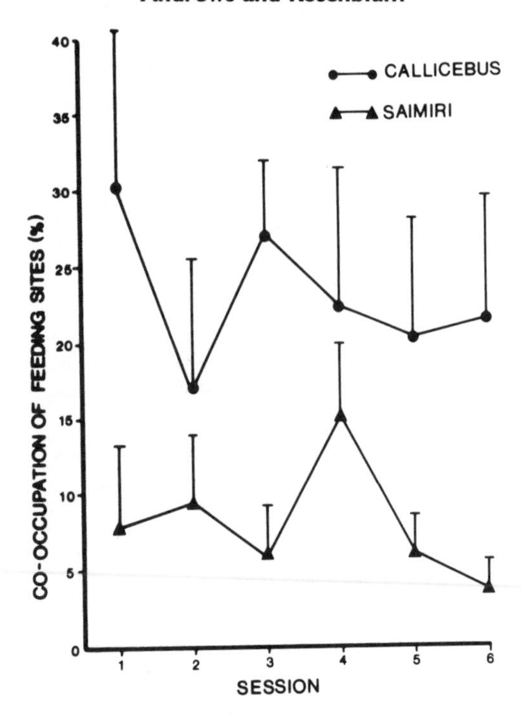

Fig. 6. Co-occupation of feeding sites as the mean percentage of all sites occupied by members of *Callicebus* and *Saimiri* pairs.

pairmates were together on their respective first visits; values are expressed as the percentage of the total number of feeding stations visited in a session by either one or both members of the pair. The percentage of feeding sites co-occupied by *Callicebus* pairmates was 23.0%, whereas the value was only 8.0% for *Saimiri,* and this difference was also significant [$F(1,14) = 5.77$, $P < .05$].

The data on sex differences in food acquisition within species provides one additional piece of evidence that species differences in social attraction resulted in a differential impact of social attraction on foraging success for *Callicebus* and *Saimiri*. Here again, the data show that *Callicebus* regularly fed together, whereas *Saimiri* pairmates tended to be competitive in their foraging patterns. *Callicebus* males and females obtained 25.8% and 36.8% of the available food, respectively, and the difference between the sexes was not significant. In contrast, *Saimiri* males and females obtained 55.3% and 28.5% of the available food, respectively, and the difference between the sexes was highly significant [$F(1,14) = 11.32$, $P < .01$]. The advantage for

Saimiri males is likely due to a difference in size between males and females that favors the males (1.0 kg vs. 0.7 kg). No sexual dimorphism in size is evident for the slightly larger *Callicebus* (1.2 kg). In fact, the physical and behavioral characteristics of male and female *Callicebus* appear to form the basis of a true partnership. Attending to the partnership while foraging may influence the quality of the foraging activity and, consequently, result in a reduced foraging efficiency for each of the individuals involved. Although other factors undoubtedly contribute to the differences in foraging success between *Callicebus* and *Saimiri* pairs, there is strong support for the hypothesis that foraging patterns that incorporate positive social referencing, as in *Callicebus,* may result in short-term deficits in foraging efficiency.

DISCUSSION

It is clear that we cannot expect to understand the variety of social factors that can influence the foraging patterns of primates in social groups by focusing exclusively on the competition between individuals. Overt, proximal competition for resources does not appear to play an important role in the relationships of mother macaques and their young infants, nor between the monogamous adult male-female dyads in *Callicebus*. Positive, affiliative attraction or bonding between individuals, and perhaps between individuals and particular subgroups, draws the individual away from a more complete focus on the physical dimensions of the foraging tasks that it confronts. Further complicating the task of the social forager, and our efforts to explicate the sources of variation in the behavior we observe, is the fact that social and foraging goals are neither mutually exclusive nor irrelevant to one another. Through strategic adjustments in behavior, individuals may pursue these objectives simultaneously. It is likely that the behavior of an animal dealing with two tasks at the same time will not represent optimal performance on either task considered alone. Variations in the degree to which effort is focused on one of the tasks, furthermore, will likely be reflected in the quality of performance on the other. The results of the studies presented above certainly support this position.

For bonnet macaque mother-infant dyads, when it is more difficult for the mother to obtain her daily ration of food, there appears to be a qualitative change in the infant-mother interaction that adversely affects the relationship and, in consequence, the course of infant development. It is not possible at this time to identify the exact nature of the change, which can become evident in the behavior of the infant only as the result of special testing. It would have to be argued, in fact, that the mother is a strong buffer between her infant and the environment, and that, to a large extent, the mother is able to adapt her relationship with her infant to a wide range of environmental conditions. When the relationship is suddenly challenged by an environmental change, however, such as relocation to a novel environment, characteristics of the relationship that were not previously evident may be revealed. Such a challenge may also reveal differences among relationships that have developed under different environmental conditions. Furthermore, the data suggest that in some circumstances an unstable environment may render it more difficult for either member of the mother-infant dyad to alter interactive and foraging activities in a strategically adaptive manner. In any event, it is clear that more time invested in foraging does not necessarily result in an obvious reduction in time invested in infant care. It may be more appropriate to address changes in the relationship in terms of psychological investment. A mother in a demanding situation may be spending time with her infant and expending energy on its care (e.g., carrying it), but the care may be less satisfactory for the infant than it would be if the mother was in a less demanding situation where she could attend more closely to the state of her infant. In particular, the mother may be less responsive to subtle changes in the infant's state and to the infant's emerging communicative repertoire. The consequences of this reduction in the infant's capacity to elicit caretaking may lead to less "security" in the attachment, heightened infant disturbance in interactions outside the dyad and diminished capacity for, or slower development of, independent functioning.

The measures taken in our studies with the bonnet macaque dyads have illuminated the influence that foraging tasks may have on the qualitative performance of a critical social task, i.e., maternal behavior. In contrast, the measures taken on the *Callicebus* pairs highlighted the influence that a social task, i.e., maintaining a close association with the pairmate, may have on the efficiency of performance of a foraging task. Despite the food motivation and the extent of travel within the foraging area, which were comparable between species, *Callicebus* pairs were much less successful in the foraging task than were *Saimiri* pairs. This suggests that the *Callicebus* pairs were less focused on the foraging activity than were the *Saimiri* pairs. In light of the evidence of the strong attraction between pairmates during the foraging sessions, it is reasonable to conclude that the lower success of *Callicebus* relative to *Saimiri* was due, at least in part, to the greater involvement of *Callicebus* with the pairmate. In fact, observations revealed that it was not uncommon during the testing sessions for a *Callicebus* subject to deviate from a potentially successful foraging path in a direction that resulted in approach to its pairmate that was following a different, and often less fruitful path. In contrast to *Callicebus* individuals, *Saimiri* searched for food separately, with *Saimiri* females apparently avoiding potential conflict with males over food. Unlike the *Callicebus, Saimiri* subjects in this situation are not confronted with a possible conflict between approaching the pairmate at one location or approaching a potential food source at another location.

Study of bonnet mother-infant dyads and *Callicebus* male-female pairs has been fruitful in suggesting a modification of our perspectives regarding the behavior of social foragers, i.e., the need to incorporate consideration of the strong, positive bonds that are essential ingredients of these relationships. However, the view proposed here is by no means restricted to these particular relationships. There is ample evidence that there are other species in which social relationships develop that are as strong as those examined here. Furthermore, it is likely that in species in which bonds are not as strong, animals still must adopt strategies to deal simultaneously with foraging tasks and positive social tasks. For example, Pitcher and Magurran [1983] have suggested that a goldfish (*Carassius auratus*) ''informed'' by experience of food-patch profitabilities is drawn to less profitable patches when other uninformed fish are foraging there. Focusing upon primates, other studies may be noted that support the general view that in gregarious species, foraging patterns of individuals may be affected by the activities of others with which there is some bond, i.e., the effects of the positive affiliative locus (PAL).

The work of Menzel and his associates has been quite sensitive in demonstrating the interactive role of social and foraging variables in two widely divergent primate species. In studying the food-searching behavior of young chimpanzees, Menzel [1971, 1973] showed that the emotional dependence of young, group-living chimpanzees on one another precluded meaningful testing of food searching patterns when an animal was separated from its group to be tested alone. When only one chimpanzee had knowledge of food locations and the other members of the group were reluctant to follow, rather than disregard its companions or the potential food source—i.e., budget separate time periods for foraging and social activities—the knowledgeable chimpanzee would go to great lengths to get the others to follow it to the food source and would often have a tantrum when efforts proved unsuccessful. As a further reflection of the role of relatively subtle social variables on foraging activities, Menzel also found that the probability of individuals following a knowledgeable leader to a food location was influenced by preference for the leader.

A somewhat different foraging style was observed in another situation with group-living marmosets [*Saguinus fuscicollis:* Menzel and Juno, 1982, 1985]. When the group entered a room in which food had been hidden, the marmosets tended to fan out, with subgroups of two animals often moving together. Although one animal could easily have consumed all the food located at a provisioned site, discovery of food did not result in monopolization. Instead, successful subjects emitted food calls that attracted the others and resulted in most group members getting at least some of the food. These marmosets appeared to have the efficient strategy of relaxing social cohesion until the discovery of food and then reforming the group through food calls. Clearly, however, the marmosets did not cease to be social animals while engaged in food gathering activities; social and foraging activities were managed together.

In addition to the focused efforts of Menzel and a few others, researchers are often compelled by their observations to make passing references to the necessity for an animal to deal simultaneously with social and foraging tasks. For example, Hall [1963] noted that even on those occasions when a baboon group was observed to divide temporarily into several smaller subgroups while foraging, each subgroup attended repeatedly to the "lead" subgroup, often shrieking in the direction of the lead subgroup; gradually all subgroups converged on the lead subgroup. Cambefort [1981] also pointed out that animals must deal simultaneously with social and foraging tasks. In introducing his study of feeding habit propagation in baboons (*Papio ursinus*) and vervets (*Cercopithecus aethiops*), he asserted that an individual is forced to keep other troop members under constant surveillance because of a dependence upon the troop. In general, it seems reasonable to suppose that an animal cannot afford the luxury of attending solely to one task, such as foraging, while neglecting all others.

In conclusion, it is not suggested that views on competition and time and energy budgeting cannot be usefully applied to the behavior of social foragers. What is proposed is that these views alone may provide only a partial understanding of the ways in which social and foraging variables interact to shape individual behavior. Furthermore, the view that an animal may often have to deal simultaneously with social (affiliative, as well as competitive) and foraging tasks is crucial to a full appreciation of the often subtle and complex strategic adjustments in behavior that must be made by the social forager.

ACKNOWLEDGMENTS

The research reported in this paper was supported in part by USPHS grants MH15965 and RR00169 and funds from the H.F. Guggenheim Foundation, the State University of New York, and the Department of Psychology, University of California, Davis.

REFERENCES

Ainsworth MDS, Wittig BA (1969): Attachment and exploratory behavior of one-year-olds in a strange situation. In Foss BM (eds): "Determinants of Infant Behaviour IV." London: Methuen & Co. Ltd., pp 111–136.

Alexander RD (1974): The evolution of social behavior. Ann Rev Ecol Syst 5:325–383.

Andrews MW (1984): Comparative use of space by two species of New World monkeys with special reference to foraging behavior. Doctoral Dissertation University of California, Davis.

Andrews MW (1986): Contrasting approaches to spatially distributed resources by *Saimiri* and *Callicebus*. In Else JG, Lee PC (eds): "Primate Ontogeny, Cognition and Social Behaviour." Cambridge: Cambridge Univ. Press, pp 79–86.

Baker MC, Belcher CS, Deutsch LC, Sherman GL, Thompson DB (1981): Foraging success in junco flocks and the effects of social hierarchy. Anim Behav 29:137–142.

Baldwin JD, Baldwin JI (1972): The ecology and behavior of squirrel monkeys (*Saimiri oerstedi*) in a natural forest in western Panama. Folia Primatol (Basel) 18:161–184.

Baldwin JD, Baldwin JI (1976): Effects of food ecology on social play: A laboratory simulation. Z Tierpsychologie 40:1–14.

Bernstein IS (1981): Dominance: The baby and the bathwater. Behav Brain Sci 4:419–457.

Bertram BCR (1978): Living in groups: Predators and prey. In Krebs JR, Davies NB (eds): "Behavioural Ecology." Oxford: Blackwell Scientific Publications, pp 64–96.

Cambefort JP (1981): A comparative study of culturally transmitted patterns of feeding habits in the chacma baboon *Papio ursinus* and the vervet monkey *Cercopithecus aethiops*. Folia Primatol (Basel) 36:243–263.

Crook JH, Gartlan JS (1966): Evolution of primate societies. Nature 210:1200–1203.

Cubicciotti D III, Mason WA (1975): Comparative studies of social behavior in *Callicebus* and *Saimiri:* Male-female emotional attachments. Behav Biol 16:185–197.

de Waal FBM (1984): Coping with social tension: Sex differences in the effect of food provision to small rhesus monkey groups. Anim Behav 32:765–773.

Galdikas BMF, Teleki G (1981): Variations in subsistence activities of female and male pongids: New perspectives on the origins of hominid labor division. Curr Anthropol 22:241–247.

Hall KRL (1963): Variations in the ecology of the chacma baboon, *Papio ursinus*. Symp Zool Soc Lond 10:1–28.

Hamilton WJ III, Busse C (1982): Social dominance and predatory behavior of chacma baboons. J Hum Evol 11:567–573.

Jaeger RG, Joseph RG, Barnard DE (1981): Foraging tactics of a terrestrial salamander: Sustained yield in territories. Anim Behav 29:1100–1105.

Jaeger RG, Nishikawa KCB, Barnard DE (1983): Foraging tactics of a terrestrial salamander: Costs of territorial defence. Anim Behav 31:191–198.

Kacelnik A, Houston AI, Krebs JR (1981): Optimal foraging and territorial defence in the Great Tit (*Parus major*). Behav Ecol Sociobiol 8:35–40.

Krebs JR (1974): Colonial nesting and social feeding as strategies for exploiting food resources in the Great Blue Heron (*Ardea herodias*). Behaviour 50:99–134.

Lewis M, Goldberg S (1969): Perceptual-cognitive development in infancy: a generalized expectancy model as a function of mother-infant interaction. Merrill-Palmer 15:81–100.

Loy J (1970): Behavioral responses of free-ranging rhesus monkeys to food shortage. Am J Phy Anthropol 33:263–272.

Mason WA (1974): Comparative studies of social behavior in *Callicebus* and *Saimiri:* Behavior of male-female pairs. Folia Primatol (Basel) 22:1–8.

Mason WA (1975): Comparative studies of social behavior in *Callicebus* and *Saimiri:* Strength and specificity of attraction between male-female cagemates. Folia Primatol (Basel) 23:113–123.

Menzel EW (1971): Communication about the environment in a group of young chimpanzees. Folia Primatol (Basel) 15:220–232.

Menzel EW (1973): Chimpanzee spatial memory organization. Science 182:943–945.

Menzel EW Jr, Juno C (1982): Marmosets (*Saguinus fuscicollis*): are learning sets learned? Science 217:750–752.

Menzel EW Jr, Juno C (1985): Social foraging in marmoset monkeys and the question of intelligence. Philos Trans R Soc Lond [Biol] 308:145–158.

Milinski M (1982): Optimal foraging: The influence of intraspecific competition on diet selection. Behav Ecol Sociobiol 11:109–115.

Phillips MJ, Mason WA (1976): Comparative studies of social behavior in *Callicebus* and *Saimiri:* Social looking in male-female pairs. Bull Psychonomic Soc 7:55–56.

Pitcher TJ, Magurran AE (1983): Shoal size, patch profitability and information exchange in foraging goldfish. Anim Behav 31:546–555.

Plimpton E, Rosenblum L (1983): The ecological context of infant maltreatment in primates. In Reite M, Caine NG (eds): "Child Abuse: The Nonhuman Primate Data." New York: Alan R. Liss, Inc., pp 103–117.

Plimpton EH, Swartz KB, Rosenblum LA (1981): The effects of foraging demand on social interactions in a laboratory group of bonnet macaques. Int J Primatol 2:175–185.

Post DG, Hausfater G, McCuskey SA (1980): Feeding behavior of yellow baboons (*Papio cynocephalus*): Relationship to age, gender and dominance rank. Folia Primatol (Basel) 34:170–195.

Richard A (1974): Intra-specific variation in the social organization and ecology of *Propithecus verreauxi*. Folia Primatol (Basel) 22:178–207.

Rodman PS (1977): Feeding behaviour of orang-utans of the Kutai Nature Reserve, East Kalimantan. In Clutton-Brock TH (ed): "Primate Ecology: Studies of Feeding and Ranging Behaviour in Lemurs, Monkeys and Apes." London: Academic Press, pp 383–413.

Rosenblum L, Youngstein K (1974): Developmental changes in compensatory dyadic responses in mother and infant monkeys. In Lewis M, Rosenblum L (eds): "The Effect of the Infant on Its Caregiver: The Origins of Behavior, Vol. 1." New York: John Wiley & Sons, pp 141–162.

Rosenblum LA, Sunderland G (1982): Feeding ecology and mother-infant relations. In Hoff LW, Gandelman R, Schiffman HR (eds): "Parenting: Its Causes and Consequences." Hillsdale, NJ: Lawrence Erlbaum Assoc., pp 75–110.

Rosenblum LA, Paully GS (1984): The effects of varying environmental demands on maternal and infant behavior. Child Dev 55:305–314.

Rosenblum LA, Clark RW, Kaufman IC (1964): Diurnal variations in mother-infant separation and sleep in two species of macaque. J Comp Physiol Psych 58:330–332.

Rosenblum LA, Kaufman IC, Stynes AJ (1969): Interspecific variations in the effects of hunger on diurnally varying behavior elements in macaques. Brain Behav Evol 2:119–131.

Slatkin M, Hausfater G (1976): A note on the activities of a solitary male baboon. Primates 17:311–322.

Southwick CH (1967): An experimental study of intragroup agonistic behavior in rhesus monkeys (*Macaca mulatta*). Behaviour 28:182–209.

Struhsaker TT (1974): Correlates of ranging behavior in a group of red colobus monkeys (*Colobus badius tephrosceles*). Am Zool 14:177–184.

Tilson RL, Hamilton WJ III (1984): Social dominance and feeding patterns of spotted hyaenas. Anim Behav 32:715–724.

Wrangham RW (1974): Artificial feeding of chimpanzees and baboons in their natural habitat. Anim Behav 22:83–93.

Ecology and Behavior of Food-Enhanced Primate Groups, pages 269–296

13

Demography and Mother-Infant Relationships: Implications for Group Structure

Carol M. Berman

Caribbean Primate Research Center, Punta Santiago, Puerto Rico, 00749; and Department of Anthropology, State University of New York at Buffalo, Buffalo, New York, 14261

INTRODUCTION

There is a growing consensus among behavioral ecologists that demographic processes play a crucial role in determining social structure among primates through their effects on individual social relationships. It is also believed that one of the major ways in which demography may operate is through developmental processes. Altmann and Altmann (1979) describe a theoretical cycle of effects of 1) behavior on demographic processes, 2) demographic processes on the size and composition of the group, and, finally, 3) group size and composition on behavior. Although there has been some theoretical and empirical research on steps 1 and 2 of the cycle (e.g., Cohen, 1969, 1972; Keiding, 1977; Dunbar, 1985), fewer data are available on the effects of group size and composition on individual social relationships and on social organization in naturally organized groups. Hausfater (1981) has used simulation modeling to illustrate possible relationships between demographic variation and observed variation in 1) adult male reproductive strategies in multimale groups, 2) the occurrence of infanticide in one-male groups, and 3) patterns of rank acquisition and maintenance among adult females. The models suggest that observed variation in behavior and group organization in small groups is as likely to be the product of demographic variation as it is of adaptation to ecological conditions. There is some empirical evidence that wild long-tailed macaques *(Macaca fascicularis)* in large social groups engage in less grooming and more agonistic interaction than those in small groups (Van Schaik et al., 1983). In addition, the size of one-male units in gelada baboons *(Theropithecus gelada)* is crucial in determining grooming patterns among females and between the male and the females (Dunbar and Dunbar, 1976; Dunbar, 1979, 1984). Almost no information is available, however, on the influence of group size and composition on infant development in naturally organized groups.

One difficulty with understanding the roles of demography in social behavior is, as Altmann and Altmann (1979) point out, that other environmental factors also affect behavior both directly (e.g., Lee, 1983; Johnson and Southwick, 1987) and indirectly through delayed effects on the size and composition of the group. Hence it is difficult to separate the effects of demography per se from those of environmental factors and environmental change. One way of overcoming this difficulty is to observe behavior in free-ranging, food-enhanced populations where animals are relatively unmanipulated but where major environmental influences such as predation and resource availability and distribution can be controlled.

In this chapter three examples of research on infant social development among free-ranging rhesus monkeys on Cayo Santiago, Puerto Rico are reviewed. The research took place between 1974 and 1984, an 11-year period during which the physical environment and amounts of food available per monkey remained virtually constant. The population and study group grew steadily and dramatically. In each example, group size and/or composition appear to affect early infant social relationships in ways which have long-term implications for social organization. The first example is concerned with the presence or absence of large kinship clusters. The second is concerned with the specific impact of immature siblings. The third example discusses the impact of changes in group size and composition as the group expanded over the years.

THE STUDY SITE AND STUDY GROUP

Cayo Santiago is a 15.2 ha island about 1 km off the east coast of Puerto Rico which supports a free-ranging, food-enhanced population of rhesus monkeys *(Macaca mulatta).* Clarence Carpenter established the colony in 1939 with the introduction of 409 rhesus monkeys from India. Since then no monkeys have been added to the population except by birth, although a number have been removed over the years. The population has been censused daily since 1956 and hence is well-habituated and made up of known individuals with known histories and maternal kinship relationships. (See Carpenter, 1942a, b; Sade et al., 1977; Rawlins and Kessler, 1983; and Sade et al., 1985 for detailed descriptions of the history, terrain, population, and management.)

Between 1972 and 1984, management practices were designed to minimize human interference and were virtually constant. The monkeys were provisioned with high protein commercial monkey chow at the liberal rate of approximately 0.5 pound (0.23 kg) per monkey per day from hog feeders located in three 0.25 acre corrals. In addition, they foraged on vegetation and soil (Sultana and Marriott, 1982) and were provided with rainwater ad libitum. Manipulation has been limited to noninvasive measurements during an annual trapping period in January and February.

Under this management scheme the population expanded at a steady geometric rate of about 13% per year (Rawlins and Kessler, 1986) from about 300 individuals in the beginning of 1974 to nearly 1,200 in 1983. A cull of 3 whole social groups in early 1984 reduced the population to about 700. No other monkeys were removed from the island during the study period.

The study group (group I), which was not directly involved in the cull, grew at a geometric rate of approximately 16% per year (Rawlins and Kessler, 1986), a rate which exceeded that of the population as a whole. Group I was formed naturally by fissioning in 1961 and was organized in the manner typical of multimale macaque groups. A number of females and their immature offspring made up the permanent core of the group along with a more transient set of adult males. Male offspring generally left their natal group at puberty and joined other groups. Adult daughters, on the other hand, remained in their natal group for life and continued to associate with their mothers and maternal kin. In this way, large maternal lineages which were also subunits of social organization developed over time.

Dominance hierarchies were constructed from the directions of fear grins and cowers between pairs of individuals. Dominance relationships among females were generally stable and linear during the study period. The dominance structure of the group was closely linked with the kinship structure: adult daughters took on ranks similar to their mothers, and lineages, as a whole, could be ranked in the same order as their matriarchs. In group I there were 3 extended lineages with members spanning as many as 4 generations. The total membership of the group varied from 53 at the beginning of 1974 to about 321 at the end of 1984. The group did not fission during this period, although it did so in both 1985 and 1986.

When groups fission, they typically split along lineage lines—that is, the lineages usually remain intact, and one or more lineages separate from the rest of the group and form a daughter group. In their study of group fissioning on Cayo Santiago, Chepko-Sade and Sade (1979) found only a few exceptions to this rule. In most of those, the lineage split along sublineage lines with the eldest daughter of the matriarch and her descendants splitting off from the rest of the lineage. The probability of fissioning was related to the size of the group or lineage, but a better predictor was the mean degree of relatedness among members (Chepko-Sade and Olivier, 1979). When this became very small, the group or lineage became likely to split.

Although social groups competed for access to the feeders, it is unlikely that this limited the total amounts of food available to individual animals in group I. Group I has consistently ranked second among the 5 to 6 social groups on Cayo Santiago over the course of the study. As a result the group has maintained priority of access to at least one feeding corral.

THE INFANT SAMPLE

Twenty infants who were born into group I during the 1974 and 1975 birth seasons were observed from birth to at least 30 weeks of age. They came from all 3 lineages. Ten (50%) were male and 10 (50%) were female. Six (30%) had primiparous mothers and 14 (70%) had multiparous mothers. These infants made up the sample for the first two examples (see Results). The sample for the third example included these infants plus 49 additional infants: one additional 1975 infant, fifteen 1979 infants, seventeen 1983 infants, and sixteen 1984 infants. The 1983 and 1984 infants were the offspring of previously observed mothers and/or of previously observed female infants. The total sample for the third example (n = 69) consisted of 35 (50.7%) males, 34 (49.3%) females, 16 (23.2%) offspring of primiparous mothers, and 53 (76.8%) offspring of multiparous mothers. The later infants were observed using essentially the same methods as the earlier infants.

DATA COLLECTION

Observation conditions were excellent; it was possible to observe intimate details of interaction almost continuously without the use of binoculars by following infants at a distance of a few meters. Focal-animal sampling methods (Altmann, 1974) were used to record the social interaction of infants with their mothers and with other members of the group. Each infant was observed individually for 3 hours during each fortnight age period. All interactions involving the infant were listed chronologically on checksheets along with the identities of the initiator and the recipient of the interaction and the time it occurred. At 2-minute intervals, point (instantaneous) time samples were taken by tape recorder identifying the individuals in contact with the infant, within 60 cm, and between 60 cm and 5 m. All observations were made between 7:00 a.m. and 12 noon AST, hours which included all major group activities: feeding at the food bins, foraging, traveling, socializing, and resting. However, time spent near the feeder bins was not included in the observations. The sequence of infants observed was determined by a randomized order arranged at the beginning of each week.

MEASURES

The following measures were used in the analyses. Each measure was tabulated separately for each mother-infant pair for each fortnight age period:

1) *time off*—the percentage of time mothers and infants were neither in ventro-ventral contact or in nipple contact was estimated from the percentage of 2-minute point time samples out of 90; 2) *time > 5 m*—the percentage of time off spent more than 5 m from the mother was estimated from point time samples; 3) *time > 60 cm*—the percentage of time off spent at a distance of more than 60 cm from the mother was estimated from point time samples; 4) *time on ventrum*—the percentage of time mother-infant pairs spent in ventro-ventral contact was estimated from the percentage of 2-minute point time samples out of 90 during which they were observed in ventral contact; 5) *relative rejections*—the proportion of all attempted nipple contacts, including those initiated by the mother, that were prevented by the mother; 6) *maternally initiated contact*—the proportion of successful nipple contacts which were initiated by the mother, and 7) *the proximity index*—a measure of the infant's relative role in maintaining proximity within 60 cm of the mother; it is the percentage of approaches made by the infant to the mother over a 60 cm limit minus the percentage of departures made by the infant from the mother over a 60 cm limit. (The index ranges from -100% to +100%; negative values indicate that the mother is taking the primary role in maintaining proximity, whereas positive values indicate that the infant is taking the primary role. See Hinde and Atkinson, 1970, for a detailed description of the properties of this index.)

RESULTS AND DISCUSSION

The General Course of Infant Development

The general course of development of mother-infant relationships in rhesus monkeys has been described by Harlow and his coworkers (e.g., Harlow and Harlow, 1965) and by Hinde and his coworkers (e.g., Hinde and Spencer-Booth, 1967). Very briefly, mothers and infants initially spend nearly all their time in ventro-ventral contact. Mothers rarely reject their infants' attempts to contact them and are primarily responsible for maintaining contact and proximity with them (as shown by a negative proximity index). At this stage they are relatively intolerant of the attention paid to infants by other group members and they frequently restrain infants from interacting with them. Gradually, however, mothers and infants spend less time in contact and in close proximity. Mothers begin to reject infants' attempts to regain close contact, and infants become primarily responsible for maintaining contact and proximity with the mother (as shown by a positive proximity index). Mothers eventually cease restraining their infants and interfere less frequently in their interactions with other group members.

Although this general pattern is relatively consistent across populations (e.g., Seay and Gottfried, 1975), whether in captivity (e.g., Hinde and Spencer-Booth, 1967), in free-ranging, food-enhanced colonies (Berman, 1980), or in the wild (e.g., Johnson and Southwick, 1984, 1987), there is wide variation within social groups. A small amount of inter-individual variation in mother-infant relationships appears to be due to variation in infants' sex, mothers' parity, and mothers' dominance. However, the particular ways in which these factors appear to affect mother-infant interaction are not consistent across studies and may depend on the way these factors interact with other factors associated with the various social and physical environments in which they have been examined (see review in Berman, 1984).

Example I: The Presence of Kinship Lineages

As with nearly all aspects of rhesus monkey social behavior (e.g., Sade, 1972, 1977), the presence or absence of maternal kin appears to influence infant social development. The first indication of this came from a comparison of measures of mother-infant interaction on Cayo Santiago with those in the captive socially living colony of rhesus monkeys at Madingley, England (Berman, 1980). The results of this comparison are summarized briefly here. At Madingley, monkeys lived in 6 permanent social groups of about 8 to 10 monkeys—1 adult male, 2 to 4 adult females, and their immature offspring. The original monkeys forming the colony in the early 1960s were not known to be related. Thus these groups lacked the elaborate lineage structure of the Cayo Santiago population. However, by 1974 small kinship clusters had begun to develop in most of the pens.

The general course of development of mother-infant relationships was remarkably similar across the 2 colonies both qualitatively and quantitatively. Nevertheless, there were small, consistent differences between free-ranging and captive pairs. It was possible to compare directly three measures of mother-infant interaction: time off, relative rejections, and the proximity index. Free-ranging pairs spent somewhat more time off than captive pairs in 13 of 15 fortnight age periods (Fig. 1). Free-ranging mothers also rejected their infants more frequently in all 15 age periods. Finally, free-ranging infants took slightly larger roles in maintaining proximity to their mothers in all but one age period (Fig. 2). These differences are consistent with the conclusion that free-ranging mothers (rather than infants) were seeking contact and proximity with their infants less than captive mothers, and in this sense, could be described as more rejecting and more encouraging of independence (see Hinde, 1969, 1974).

Obviously, many aspects of the two environments differed, and it was not immediately clear which differences were responsible for the observed differences between the mothers. However, the observation that the Madin-

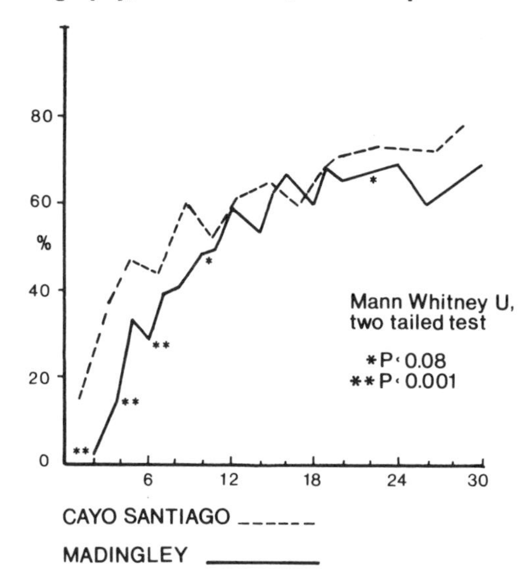

Fig. 1. Median percentages of time spent off the mother for Cayo Santiago and Madingley pairs as a function of infant age. Data are derived from 90 2-minute point time samples per infant per fortnight age period from birth to 30 weeks of age. Asterisks refer to the significance of differences between colonies. For details of the sample and methodology, see Berman (1980).

gley mothers seemed to be becoming more like free-ranging mothers over the years suggested that important differences between the 2 colonies were likely to be related to factors which were changing at Madingley over time, and not to stable differences between the colonies such as diet or climate. For example, in 1968 the proximity index at Madingley was negative for a longer period of time than in 1974, and it leveled out at a lower value (see Fig. 3). In other words, Madingley infants before 1968 began to play the primary role in maintaining proximity to the mother at a later age than subsequent Madingley infants, and even after they took on the primary role, they took less responsibility for maintaining proximity than did subsequent infants. In both cases, Madingley infants played smaller roles in maintaining proximity to their mothers than did Cayo infants.

After eliminating a number of alternative explanations (e.g., differences in environmental complexity, ages of the colonies, parities of the mothers, sexes of the infants, origins of the mothers, and amounts of human disturbance), the hypothesis remained that free-ranging mothers were more relaxed with their infants because they were able to raise them among their kin. When the Madingley colony was established in the early 1960s, none of the adult females were known to be related. By 1968, 3 of the 6 pens

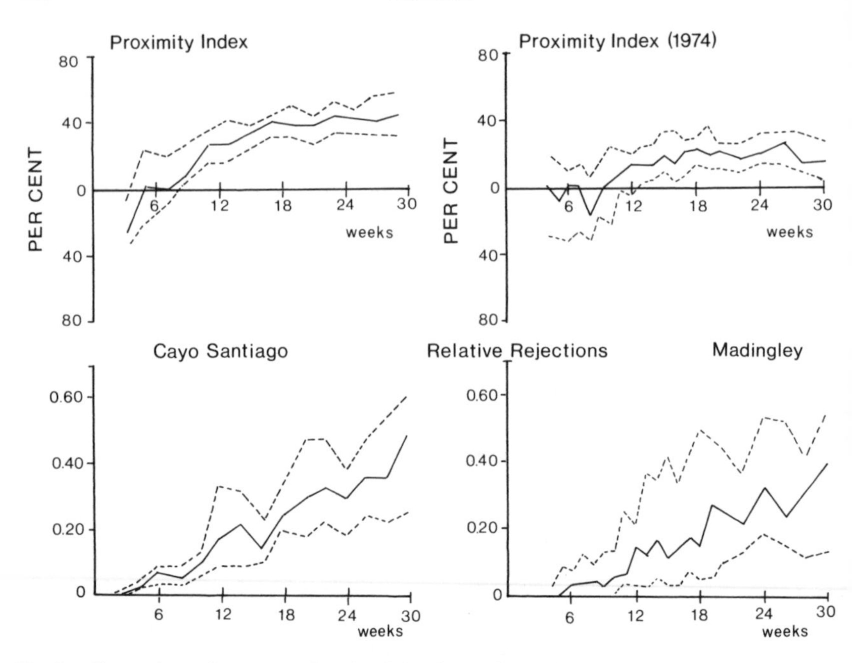

Fig. 2. Comparison of measures of mother-infant interaction at Cayo Santiago (**left**) and Madingley (**right**). Medians and interquartile ranges for the proximity index (**top**) and the relative frequency of rejections of nipple contact by the mother (**bottom**). For details of sample and methodology, see Berman (1980).

contained clusters of kin that extended beyond a single mother-infant pair. In 1974, this was the case in 5 of 6 pens. Hence, changes in the captive mothers over time appear to have paralleled the development of small kinship clusters.

The apparent tendency for mothers to develop a relaxed maternal style in the presence of matrilineal kin has important implications for understanding the development and maintenance of lineage-based group structure over generations. Infants (particularly females) continue to associate with their mothers after infancy and develop networks of relationships which mirror those of their mothers (Berman, 1982a). Mothers with few kin within the group may allow their infants fewer opportunities to form relationships with other members of the group than mothers with more kin. As a result, these infants may develop less intense or less relaxed relationships with others. Their closest relationships would be formed with kin, because these are also the mother's closest associates. The persistence of these patterns beyond infancy would lead eventually to the development of subunits of organization within the groups based on maternal lineages. The continued rearing of each

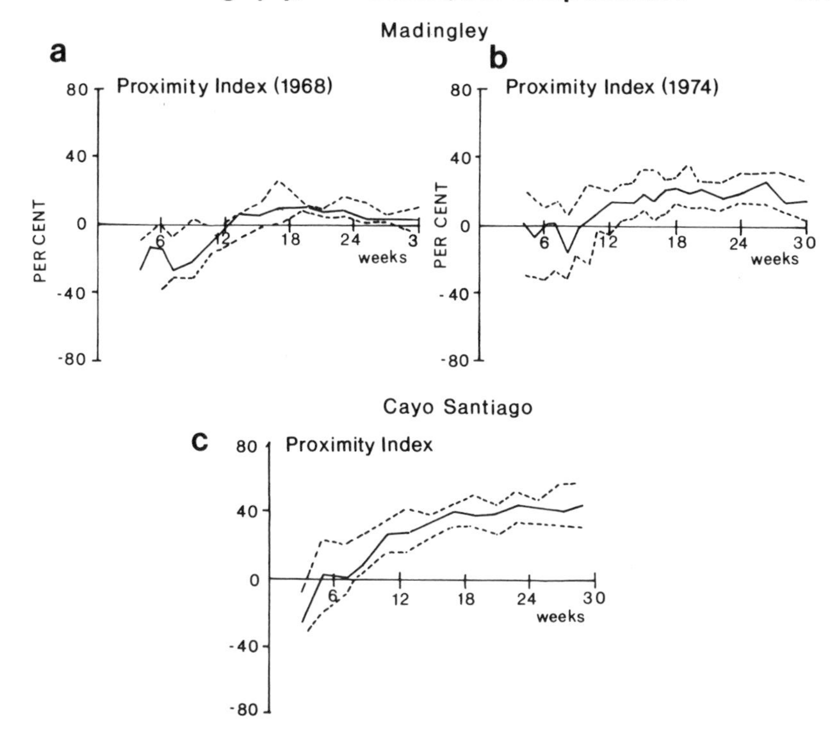

Fig. 3. Medians and interquartile ranges for the proximity index for **a**) Madingley pairs (compiled in 1968); **b**) Madingley pairs (compiled in 1974); and **c**) Cayo Santiago pairs. Details of the sample and methodology are in Berman (1980).

new infant within these units would serve to maintain their structure over generations.

Example 2: The Presence of Immature Siblings

The presence of particular kinds of kin are also likely to have important influences on maternal style. For example, Cayo Santiago mothers with several immature offspring within the group tend to be more rejecting and less solicitous of their infants than mothers with fewer immature offspring (Berman, 1984). Mothers with relatively more immature offspring within the group spend significantly less time in ventro-ventral contact with their infants and more time at a distance of at least 5 m from the infant. In addition, they initiate significantly smaller proportions of the nipple contacts, reject their infants significantly more, and take a relatively smaller role in maintaining proximity with them (Table I, row 3).

TABLE I. Kendall Rank Order Correlation Coefficients for Measures of Mother-Infant Interaction With Mother's Age, Total Maternal Experience, and Number of Immature Siblings

	Time on ventrum	Time > 5 m	Maternally initiated contact	Proximity index	Rejections
Mother's age	−0.30*	0.30*	−0.52**	0.31*	0.41**
Mother's experience	−0.34*	0.32*	−0.59**	0.39**	0.47**
No. of immature sibs	−0.34*	0.31*	−0.62**	0.42**	0.49**

**P<0.01; *P<0.05.
Total 20 mother-infant pairs.
Data are from Berman (1984).

In this expanding population, where natality is high and mortality is low, the numbers of immature young a mother has within the group is correlated with her age and maternal experience (measured as the total number of offspring she has reared whether or not they are still alive or within the group). Thus older and more experienced mothers also appear to be more rejecting and less solicitous than younger, less experienced mothers (Table I, rows 1, 2). [This is a result which is not in harmony with predictions of parent-offspring conflict theory (Trivers, 1974) and will be discussed in detail elsewhere (Berman in prep. a)]. It is possible to separate the effects of numbers of immature offspring from those of age and experience by repeating the analysis on mothers that are 6 years of age or older. On Cayo Santiago, females typically begin to reproduce at 3 or 4 years of age. Many give birth each year and by 6 years of age each female has had the opportunity to produce the maximum number of 3 immature offspring in addition to the infant on hand. Among mothers 6 years and older, neither maternal age nor experience are strongly related to measures of mother-infant interaction (Table II). However, numbers of immature siblings are significantly related to two measures: mothers with more immature offspring initiate smaller proportions of nipple contact with their infants and play a relatively smaller role in maintaining proximity with them than do mothers with fewer immatures.

These results illustrate not only that the presence of immature siblings can affect an infant's relationship with its mother, but also that the apparent relationship between maternal style and other attributes of the mother (e.g., age and maternal experience) may depend on the demographic characteristics of the group. In groups with lower natality and higher juvenile mortality, there is likely to be a different kind of relationship between maternal age and experience and numbers of immature siblings. In these groups, older mothers would not necessarily be expected to manifest more rejecting styles of

TABLE II. Kendall Rank Order Correlation Coefficients for Measures of Mother-Infant Interaction With Mother's Age, Total Maternal Experience, and Number of Immature Siblings

	Time on ventrum	Time > 5 m	Maternally initiated contact	Proximity index	Rejections
Mother's age	−0.08	0.14	−0.14	0.05	0.14
Mother's experience	−0.26	0.26	−0.32	0.20	0.09
No. of immature sibs	−0.07	0.30	−0.67**	0.60**	0.07

Sample includes mothers 6 years old and older; total 9 mother-infant pairs. Data are from Berman (1984).
**P<0.01.

mothering than younger mothers (e.g., Nicolson, 1982). Hence theories concerning inter-individual variations in maternal investment must consider the demographic characteristics of the group as well as the attributes of individual mothers.

The precise way in which immature siblings affect their mothers is under investigation. It may be that immature siblings make additional demands on mothers, leaving them less time or energy to care for infants. Alternatively, immature siblings may invest care in infants, reducing the mothers' burden of care. In addition there is evidence that the ages and sexes of siblings are important factors influencing mother-infant interaction and that these factors interact in complicated ways with the infant's sex (Hooley and Simpson, 1983): Infants with yearling siblings spend less time in contact with their mothers in the early weeks and more time playing with group companions than infants with older siblings. Mothers with male infants restrain their infants more if they have older siblings than if they have yearling siblings, probably because the older and larger siblings pose more of a threat to the mother and infant. Finally, yearling sisters appear to influence mother-infant interaction more and in different ways than yearling brothers.

Example 3: Changes in Infant Social Development and Social Organization with Population Growth

On Cayo Santiago, as in populations of wild rhesus monkeys (e.g., Malik et al., 1984) and Japanese macaques (Mori, 1979), food enhancement has led to high natality, low mortality, and rapid population growth. Between 1974 and 1984, the physical environment and the amount of food available per monkey were constant on Cayo Santiago, providing the opportunity to examine the consequences of population growth—in terms of changes in group composition, changes in the infants' social milieu, and subsequent changes in its social relationships—in the absence of confounding changes in resource availability. Specific changes were found in mother-infant interaction and in infant social networks which have implications for social organization, particularly for the likelihood of fissioning along lineage lines (Berman in prep. b). These findings are summarized below.

Changes in group size and composition. As stated previously, group I expanded at the rate of about 16% per year during the study period, from 53 individuals at the beginning of 1974 to 321 at the end of 1984. The proportions of the group represented by each age-sex class varied from year to year, but for the most part these changes were not in any consistent direction. The exception was for nonnatal adolescent and adult males whose proportional representation in the group increased each year from 13.1% of the group in 1974 to 23.9% in 1984.

The maternal kinship structure of the group also underwent changes. Figure 4 compares kinship structures in 1974–1975 and in 1983. All possible dyads involving infants with other group members were divided into those in which the infants' companion was a close maternal relative (sibling or grandmother), a more distant maternal relative (any other lineage member), or an unrelated monkey (member of another lineage). The distributions of each sort of dyad were compared between years using a Chi square analysis. Although each infant in 1983 had more kin (in absolute terms) than each infant in 1974 and 1975, their close maternal kin made up a significantly smaller proportion of their social companions within the group.

To summarize these changes, infants in the later years were being brought up on a more crowded island, in a much larger group, with proportionally more nonnatal males and proportionally fewer close kin than infants in the early years.

Infants' social experience. The next questions asked concerned the ways in which these demographic changes were reflected in the infant's social environment. Were infants in the later years surrounded by many more companions most of the time or by different sorts of companions than earlier infants—or were their experiences with group members regulated in some way by their mothers, themselves, or by others? An initial approach was simply to look at the mean number of social companions near the infant at any one time. The occurrence of change depended very much on the way one defined "near." For example, the mean number of group members within 60 cm (touching distance) of infants at any one time remained nearly constant at between 1 or 2 animals in spite of the increase in group membership (Fig. 5). This was not expected in view of infants' attractiveness to others (Rowell et al., 1964; Berman, 1982b) and in view of the fact that it is physically possible to have as many as 6 to 8 monkeys as close to infants as 60 cm at once. This suggested that other factors, perhaps actively involving the mother or infant, were acting to regulate close proximity with the infant.

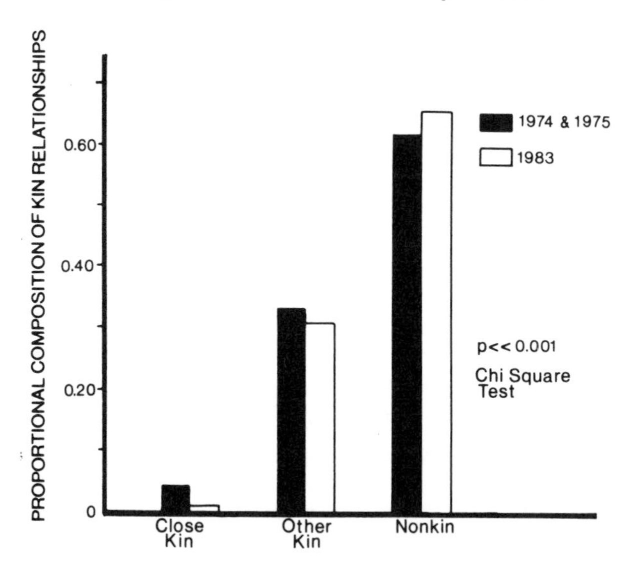

Fig. 4. The distribution of dyads involving infants with other group members in group I by the maternal kinship relationship of the infant with the other group member. Infants in 1974 and 1975 are compared with infants in 1983. Close kin = siblings and grandmothers. Other kin = other members of the infant's matrilineage. Nonkin = members of a different matrilineage. N (1974, 1975) = 21 infants. N (1983) = 17 infants.

In contrast, changes in the number and composition of monkeys at a moderate distance (<5 m) from the infant mirrored changes in the group as a whole. (Five meters is roughly the area making up the size of a cluster of monkeys at rest.) The mean number of monkeys within 5 m of infants at any one time increased steadily and markedly over the years (Fig. 6). Increases in all age-sex-kinship categories were marked. Moreover, there was a proportional increase in numbers of unrelated animals and a proportional decrease in the numbers of close maternal kin (see Table III). In other words, infants in later years were not only surrounded by many more monkeys but also by monkeys made up of a larger proportion of nonkin, and hence by less familiar and potentially more frightening individuals.

Mother-infant relationships. With such marked changes in the infants' social environment, certain compensatory changes in mother-infant relationships would be expected. In a species such as rhesus where infants are attractive to a variety of group members and where mothers are relatively intolerant of the attention paid to their infants, one would predict that mothers would respond to such a situation with increased protective behavior toward their infants. In this case one would predict specifically that mothers in the

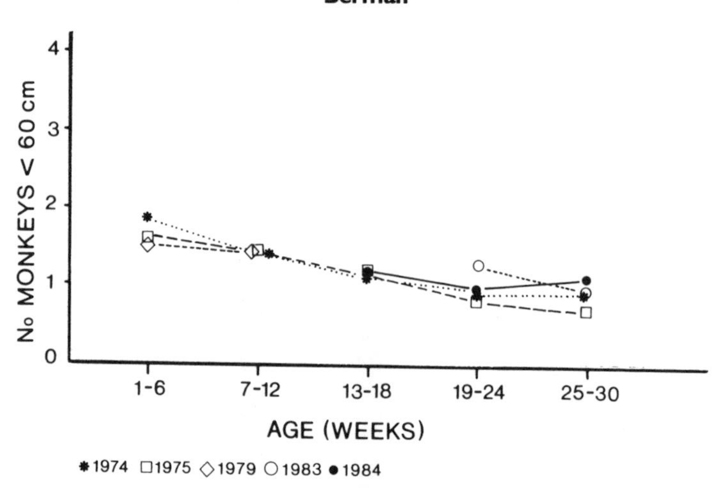

Fig. 5. Mean numbers of group members within 60 cm of the infant at any one time as a function of infant age and year of observation. Data are derived from point time (instantaneous) samples. N (1974) = 9 infants; N (1975) = 12 infants; N (1979) = 15 infants; N (1983) = 17 infants, N (1984) = 16 infants.

later years of the study would seek proximity with their infants more—that is, they would spend more time near them and take a larger role in maintaining proximity with them. This is indeed what the data showed.

Infants spent significantly less time at a distance of at least 5 m from the mother with each year of observation, and this was significant in most age periods (see Fig. 7). Infants also spent less time beyond 60 cm of mothers with each year of observation, but these results were less marked. Figure 8 shows that the proximity index also decreased with year of observation. Hence mothers in the later years not only spent more time near their infants, they also took a relatively greater role in maintaining proximity with them. This pattern of changes is consistent with the conclusion that changes in mothers (rather than infants) were primarily responsible for changes in mother-infant relationships over the years (see Hinde, 1969, 1974), and more specifically that mothers in the later years were more protective than mothers in the earlier years of the study. Their increased proximity-seeking probably functioned to keep down the number of monkeys within touching distance of their infants.

Interestingly, there were few changes in measures of *contact* between mothers and infants. There were no changes in the amount of time spent in contact, in the rate at which mothers rejected infants' attempts to get on the nipple, or in the proportion of nipple contacts initiated by mothers. This may have been because there had been no change in the number of other monkeys

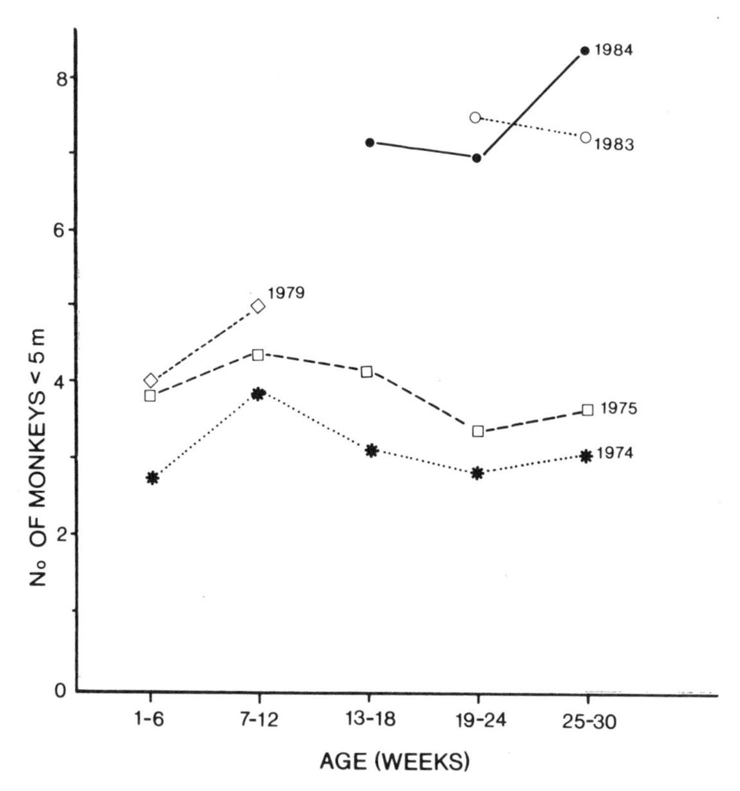

Fig. 6. Mean numbers of group members within 5 m of the infant at any one time as a function of infant age and year of observation. Data are derived from point time (instantaneous) samples. Sample sizes are as in Figure 5.

within touching distance of infants. Hence, there was no increased need to protect infants from contact with others by holding onto them more.

In order to test whether the mothers' increased protectiveness was actually in response to the increased presence of group members within 5 m and not in response to other factors, amounts of time individual 1983 mothers and infants were more than 5 m from one another were examined as a function of the number of other monkeys near the infant at the time of observation. Observed values were compared with those expected based on the null hypothesis that mothers and infants distributed their time apart from one another without respect to the numbers of other monkeys within 5 m of the infant.

For example, suppose that 100 point (instantaneous) time samples were recorded for an infant in which the identity of all other monkeys within 5 m

TABLE III. Proportions of Group Members Observed Within 5 m of Infants Who Were Close Kin (Siblings and Grandmothers), Other Kin (Other Members of the Infant's Lineage), and Nonkin (Members of Other Lineages): Directions of Differences Between Infants Born in 1974 and 1975 and Those Born in 1983

	Close kin	Other kin	Nonkin
A♀♀	−	+	−
J♀♀	−	−	+
Y♀♀	−	=	+
J & Y♂♂	−	+	+
Infants		=	+

+: '83 > '74, '75.
−: '83 < '74, '75.
=: '83 = 74, '75.
A♀♀ = Adult females (4 years and older).
J♀♀ = Juvenile females (2 and 3 years).
Y♀♀ = Yearling females (1 year).
J & Y♂♂ = Juvenile and yearling males (1–3 years).
Infants = Individuals born during the same birth season as the subjects.
N (1974, 1975) = 20 infants.
N (1983) = 17 infants.

of the infant was noted. Suppose again that the mother was observed to be beyond 5 m of the infant in 50 or half of those samples. If the infant was recorded within 5 m of 0–4 other individuals in 40 samples, within 5 m of 5–9 other individuals in 25 samples, within 5 m of 10–14 other individuals in 20 samples, and within 5 m of 15 or more other individuals in the remaining 15 samples, one would expect to find the mother beyond 5 m of the infant in 50% × 40 = 20 samples during which the infant was near 0–4 other individuals, and in 50% × 25 = 12.5 samples during which the infant was near 5–9 other individuals. The comparable figures for 10–14 individuals and 15 or more individuals would be 50% × 20 = 10 and 50% × 15 = 7.5, respectively.

Table IV shows the directions of the differences between observed and expected values for 8 individual 1983 infants between the ages of 27 to 30 weeks. In all 8 cases, mothers were more likely than expected to be more than 5 m from their infants when there were few other monkeys within 5 m of the infants and less likely than expected to be at a distance from infants when there were many other monkeys within 5 m of them. These results reached significance in 4 out of the 8 cases.

There is some evidence that mothers were reacting specifically or particularly to increases in particular classes of animal <5 m from infants rather than to the increase in the total number of monkeys <5 m from them. Table V shows a similar analysis for nonnatal adult and adolescent males

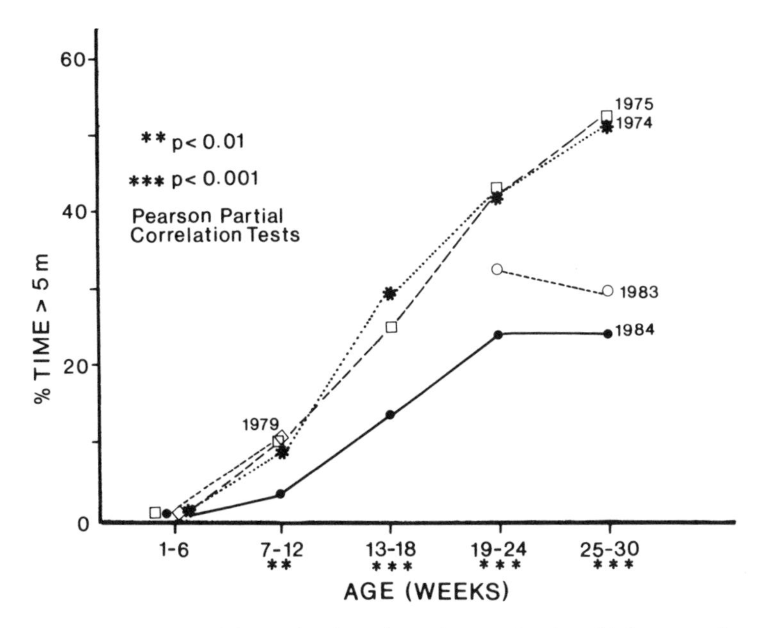

Fig. 7. Median percentage of time > 5 m from the mother as a function of infant age and year of observation. Asterisks refer to significant results of Pearson partial correlation tests for % time > 5 m with year of observation controlling simultaneously for mother's age, mother's total maternal experience (number of infants reared), and number of immature siblings within the group. Separate analyses in which variations in infant's sex, mother's dominance status, and mother's parity were controlled produced similar results. Sample sizes are as in Figure 5.

near infants. In this analysis it was necessary to control for the total number of monkeys <5 m from infants because there was a tendency for more males to be <5 m from infants when more monkeys of any kind were <5 m from them. This was done by examining only those samples in which 5 to 7 monkeys (approximately the median) were within 5 m of the infant and by expanding the age range of the infant data to 23 to 34 weeks (to insure a reasonable sample size per infant). All 8 infants were at a distance from their mothers more than expected when no male was <5 m and less than expected when one or more males were <5 m ($p < 0.008$, Sign test). However, differences between expected and observed values were small; individual Chi square tests fell short of significance for all but one infant.

To summarize changes in mother-infant relationships, different processes appeared to be governing the exposure infants had to other group members at different distances. Some process or processes not involving increased mother-infant *contact* appeared to be regulating the mean number of monkeys within touching distance of infants. The number of individuals

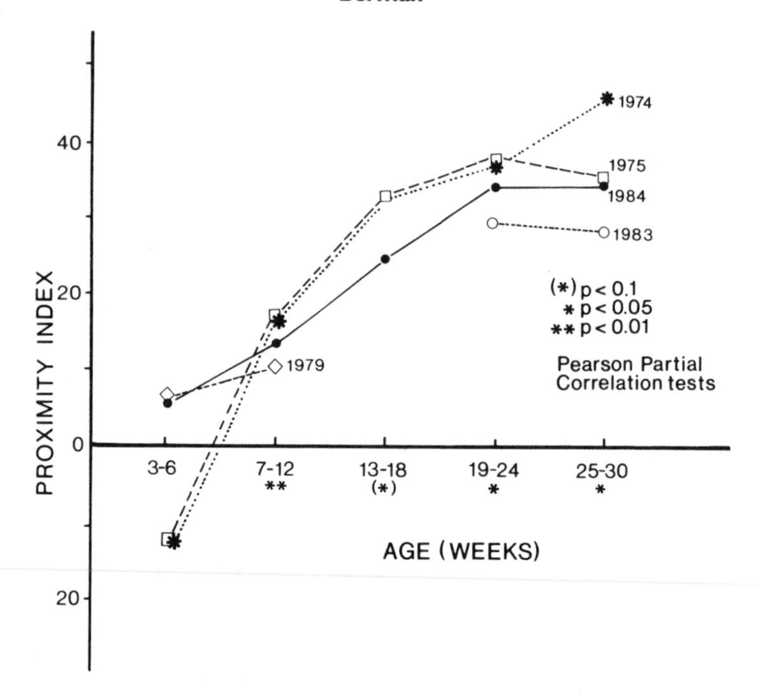

Fig. 8. Median values for the proximity index as a function of infant age and year of observation. Asterisks refer to significant results for Pearson partial correlation tests of the index with year of observation, controlling simultaneously for mother's age, mother's total maternal experience, and numbers of immature siblings within the group. Separate analyses in which variations in infant's sex, mother's dominance status, and mother's parity were controlled produced similar results. Sample sizes are as in Figure 5.

within 5 m did not appear to be as tightly regulated. Changes in this mirrored changes in the size and kin composition of the group in general. Mothers in the later years appeared to be seeking more proximity with their infants in response to these changes, and it is possible that this served to regulate the numbers of monkeys coming within touching distance of infants.

Changes in the infants' social network. In a final set of analyses, the amounts of time specific kinds of companions spent within 5 m of infants were examined in order to get a more detailed picture about how these changes were likely to have affected the infants' social network and in order to get an idea of the processes which may have governed these changes. Were changes in the distribution of monkeys within 5 m of infants due solely and directly to changes in group composition over the years and not to other factors, such as changes in the overall dispersion of the group or specific changes in the behavior of individuals? (Changes in dispersion and individual

TABLE IV. Directions of Differences Between Observed and Expected Amounts of Time That Individual Infants Were Beyond 5 m of Their Mothers as a Function of the Number of Other Group Members Within 5 m of the Infant at the Time

Infant ID	No. of monkeys < 5 m from infants				n	Significance of χ^2 test
	0–4	5–9	10–14	15+		
G51	+	+	0	−	34	0.01
G06	+	0	−	−	13	ns
G73	+	+	−	−	35	0.01
G10	+	0	−	−	18	0.001
G68	+	+	−	−	48	0.02
G46	+	+	−	−	30	ns
G21	+	0	0	−	17	ns
G80	+	−	−	−	24	ns

Expected values are based on the null hypothesis that mothers and infants distribute their time at a distance from each other without regard to the number of individuals within 5 m of infants. For an example of the calculations of expected values, see text.
+: observed values greater than expected values.
− : observed values less than expected values.
0 : observed values equal to expected values.

TABLE V. Directions of Differences Between Observed and Expected Amounts of Time That Individual Infants Were Beyond 5 m of Their Mothers as a Function of the Number of Nonnatal Adult Males That Were Within 5 m of the Infant at the Time

Infant ID	No. of monkeys < 5 m from infants			Significance of χ^2 test
	0	1+	n	
G51	+	−	32	ns
G06	+	−	20	ns
G73	+	−	40	ns
G10	+	−	21	0.05
G68	+	−	29	ns
G46	+	−	32	ns
G21	+	−	9	
G80	+	−	19	ns

Overall $\chi^2 = 7.95$, df = 6, $p < 0.30$.
Expected values are based on the null hypothesis that mothers and infants distribute their time at a distance from each other without regard to the number of adult males within 5 m of infants. For an example of the calculations of expected values, see text. Data are based on instantaneous samples where 5–7 animals other than the mother were within 5 m of the infant.
+, −, 0: as in Table IV.

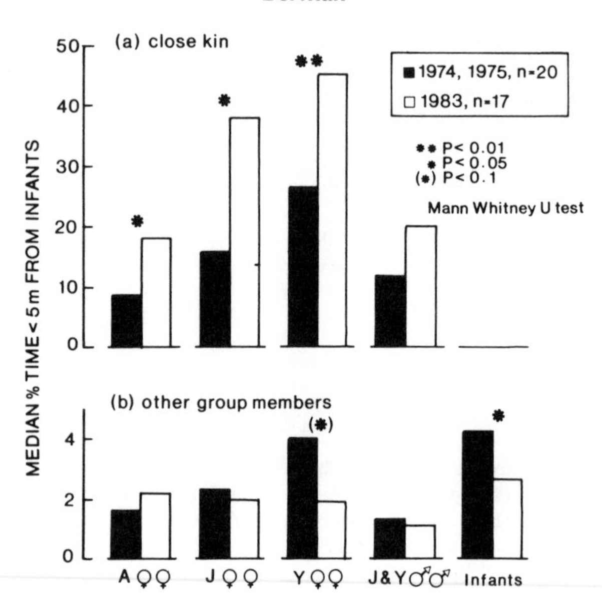

Fig. 9. Median percentages of time infants spent within 5 m of *each* of their close kin (siblings and grandmothers) and *each* other group member by the age-sex class of the group member. Infants in 1974 and 1975 are compared with infants in 1983. A♀♀ = adult females (4 years and over); J♀♀ = juvenile females (2 & 3 years); Y♀♀ = yearling females (1 year); J & Y♂♂ = juvenile and yearling males (1–3 years); infants = other individuals born during the same birth season as the subjects.

behavior could of course be mediated by changes in group composition.) To do this, the null hypothesis that the amounts of time that particular *individuals* were within 5 m of infants did not change over the years was tested. The null hypothesis predicted, for example, that individual juvenile sisters in 1983 would have been observed within 5 m of their infant sibs the same amount as individual juvenile sisters had been in the earlier years.

Figure 9 shows the median percentage of time infants in 1983 and in 1974 and 1975 spent within 5 m of *each* individual belonging to each age-sex class. Close maternal kin (sibs and grandmothers) and other group members are shown separately. Infants in 1983 tended to spend more time within 5 m of each of their close kin than did infants in 1974 and 1975. This was the case for all kinds of close kin: adult females, juvenile females, yearling females, and yearling and juvenile males. The differences reached significance in all but one age-sex class. There was also a tendency to spend less time with less closely related individuals belonging to 4 out of 5 age-sex classes. However, these tendencies were marked only for infant companions. Infants were within 5 m of each of their peers less in 1983 than in earlier years.

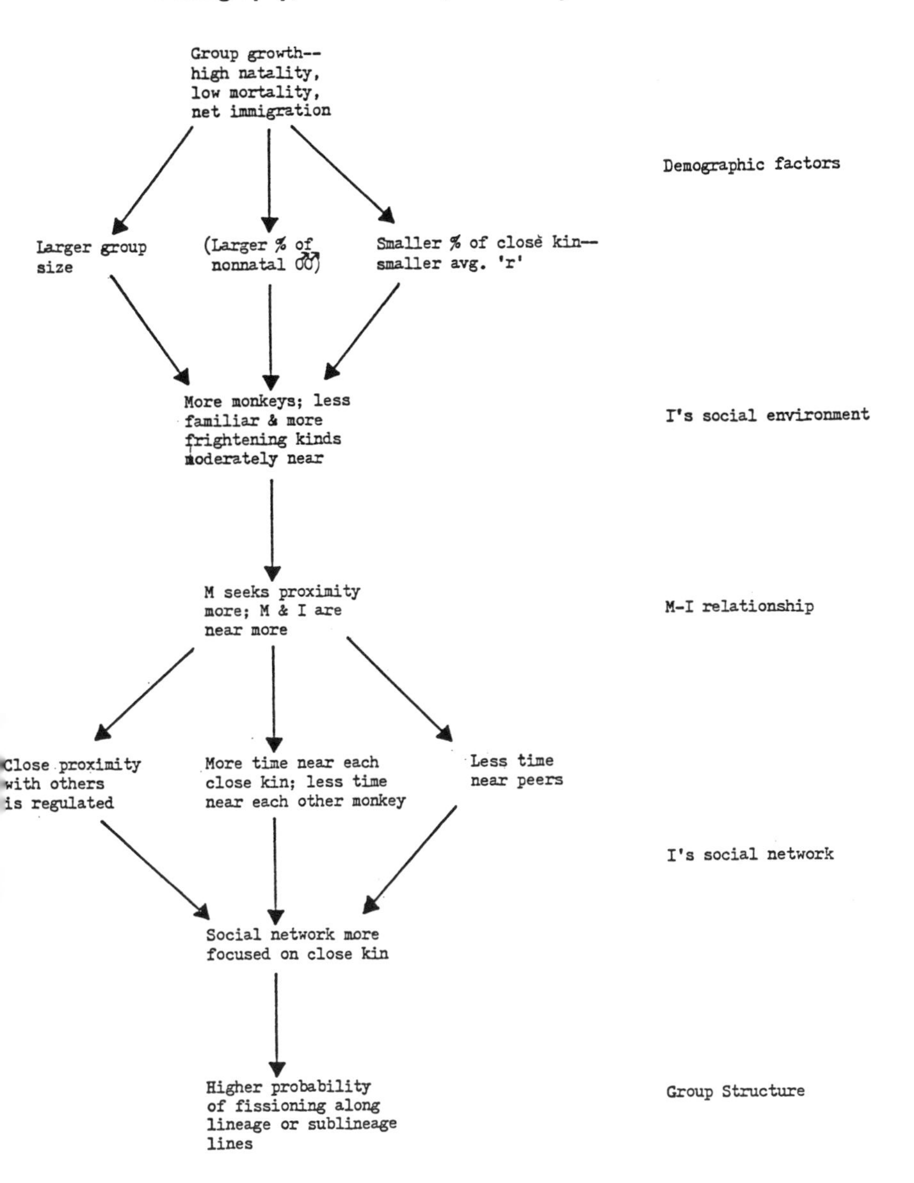

Fig. 10. Hypothetical links between group growth, the infant's social environment, its relationship with its mother, its social network, and group structure.

It is possible to explain these results as consequences of infants being kept near mothers more. Close kin spend a great deal of time near the mother; thus infants who are near mothers more are likely to encounter their kin more. Infant-infant pairs (none of whom are siblings) probably had fewer opportunities to interact because each infant was kept near its own mother more.

At this point it is not known whether these changes in proximity relationships between infants and group members are also reflected in measures of more intimate affiliative interaction (e.g., contact and play). However, past analyses (Berman, 1978, 1982a) indicate that measures of more intimate interaction correlate very closely with proximity measures, and that one can map out infants' affiliative social network very accurately with proximity measures. Given this, it is reasonable to describe these changes as a narrowing of infant networks of relationships within the group through an increased emphasis on close maternal kin and a decreased emphasis on relationships with peers and less closely related monkeys of all kinds. A focusing of networks onto close maternal kin might be expected to lead eventually to a high probability of group fissioning along lineage or even sublineage lines.

Figure 10 summarizes the argument presented in this example. The figure shows hypothetical links between demographic factors affecting group size and composition, the infants' social environment, the mother-infant relationship, the infants' social network, and, finally, group structure. Briefly, group growth (which in this case is a result of high natality, low mortality, and net immigration into the group) appears to lead to specific kinds of changes in group composition—a larger group and one in which each individual has a smaller percentage of close maternal kin. This is equivalent to saying that the average degree of relatedness decreases. In group I, a larger proportion of nonnatal males also followed, but as this was not general in other social groups of the population, this is in parentheses in the figure and would probably not be included in a more general model. All these changes are followed by changes in the infants' social environment: infants are surrounded by more monkeys and a larger proportion of those monkeys are likely to be unfamiliar and potentially more frightening to them and to their mothers. As a result, mothers seek proximity with their infants more, and infants spend more time near their mothers. This in turn has implications for the development of the infants' social network of relationships among the group. The infants' close proximity is regulated so that they have no more opportunity to associate intimately with other monkeys than did previous infants. They spend more time with each of their close maternal relatives and less time with each more distantly related or unrelated monkey. In particular, they spend less time with each of their peers. This constitutes a focusing of infants' social networks more narrowly onto close kin and away from others.

If this focusing of social networks persists as infants mature, and if it continues to develop in subsequent crops of infants within the group, one would expect a gradual weakening of bonds between lineages within the group and possibly between sublineages within lineages. This should lead eventually to a higher probability of group fissioning along lineage or sublineage lines.

The figure should be qualified in two ways. First, the links proposed here are not all necessarily meant to be causal in nature. The figure might be best thought of as a chronology of events at this point. Second, the processes described are not meant to be exhaustive of what occurs in group fissioning. They are meant merely to suggest some of the developmental components of the process. Although an empirical test of the model would be difficult in the field, certain aspects of it can and will be tested. As mentioned earlier, group I fissioned in 1985 and again in 1986. It will be possible to look at infant relationships with mothers, close kin, and other monkeys within the daughter groups to determine whether they are in harmony with the hypotheses suggested here.

Group fissioning among gelada baboons also appears to be associated with a narrowing of social networks toward close kin (Dunbar, 1984). As the one-male units in which they live increase in size, females tend to groom their close female relatives progressively more to the near exclusion of others. Dunbar suggested that the lack of cohesiveness that results among females makes the unit particularly vulnerable to fissioning along lineage lines when a follower male joins. The extent to which adult female rhesus monkeys also focus their social networks on kin prior to group fissioning is under investigation.

As Dunbar pointed out, these phenomena do not constitute an example of the traditional Christian-Calhoun syndrome of behavioral effects when population density increases and migration is prevented (Calhoun, 1962; Christian, 1961). In early 1984 the population density on Cayo Santiago decreased with the cull of 3 social groups. Nevertheless, the trends in mother-infant interaction and in the infants' social environment continued unabated as the group size increased and the mean degree of relatedness decreased within the study group.

GENERAL DISCUSSION

In nearly every environment in which they have been observed, macaque mothers are initially protective of infants. Gradually they encourage more independence. The manner and extent to which they encourage independence, however, vary considerably with several aspects of the physical and social environment. In the three examples above, the availability of kin, in terms of numbers, age, sex, degree of relatedness, and relative representation in the group, were shown to be particularly important aspects of that environment. These demographic factors appear to be responsible for variations in mother-infant relationships between colonies, among individuals, and over several years of group expansion. Although they are certainly not the only sources of variation among mother-infant pairs, they appear to account for more variation than several other more widely cited factors (Berman, 1984). The fact that these observations have been made in a free-ranging colony in which amounts of food per monkey (and other environmental resources) have been virtually stable suggests that they are not mere artifacts of variations in resource availability.

The three examples also illustrate that, at least in expanding populations, demographic factors not only appear to exert short-term variations in relationships, but also appear to have long-term implications for the development, maintenance, and elaboration of matrilineally based group structure, and eventually for patterns of fissioning. Mothers, through their control of their infants, pass on their networks of relationships to them (Berman, 1982a). Infants retain these networks beyond the age when they are directly controlled by mothers. Individuals with whom mothers allow free interaction (and these tend to be close kin, females and younger immatures) become close associates of infants, and those whom mothers avoid, apparently perceive as dangerous, or to whom they cannot themselves gain access do not become close associates of their infants. This allows mothers' apparently sensitive responses to the social characteristics of the group at the time infants are reared to be translated by offspring into longer-term modifications in social networks. If the mothers' responses persist or progress with each subsequent infant, long-term changes in social structure are expected. This is not to say that older offspring and adults may not also respond to the social environment by modifying their social networks. The extent to which they do so and the extent to which they retain their very early networks into adulthood are under investigation.

An obvious conclusion of the findings presented here is that descriptions of species-specific behavior and group structure must place more emphasis on demographic effects. Such descriptions must be of a dynamic nature, incorporating the range of variation in behavior and organization in a variety of demographic conditions. Altmann and Altmann (1979) pointed out that the availability of kin and other kinds of group companions can be expected to vary drastically depending on whether a population is expanding, stable, or declining. As all three conditions are found in nature (Dunbar, 1979; Struhsaker, 1976; Dittus, 1980; Southwick et al., 1980; Jolly et al., 1982a, b), none can be ignored. Since food availability clearly affects aspects of social structure through its effects on spatial relationships and demography, descriptions of the type, distribution, and amount of food available are also essential in all studies of social behavior. This is equally important for groups which are and are not food enhanced.

SUMMARY

The effects of demographic processes on social relationships and on social structure are difficult to separate from the effects of environmental change in most primate populations. This difficulty can be overcome, however, in free-ranging, food-enhanced populations where major environmental influences can be controlled. The influence of demographic variables and of population growth on mother-infant relationships and infant social networks was examined here among free-ranging rhesus monkeys on Cayo Santiago. Data are from 1974 to 1984, a period during which the physical environment and amounts of food per monkey remained virtually constant. Three examples of research were reviewed in which mothers were found to adjust the degree to which they encourage independence in their infants in relation to group size and/or aspects of group composition. The availability of kin, in terms of numbers, degree of relatedness, age, sex, and relative representation in the group were found to be associated with variations in mother-infant relationships between colonies, among pairs within a group at any one time and over several years of group expansion. Group growth in particular led to increased maternal protectiveness and a subsequent focusing of infant social networks more narrowly onto close kin and away from peers and other group members. It is suggested that mothers' apparently sensitive responses to the social characteristics of the group at the time infants are reared have long-term influences on offspring social networks. Their influence on offspring networks in turn has long-term implications for social structure, particularly for the development, elaboration, and eventual fissioning of lineages.

ACKNOWLEDGMENTS

I am grateful to Donald Sade, Richard Rawlins, Gilbert Meier, Matt Kessler, and Dell Collins for permission to carry out research on Cayo Santiago. Cayo Santiago was supported over this 11-year period by contracts NIH DRR 71-2003 and NIH RR-7-2115 and by grant RR-01293 to the University of Puerto Rico. I received support from the Wenner Gren Foundation for Anthropological Research, the Explorer's Club, Sigma Xi Society, NIMH predoctoral fellowship MH05195, NSF postdoctoral fellowship SPI-7815548, and NIMH grant MH 38647. I am most grateful to Professor Robert A. Hinde for guidance during the early years of the study, to Laureen Busacca, Anne Homer, and Leland Smith for assistance in data collection during the later years of the study, and to Bart Brown for typing the manuscript. Finally, thanks go to the staff of the Caribbean Primate Research Center, particularly A. Figueroa, E. Davila, and J. Berard for their persistent and careful census-taking.

REFERENCES

Altmann J (1974): Observational study of behaviour: Sampling methods. Behaviour 49:227–267.

Altmann J, Altmann SA (1979): Demographic constraints on behavior and social organization. In Bernstein I, Smith EO (eds): "Primate Ecology and Human Evolution." New York: Garland STMP Press, pp 47–64.

Berman CM (1978): Social relationships among free-ranging infant rhesus monkeys. Ph.D. thesis. U.K.: University of Cambridge.

Berman CM (1980): Mother-infant relationships among free-ranging rhesus monkeys on Cayo Santiago: A comparison with captive pairs. Anim Behav 28:860–873.

Berman CM (1982a): The ontogeny of social relationships with group companions among free-ranging infant rhesus monkeys. I. Social networks and differentiation. Anim Behav 30:149–162.

Berman CM (1982b): The ontogeny of social relationships with group companions among free-ranging infant rhesus monkeys. II. Differentiation and attractiveness. Anim Behav 30:163–170.

Berman CM (1984): Variation in mother-infant relationships: Traditional and nontraditional factors. In Small M (ed): "Female Primates—Studies by Women Primatologists." New York: A.R. Liss, Inc., pp 17–36.

Berman CM (in prep. a): The influence of maternal age, maternal experience and siblings on maternal behavior and infant development: Data from free-ranging rhesus monkeys.

Berman CM (in prep. b): Links between population growth, social structure and social development among free-ranging rhesus monkeys on Cayo Santiago.

Calhoun JB (1962): Population density and social pathology. Sci Am 206:139–148.

Carpenter CR (1942a): Characteristics of social behavior in nonhuman primates. Trans NY Acad Sci 4:248–258.

Carpenter CR (1942b): Sexual behavior of free-ranging rhesus monkeys (*Macaca mulatta*): Specimens, procedures, and behavioral characters of estrus. J Comp Psychol 33: 143–162.

Chepko-Sade BD, Sade DS (1979): Patterns of group splitting within matrilineal kinship groups. Behav Ecol Sociobiol 5:67–86.

Chepko-Sade BD, Olivier TJ (1979): Coefficient of genetic relationship and the probability of intragenealogical fission in *Macaca mulatta*. Behav Ecol Sociobiol 5:263–278.

Christian JJ (1961): Phenomena associated with population density. Proc Natl Acad Sci 47:428–449.

Cohen JE (1969): Natural primate troops and a stochastic population model. Am Nat 103:455–477.

Cohen JE (1972): Markov population processes as models of primate social and population dynamics. Theor Pop Biol 3:119–134.

Dittus WPJ (1980): The social regulation of primate populations. In Lindburg D (ed): "The Macaques: Studies in Ecology, Behavior and Evolution." New York: Van Nostrand.

Dunbar RIM (1979): Population demography, social organization, and mating strategies. In Bernstein I, Smith EO (eds): "Primate Ecology and Human Origins." New York: Garland STMP Press.

Dunbar RIM (1984): "Reproductive Decisions: An Economic Analysis of Gelada Baboon Social Strategies." Princeton, NJ: Princeton University Press.

Dunbar RIM (1985): Population consequences of social structure. In Sibley RM, Smith RH (eds): "Behavioural Ecology: Ecological Consequences of Adaptive Behaviour." London: Blackwell Scientific Publications.

Dunbar RIM, Dunbar EP (1976): Contrast in social structure among black-and-white Colobus monkey groups. Anim Behav 24:84–92.

Harlow HF, Harlow MK (1965): The affectional system. In Schrier AM, Harlow HF, Stollnitz F (eds): "Behaviour of Nonhuman Primates, Vol. 2." New York: Academic Press.

Hausfater G (1981): Computer models of primate life-histories. In Alexander RD, Tinkle D (eds): "Natural Selection and Social Behavior." New York: Chiron, pp 345–362.

Hinde RA (1969): Analysing the roles of partners in a behavioural interaction—Mother-infant relations in rhesus macaques. Ann NY Acad Sci 159:651–667.

Hinde RA (1974): "Biological Bases of Human Social Behaviour." New York: McGraw Hill.

Hinde RA, Atkinson S (1970): Assessing the roles of social partners in maintaining mutual proximity as exemplified by mother-infant relations in rhesus monkeys. Anim Behav 18:169–176.

Hinde RA, Spencer-Booth Y (1967): The behaviour of socially-living rhesus monkeys in their first two and a half years. Anim Behav 15:169–196.

Hooley JM, Simpson MJA (1983): Influence of siblings on the infant's relationships with the mother and others. In Hinde RA (ed): "Primate Social Relationships: An Integrated Approach." Oxford: Blackwells Scientific Publications, pp 139–142.

Jolly A, Gustafson H, Oliver WLR, O'Connor SM (1982a): *Propithecus verreauxi* population and ranging at Berenty, Madagascar, 1975 and 1980. Folia Primatol 39:124–144.

Jolly A, Oliver WLR, O'Connor SM (1982b): Population and troop ranges of *Lemur catta* and *Lemur fulvus* at Berenty, Madagascar: 1980 census. Folia Primatol 39:115–123.

Johnson RL, Southwick CH (1984): Structural diversity and mother-infant relations among rhesus monkeys in India and Nepal. Folia Primatol 43:198–215.

Johnson RL, Southwick CH (1987): Ecological constraints on the development of infant independence in rhesus monkeys. Am J Primatol 13:103–118.

Keiding N (1977): Statistical comments on Cohen's application of a simple stochastic population model to natural primate troops. Am Nat 111:1211–1219.

Lee PC (1983): Ecological influences on relationships and social structures. In Hinde RA (ed): "Primate Social Relationships: An Integrated Approach." London: Blackwells Scientific Publishers, pp 225–229.

Malik I, Seth PK, Southwick CH (1984): Population growth of free-ranging rhesus monkeys at Tughlaqabad, northern India. Am J Primatol 7:311–321.

Mori A (1979): Analysis of population changes by measurement of body weight in the Koshima troop of Japanese monkeys. Primates 20:371–397.

Nicolson NA (1982): Weaning and the development of independence in olive baboons. Ph.D. thesis. Cambridge, MA: Harvard University.

Rawlins RG, Kessler MJ (1983): Congenital and hereditary abnormalities in the rhesus monkeys (Macaca mulatta) of Cayo Santiago. Teratology 28:169–174.

Rawlins RG, Kessler MJ (1986): Demography of free-ranging Cayo Santiago macaques, 1976-1983. In Rawlins RG, Kessler MS (eds): "The Cayo Santiago Macaques: History, Biology and Behavior." Albany, NY: SUNY Press, pp 47–72.

Rowell TE, Hinde RA, Spencer-Booth Y (1964): "Aunt"-infant interaction in captive rhesus monkeys. Anim Behav 12:219–226.

Sade DS (1972): A longitudinal study of social behavior of rhesus monkeys. In Tuttle R (ed): "Functional and Evolutionary Biology of Primates." Chicago: Aldine-Atherton, pp 378–398.

Sade DS, Cushing K, Cushing P, Dunaif J, Figueroa A, Kaplan JR, Lauer C, Rhodes D, Schneider J (1977): Population dynamics in relation to social structure on Cayo Santiago. Yearbook Phys Anthropol 20:253–262.

Sade DS, Chepko-Sade BD, Schneider JM, Roberts SS, Richtsmeier JT (1985): "Basic Demographic Observations on Free-ranging Rhesus Monkeys." New Haven, CT: HRAF Inc.

Seay B, Gottfried NW (1975): A phylogenetic perspective for social behavior in primates. J Gen Psychol 92:5–17.

Southwick CH, Richie T, Taylor H, Teas J, Siddiqi MF (1980): Rhesus monkey populations in India and Nepal: Patterns of growth, decline and natural regulation. In Cohen MN, Malpass RS, Klein HG (eds): "Biosocial Mechanisms of Population Regulation." New Haven: Yale Univ. Press, pp 157–170.

Struhsaker TT (1976): A further decline in numbers of Amboseli vervets. Biotropica 8:211–214.

Sultana CJ, Marriott BM (1982): Geophagia and related behaviors of rhesus monkeys (*Macaca mulatta*) on Cayo Santiago Island, Puerto Rico. Int J Primatol 3:338.

Trivers RL (1974): Parent-offspring conflict. Am Zool 14:249–264.

Van Schaik CP, Van Noordwijk MA, deBoer R, den Tonkelaar I (1983): The effect of group size on time budgets and social behaviour in wild long-tailed macaques (*Macaca fascicularis*). Behav Ecol Sociobiol 13:173–181.

Ecology and Behavior of Food-Enhanced Primate Groups, pages 297–311

14

Ecological Constraints and Opportunities: Interactions, Relationships, and Social Organization of Primates

Phyllis C. Lee

Sub-Department of Animal Behaviour, University of Cambridge, Cambridge CB3 8AA, United Kingdom

INTRODUCTION

Primatologists have long sought to explain the variability in primate social systems in terms of ecological factors, typically the distribution, abundance, and quality of foods exploited by groups in different habitats (e.g., Hall, 1965). Many past categorizations have been based on a deterministic "eco-correlates" approach using relatively simple correlations between general feeding habits (e.g., Crook and Gartlan, 1966) or rainfall (e.g., Popp, 1983). More complex interactions between habitat, diet, and energetic requirements (Clutton-Brock and Harvey, 1977a,b; Eisenberg et al., 1972; Harvey and Clutton-Brock, 1981; Sailer et al., 1985) have greater explanatory power, tending toward the ultimate evolutionary view that groups are alliances of individuals cooperating to maximize access to critical resources (Wrangham, 1980, 1983).

In asking questions about the social structure of a primate group, we are asking about dynamic processes of relationships between the individuals in that group (Hinde, 1976, 1983; Richard, 1985; Seyfarth et al., 1978). However, what is typically measured are the rates at which individuals interact and the context in which those interactions take place. Additional information on the nature of a relationship comes from examining the consistency of interactions and their patterning through time (see Hinde and Stevenson-Hinde, 1976). A distinction between measurement of frequencies of interactions and the interpretation of the nature of the relationship, which includes information about context, patterning, and consistency, is relevant in discussions of food-enhanced primate groups. Although the rates at which interactions occur among members of such groups may differ from those between members of "natural" groups, the contexts of the interactions and their patterning through time, and thus the nature of the relationships, may

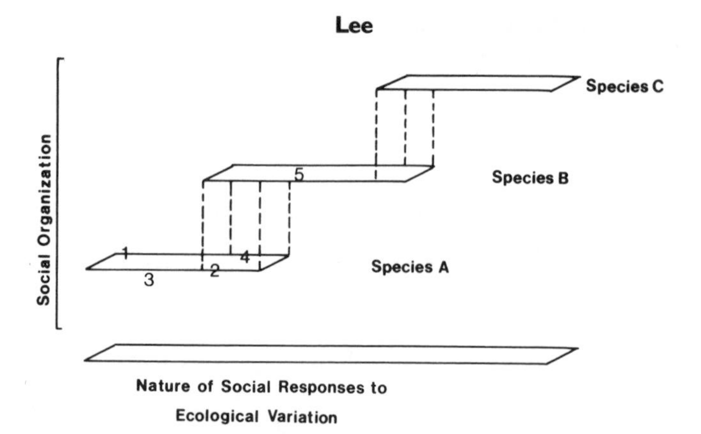

Fig. 1. A schematic representation of the range of social variability for 3 hypothetical species of primates. The points represent different food conditions for groups in different seasons or different habitats. The social variation can be seen as having at least 2 dimensions.

be similar. In order to understand primate social structure, the range of variability in interactions and relationships and the food-related factors affecting that variability need to be determined.

A number of studies on species such as lemurs (Richard, 1974), squirrel monkeys (Baldwin and Baldwin, 1976), mangabeys (Chalmers, 1968), toque macaques (Dittus, 1979), and baboons and vervets (Lee, 1983a) have shown that variation in rates and the nature of social interactions within a group can be related to seasonal changes in the nature of the food supply. The contrasts between "good" seasons of ecological opportunities and "poor" seasons of ecological constraints allow for interpretations of variability in social structure with increased accuracy and validity. Furthermore, since the "poor" season limits the range of behavioral options open to selection (e.g., Boyce, 1979), fundamental principles underlying social structure may be derived by examining variability in behavior as a result of seasonal contrasts. Each season may have its specific constraints, but these may best be revealed by understanding the changes occurring between seasonal periods.

To accomplish this, variability in social responses to ecological contrasts between seasons within a group, between groups in different habitats, and between different species of primates can be schematically diagrammed as in Figure 1. We can postulate that seasonal limitations represent one point (1) along the line for Species A, while seasonal opportunities are another point (2). Two groups of Species A in different habitats can be represented at points 3 and 4. The range of social responses shown by different species in the context of different social organizations can overlap. When seasonal habitat conditions are at the extreme end of the range for a given species (A), it may show social responses more typical of Species B.

The conditions accompanying food enhancement, such as spatial localization of foods or constraints on inter-individual distances and dispersal, may be as important in influencing social structure as are the changes in food abundance and quality. For food-enhanced groups, we need to determine if the social variability will be represented by a point still within the species' range (A), or if provisioning can result in social responses (point 5) which lie along the line typical for Species B. Among food-enhanced groups, we might expect either the lower range of social variability to be truncated, or a social structure of a different general type. The contrasts between food abundance and deprivation, and the resulting variation in the nature and patterning of interactions and relationships will be explored here.

SEASONALITY AND PRIMATE SOCIAL STRUCTURE

A direct, causal link between the energy available from the environment and the behavior of individuals, expressed by the daily metabolic requirements of an individual (based on size and activity) and its daily intake of nutrients, is demonstrated by the papers in Section II of this volume. Among populations of primates inhabiting environments with seasonal extremes in nutrient and energy availability, patterns of behavior change seasonally to reflect these extremes. These changes in behavior, in turn, may affect social relationships by altering the frequency and patterning of interactions. Changes in the demographic composition and structure of groups, as a result of different levels of nutrition acting on reproduction and survival (see papers in Section III), also affect individual behavior and relationships. This paper will attempt to describe some of these inter-relations, focusing on changes in behavior, in relationships and social structure in environments with seasonal contrasts.

Seasonal Changes in Activities

Activities change in relation to seasonal limits on food quality and availability. The magnitude and direction of those changes need to be assessed for their functional variability for a wider variety of habitats. Overall, the relations between nutritional quality and the time and energy costs of procuring the foods (based on distribution and abundance) can be used to explain some of the seasonal patterns of activity.

When the energy available and nutrient intake per mouthful increase, the time it takes to ingest daily metabolic requirements decreases (see Homewood, 1978; Iwamoto and Dunbar, 1983). Thus in some studies, as the quality of food increases seasonally, the time spent feeding decreases (Altmann and Altmann, 1970; Kummer, 1968; Lee, 1984). Alternatively, when an increase in food quality and availability is associated with an

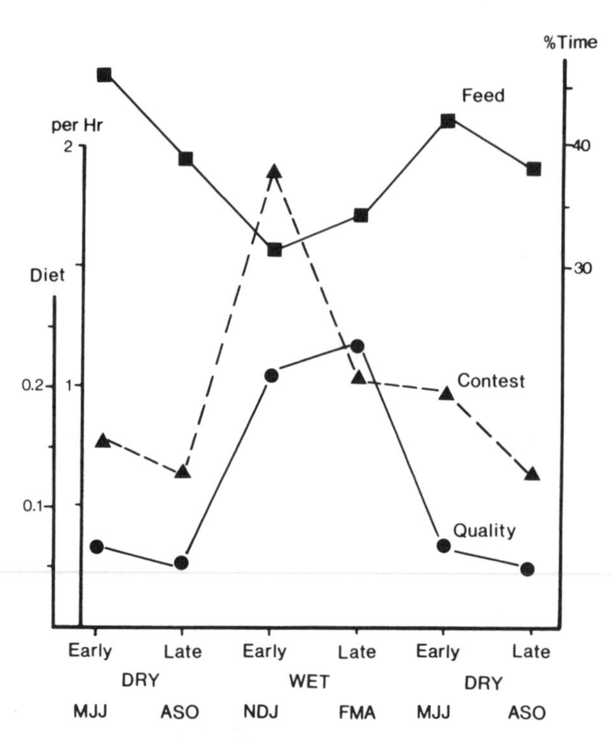

Fig. 2. A plot of the mean percent of time spent feeding, and hourly rates of competition and aggression (contests), against estimates of dietary quality in different seasons for free-ranging vervet monkeys. (Estimates of dietary quality are explained in Lee, 1984.)

increase in time spent feeding (e.g., Richard, 1977; Wrangham, 1977; Homewood, 1978), minimizing energy expenditure during the poor season might limit the time available for feeding. When these energetic constraints are lifted, the animals can feed for longer. Both such effects can occur and examples are presented in Figure 2. Using data from a study of free-ranging immature vervet monkeys (*Cercopithecus aethiops*) (ranging in age from 1–5 years), Figure 2 shows that seasonal changes in estimated dietary quality were associated with corresponding changes in the time spent feeding. As dietary quality declined in the late dry (poor) season, feeding-time dropped, possibly to minimize energy expenditure. When quality rose dramatically during the wet season, feeding time declined even further as the energy available per unit time increased. The time spent feeding then increased with a further rise in quality, possibly to take advantage of opportunities for fat storage, or as a result of changes in the dispersion of foods. The distribution of the foods may have meant longer travel times for the same energetic return of ingestion, and time spent feeding increased to account for the additional procurement costs.

Other factors such as digestive capacity and perhaps toxins (McKey et al., 1981; Oates et al., 1980) also affect the time spent feeding. Furthermore, the distribution of foods of different quality affects travel time and range use. In some simple systems, where poor season foods are of low quality and dispersed, then travel time, foraging time, and day range increase (Kummer, 1968; Lee, 1983a). In more complex systems, where low quality resources are clumped (e.g., red colobus *(Procolobus badius);* Clutton-Brock, 1977), or when renewal rates of evenly distributed resources vary seasonally (e.g., mangabeys *(Cercocebus albigena);* Waser and Homewood, 1979), simple relations between food quality and activity patterns can not be expected. As Harvey and Clutton-Brock (1981) have demonstrated, the combined metabolic requirements of a group (a function of individual requirements based on body size and the number of animals in a group) also affect day range and home range size differently for different distributions of resources.

Two further generalizations can be made about seasonal influences on activities. First, in species where the rate of digestion is not limiting, resting time tends to be inversely correlated with feeding and travel time. Thus, in species where the time spent in activities throughout a day is limited by energy, resting time declines as feeding time increases. The exceptions are the fermenting species, such as colobines (Oates, 1977; McKey et al., 1981), where increases in resting time are related to the time needed to ferment foods of decreasing quality. Second, the time spent in social activities tends to decrease with increases in time spent procuring energetic requirements. An interesting exception to this pattern is a relatively small seasonal change in time spent in grooming (Lee, 1983a). Dunbar and Sharman (1984) have shown that as feeding time increases, the time used for grooming is taken from resting time, which declines in relation to feeding time plus grooming time.

In the most simplified model of an ecological system, the following assumptions about seasonal contrasts in activities can be made. During the good season, high quality foods are relatively abundant and may be either uniformly distributed or occur in large clumps and are thus relatively inexpensive to exploit. Poor seasons represent periods of low food abundance and quality, with higher energy and time costs for its procurement. During poor seasons, foods may be widely dispersed (or if clumped, in dispersed small clumps) and low in quality. These foods are used because low quality foods tend to be relatively more abundant than high quality foods and may be all that is available, but they are costly to exploit in terms of foraging time. In the poor season, food choice switches to widely available foods with a lower energetic yield and animals spend more time foraging and less time in other activities. The summation of energetic yield, distribution, and availability of seasonal foods should be the major predictor of seasonal changes in activities.

Seasonal Changes in Interactions

In this simple system, low-energy foods require more time to procure than high-energy foods, and a reduction in social time is apparent. How does this reduction in time spent in social interactions influence the distribution of interactions across individuals or within relationships, and affect social structure? In some studies, as animals exploit dispersed low-quality foods, inter-individual distances increase thus reducing opportunities for interaction (Dittus, 1979; Fairbanks and Bird, 1978; Lee, 1984, 1986; Loy, 1970). This may have dramatic consequences for frequencies of interactions as well as for the types of interactions and the choice of partners, as demonstrated for patas monkeys by Rowell and Olson (1983). Since kin tend to associate, it can be predicted that the nearest partner will be a related individual and a strong pattern of kin-biased interaction would be observed. Furthermore, since potential partners are dispersed at large distances, they must actively seek opportunities to be together in order to interact. Such interactions, while rare, may have a greater importance in maintaining or altering a relationship than those interactions which are frequent but involve little discrimination between potential partners who interact simply because they are close at hand.

Competitive interactions, which are energetically costly and may be time consuming, tend to be infrequent in severely food-constrained groups (Dittus, 1979; Lee, 1984; Loy, 1970), and increase in frequency during periods of seasonal energy abundance (Fig. 2). A straightforward relation between frequency of competition and quality of diet may not be expected, since the costs and benefits of contesting access to a food resource can be affected not only by the distribution and absolute energetic values of the resource, but also by the relative benefit gained through controlling access to a resource, and finally by the social context of the competition. If the cost of competition (e.g., risk of injury) is high, but the benefit of gaining the contested resource is also high, then there is a high probability that the interaction will take place. However, it may be rare that a contested resource can offer such high benefits and thus, competition will be relatively infrequent and limited to valued resources (see Rowell, 1966; Wrangham, 1981). Opportunities for competition resulting from proximity to other individuals, the value of the resource, the risks associated with a contest, and the probability of winning the contest all influence the rate of competitive interactions. Among food-enhanced groups, both the presence of valued resources and close proximity led to high rates of competition (Sugiyama and Ohsawa, 1982).

Of additional importance, but seldom investigated, is the mechanism by which familiar contestants increase the predictability of the outcome, or at least make a better assessment of the likelihood of defeat or victory (de Waal, 1979). In a study of juvenile olive baboons, Johnson (1984) found that most competition occurred over resources of low energetic value which were relatively easy to relocate when lost. However, the social consequences of the competition were of high value and related to the acquisition of rank. The animals were using a resource over which others were unlikely to escalate possibly to test their probability of winning a contest. Although some workers have concluded that high rates of competition are an artifact of provisioning (Rowell, 1967, 1974), the key factors are the abundance and value of the resource, the opportunities for interaction and the social context.

Aggression also appears to be more frequent when opportunities for interaction and the energy to sustain those interactions are more available (see also Figure 2). A model of rates of aggression based on ecological opportunities again seems relevant to food-enhanced groups. Groups ''naturally'' food-enhanced during good seasons can afford to interact aggressively (Lee, 1984). Some groups show increases in rates of aggression (and competition) as food declines in availability, but these then drop off as food becomes absolutely rare (Baldwin, 1971; Southwick, 1967). While high rates of aggression are frequent among food-enhanced groups (see Lee et al., 1986, and papers in this volume), frequency of aggression is probably related to the quality and distribution of resources and familiarity of contestants, but less to absolute shortage or abundance.

The quality, abundance, and distribution of foods affect the opportunities for interaction and the rates at which individuals compete and fight. But do either seasonal differences in rates within groups or habitat differences between groups predict variation in social structure? If rates decline, but the patterning of interactions across available partners remains constant, then the quality of the relationship may have changed without altering the dynamics and long-term consistency of that relationship. Thus dominance hierarchies can exist, but may not be expressed during periods of energetic constraint (see Lee, 1983a). Alternatively, if dominance relationships are expressed not through overt competition but through a low-cost behavior such as grooming (e.g., Seyfarth, 1980), alliances which may be seasonally critical in determining access to resources (food or social) can be maintained. A shift in the types of interactions observed can thus reflect either a change in the nature of the relationship or an energy conserving means of achieving the same end—enhancing individual access to resources.

The same type of shift to an energy conserving form of interaction is apparent in play between immatures. As noted for many primate populations, play is particularly sensitive to changes in dietary quality (Baldwin and

Baldwin, 1976; Lee, 1984) and becomes rare or disappears altogether under conditions of energy constraint (see also Fagen, 1981). If play is relatively costly (even if not absolutely costly; Martin, 1984), but grooming is not, then peer-peer relationships may be expressed through grooming (Lee, 1983a). As in the case of dominance relationships, peer-peer relationships appear to be expressed through interactions which change between seasonal periods, while perhaps related to the same functional ends. Thus social structure appears variable when observed in the short-term but the consistent inter-individual relationships are independent of the observed short-term variation.

Finally, ecological conditions influence maternal nutrition and hence ability to lactate, consequent infant growth, and birth spacing resulting from lactational anoestrous; all of which are components of mother-infant relationships and have a bearing on the time course of weaning. Suckling success of infants declines more rapidly under conditions of food shortage and weaning may be completed at a younger age (Lee, 1984, 1986). Infants grow slowly when nutrition is limited (Sackett et al., 1979), and may have different levels of suckling demands as a result of rapid or slow growth rates. Other components of mother-infant relationships such as frequency of rejections and spatial independence have been related to the mother's ability to meet the energetic burden of carrying infants as they increase in weight (Altmann, 1980).

Food, Demography, and Social Structure

Since high food abundance and quality appear to enhance fecundity and survival (see papers in Section III), individuals living in seasonal environments with sharp contrasts between abundance and deprivation should have lower and/or more variable reproductive rates. Changes in demographic parameters have consequences on the nature of interactions and social structure within groups, several of which have received considerable attention (see also Richard, 1985). The demographic composition of a group and fecundity have been shown to affect the number of surviving kin and thus the nature of relationships between kin (Altmann and Altmann, 1979). The number of potential partners of different ages and sex also affect the development of relationships during ontogeny (Altmann, 1979; Lee, 1986). The consistency and stability of female rank relationships are related to the number of females of different ages (Hausfater et al., 1982). The likelihood of competition between males for control of female groups in geladas (Dunbar, 1984) and among other species, the timing and probability of young males transferring between groups (Manzolillo, 1986; Packer, 1979), as well as who they accompany during transfer (Cheney and Seyfarth, 1983; Colvin, 1983) all are influenced by the composition of the group as well as that of surrounding groups.

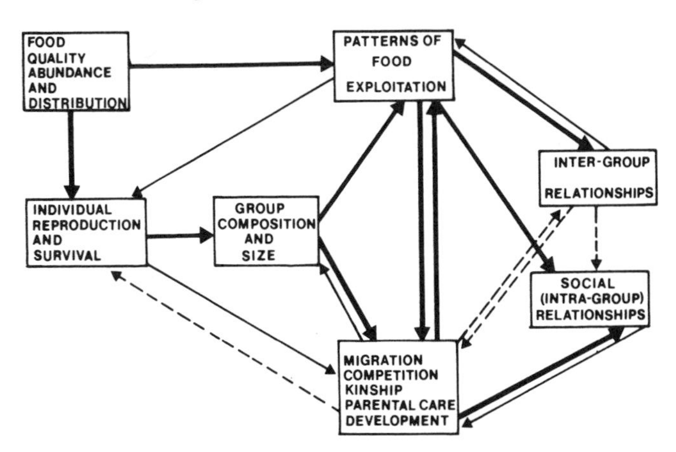

Fig. 3. Schematic effects of changes in group size and composition on relationships and social structure. The heavy lines indicate a primary effect, while the dotted lines represent a secondary effect mediated through effects of other variables.

One problem, however, is that the consequences of changes in demographic composition of a group appear to be interrelated. If birth rate increases, we can postulate concomitant increases in the size of kin networks or matrilines, in overall group size, in the complexity of relationships developed during immaturity, and changes in overall social structure as measured by relationships between individuals—alliances, competition, dispersal from natal groups, male-female friendships, etc. (Fig. 3). But no simple model yet predicts which changes will occur in relation to specific changes in demographic parameters, or how changes will interact to produce alterations in social structure.

Far more data are needed, both on responses of a single group over time and comparisons between groups of the same species under different demographic conditions. The social variables affected and the pathways for changes in social responses are interrelated and it is therefore difficult to determine ecological as opposed to demographic causality. General patterns can be proposed, as in Figure 3, which perhaps highlight the complexity and the areas of least information. In the diagram presented here, pathways for changes in social structure can be observed, as well as the points at which interaction between variables may exist. This model resembles that of Ohsawa and Dunbar (1984), but includes a greater variety of social factors.

As Hinde (1976) points out, relationships are dynamic, and affected by the relationships around them. Social structure is thus the outcome of dynamic processes of individuals interacting, forming relationships and being affected by those relationships. Specific types of relationships creating a group's social structure thus are not fixed. General patterns, both in terms of response to changes in group composition and in terms of the nature of the relationships between individuals in different groups, may be similar.

SPECULATIONS AND CONCLUSIONS

In a recent review, Cant and Temerin (1984) distinguish four aspects of foraging. These are the physical characteristics of the environment and of foods in that environment, traits of the animals such as body size, foraging problems expressed as how best to acquire food, and finally the behavioral capacities of the animal for solving foraging problems. They especially emphasize the need for perception and cognitive problem solving when the spatio-temporal distribution of food patches predicts food searching behavior and travel time.

A second level of foraging factors can also be identified, relating how animals exploit their environment to their social behavior. Returning to the original proposition that seasonally food-constrained groups represent one possibility for social responses, and ecological opportunities among seasonally food-enhanced groups give rise to different possibilities for response, can we now predict the range of variation for groups both food-deprived and food-enhanced? Using a very simple categorization (Fig. 4) of primates by three dimensions of food quality, abundance, and temporal-spatial patchiness, dietary clusters of extant primates can be found in seven of the eight categories. In this categorization, the triangle filled by constantly food-enhanced groups (e.g., captive primates) may represent the most desirable condition.

Among food-enhanced groups, a range of social responses to ecological variables should be observed, with the lower limits defined by groups receiving sporadic and unpredictable provisioning (such as temple monkeys) and the upper limits by those groups with constant, predictable access to large quantities of high-quality food (such as chimpanzees in Arnhem Zoo). If food-enhancement is rendered unpredictable by virtue of inter-individual interactions such as competition, groups should be similar in the range of social behavior to those receiving food at unpredictable times. Groups whose food is enhanced in unpredictable ways may be similar to those groups whose food is seasonally enhanced and thus variable in rates of aggression and competition (e.g., Loy, 1970), types of alliance formation and dominance relations (e.g., Datta, 1983), patterns of group fissioning (e.g., Chepko-Sade and Olivier, 1979), and even frequency of mating (e.g., Zuckerman, 1932) without major changes in the basic nature of their social relationships. Changes in the frequency with which relationships are expressed may appear

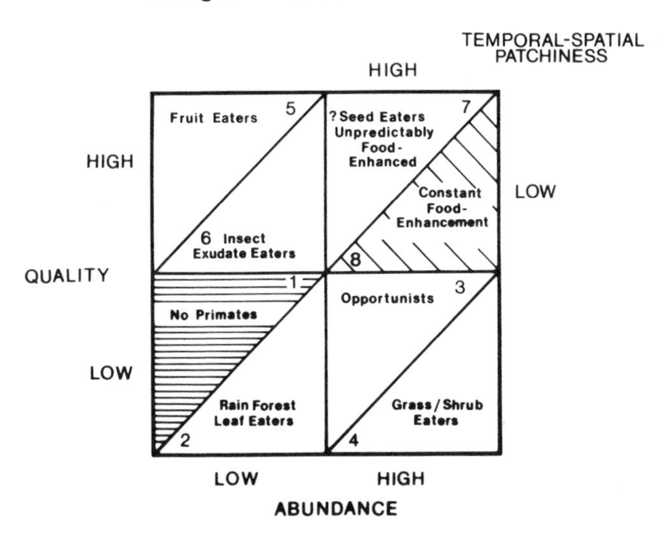

Fig. 4. A simple three-dimensional categorization of the relations between food quality, abundance, and temporal-spatial patchiness and different feeding habits of primates. Categorization of high/low indicates relative, rather than absolute, levels.

to influence social structure, but do not affect underlying principles of the group's organization. However, among primates whose cognitive skills can be turned away from the constraints and demands of foraging and toward their social relationships, the basic nature of relationships, their complexity, and their dynamic effects upon participants may alter, giving rise to variation that is no longer predictable. Examples of extreme levels of cognitive social complexity are seen primarily in groups with few foraging constraints (e.g., Bachmann and Kummer, 1980; de Waal, 1982). Thus, some food-enhanced groups represent merely a move toward increased variability upon their "species typical" line, while others do indeed represent an order of magnitude change between the ranges for different species.

Such discontinuities representing shifts between different ranges of social structure can be related to the evolution of different social organizations. Conditions of either constant food deprivation (e.g., triangle 1 in Figure 4) or constant food enhancement (triangle 8) may alter the nature of the social structure enough for different principles of organization to emerge (e.g., Figure 1). Although examples of such shifts in social organization have yet to be widely demonstrated for primates, facultative alternations between monogamy, polygamy, and polyandry in response to food abundance have been shown for avian species such as the dunnock (Davies and Lundberg, 1984). Suggestions of similar shifts in social organization among primates can be found in mountain gorillas, where stable female residence patterns have led to the atypical development of matrilines (Harcourt, 1978). Other examples of inter-group variability in the degree of cooperative polyandry (Terborgh and Goldizen, 1985) and sex ratios which are determined by birth rate (Ohsawa and Dunbar, 1984), suggest that ecological conditions and demographic conditions interact to shift inter-individual relationships over some threshold of social organization. With further data, the study of the range of variation in interactions, relationships, and resulting social structure within groups between seasons, and between groups with different ecological constraints and opportunities should lead to a greater understanding of overall patterns of social organization among primates.

SUMMARY

Past characterizations of primate social systems have tended to seek simple parameters that distinguish between apparently different types of groups. Such characteristics range from relatively simple correlations with food habits (Crook and Gartlan, 1966) or rainfall (Popp, 1982), through complex interactions between habitat, diet, and body size (Clutton-Brock and Harvey, 1977), interpretations of dynamics of inter-individual relationships (Seyfarth et al., 1978), and ultimately to the view that groups are alliances of individuals cooperating to optimize their access to resources (Wrangham, 1980, 1983). In the final case, the role of food quality, abundance, and distribution has been seen as a determinant of the costs and benefits of cooperating and competing with other individuals. The resolution of this conflict is postulated to determine the social organization.

Until recently, few studies of wild primates have examined how the patterning of relationships and the dynamics of social structure within groups are related over the short or the long-term to the availability and variability of the food supply. It is this proximate issue of the relations between food abundance and deprivation, and the observed variation in the nature and patterning of interactions and relationships that this paper explored.

ACKNOWLEDGMENTS

I thank R.W. Wrangham, D.L. Cheney, R.M. Seyfarth, M. Hauser, K. Stewart, A.H. Harcourt, and J. Altmann for long and valuable discussions. K. Lindsay made extensive comments on the draft, and R. Foley made many helpful suggestions about the models and their implications. I thank Profs. R.A. Hinde and P.P.G. Bateson for facilities and discussions while writing, and R.A. Hinde for critical comments.

REFERENCES

Altmann J (1979): Age cohorts as paternal sibships. Behav Ecol Sociobiol 6:161–169.
Altmann J (1980): ''Baboon Mothers and Infants.'' Cambridge, MA: Harvard University Press.
Altmann SA, Altmann J (1970): ''Baboon Ecology.'' Chicago: University of Chicago Press.
Altmann SA, Altmann J (1979): Demographic constraints on behavior and social organization. In Bernstein IS, Smith EO (eds): ''Primate Ecology and Human Origins.'' New York: Garland STMP, pp 47–64.
Bachmann C, Kummer H (1980): Male assessment of female choice in hamadryas baboons. Behav Ecol Sociobiol 6:315–321.
Baldwin JD (1971): The social organization of a semifree-ranging troop of squirrel monkeys *(Saimiri sciureus).* Folia Primatol 14:23–50.
Baldwin JD, Baldwin L (1976): Effects of ecology on social play: A laboratory simulation. Z Tierpsychol 40:1–14.
Boyce MS (1979): Seasonality and patterns of natural selection for life histories. Am Nat 114:569–583.
Cant JGH, Temerin LA (1984): A conceptual approach to foraging adaptation in primates. In Rodman PS, Cant JGH (eds): ''Adaptations for Foraging in Non-Human Primates.'' New York: Columbia University Press, pp 304–342.
Chalmers NR (1968): Group composition, ecology and daily activities of free-living mangabeys in Uganda. Folia Primatol 8:247–262.
Cheney DL, Seyfarth RM (1983): Nonrandom dispersal in free-ranging vervet monkeys: Social and genetic consequences. Am Nat 122:392–412.
Chepko-Sade BD, Olivier TJ (1979): Coefficients of genetic relationship and the probability of intergenealogical fission in *Macaca mulatta.* Behav Ecol Sociobiol 5:263–278.
Clutton-Brock TH (1977): Some aspects of intraspecific variation in feeding and ranging behaviour in primates. In Clutton-Brock TH (ed): ''Primate Ecology.'' London: Academic Press, pp 539–556.
Clutton-Brock TH, Harvey P (1977a): Species differences in feeding and ranging behaviour in primates. In Clutton-Brock TH (ed): ''Primate Ecology.'' London: Academic Press, pp 557–579.
Clutton-Brock TH, Harvey P (1977b): Primate ecology and social organization. J Zool Lond 183:1–39.
Colvin J (1983): Influences of the social situation on male emigration. In Hinde RA (ed): ''Primate Social Relationships.'' Oxford: Blackwells, pp 160–171.
Crook JH, Gartlan S (1966): Evolution of primate societies. Nature 210:1200–1203.
Datta SB (1983): Patterns of agonistic interference. In Hinde RA (ed): ''Primate Social Relationships.'' Oxford: Blackwells, pp 289–297.
Davies NB, Lundberg A (1984): Food distribution and a variable mating system in the dunnock *Prunella modularis.* J Anim Ecol 53:895–912.
Dittus WPJ (1979): The evolution of behaviours regulating density and age-specific sex ratios in a primate population. Behaviour 69:265–302.
Dunbar RIM (1984): ''Reproductive Decisions.'' Princeton, NJ: Princeton University Press.
Dunbar RIM, Sharman M (1984): Is social grooming altruistic? Z Tierpsychol 64:163–173.
Eisenberg JF, Muckenhirn NA, Rudran R (1972): The relations between ecology and social structure in primates. Science 176:863–874.
Fagen R (1981): ''Animal Play Behavior.'' Oxford: Oxford University Press.
Fairbanks LA, Bird J (1978): Ecological correlates of interindividual distance in the St. Kitts vervets *(Cercopithecus aethiops sabaeus).* Primates 19:605–614.

Hall KRL (1965): Social organization of Old World monkeys and apes. Symp Zool Soc Lond 14:265–289.
Harcourt AH (1978): Strategies of emigration and transfer by primates with particular reference to gorillas. Z Tierpsychol 48:401–420.
Harvey PH, Clutton-Brock TH (1981): Primate home range size and metabolic needs. Behav Ecol Sociobiol 8:151–156.
Hausfater G, Altmann J, Altmann S (1982): Long-term consistency of dominance relations among female baboons *(Papio cynocephalus)*. Science 217:752–755.
Hinde RA (1976): Interactions, relationships and social structure. Man 11:1–17.
Hinde RA (1983): General issues in describing social behaviour. In Hinde RA (ed): "Primate Social Relationships." Oxford: Blackwells, pp 17–20.
Hinde RA, Stevenson-Hinde J (1976): Towards understanding relationships: Dynamic stability. In Bateson PPG, Hinde RA (eds): "Growing Points in Ethology." Cambridge: Cambridge University Press, pp 451–479.
Homewood KM (1978): Feeding strategy of the Tana mangabey *Cercocebus galeritus galeritus*. J Zool Lond 186:375–391.
Iwamoto T, Dunbar RIM (1983): Thermoregulation, habitat quality and the behavioural ecology of gelada baboons. J Anim Ecol 52:357–366.
Johnson JA (1984): "Social Relationships of Juvenile Olive Baboons." PhD Thesis: University of Edinburgh.
Kummer H (1968): "Social Organization of Hamadryas Baboons." Chicago: University of Chicago Press.
Lee PC (1983a): Ecological influences on relationships and social structure. In Hinde RA (ed): "Primate Social Relationships." Oxford: Blackwells, pp 225–230.
Lee PC (1983b): Context-specific unpredictability in dominance interactions. In Hinde RA (ed): "Primate Social Relationships." Oxford: Blackwells, pp 35–44.
Lee PC (1984): Ecological constraints on the social development of vervet monkeys. Behaviour 91:245–262.
Lee PC (1986): Environmental influences on development: Play, weaning and social structure. In Else JG, Lee PC (eds): "Primate Ontogeny, Cognition and Social Behaviour." Cambridge: Cambridge University Press, pp 227–237.
Lee PC, Brennan EJ, Else JG, Altmann J (1986): Ecology and behaviour of vervet monkeys in a tourist lodge habitat. In Else JG, Lee PC (eds): "Primate Ecology and Conservation." Cambridge: Cambridge University Press, pp 229–235.
Loy J (1970): Behavioral responses of free-ranging rhesus monkeys to food shortage. Am J Phys Anthropol 33:263–272.
McKey DB, Gartlan SJ, Waterman PG, Choo GM (1981): Food selection by black colobus monkeys *(Colobus satanus)* in relation to plant chemistry. Biol J Linn Soc 16:115–146.
Manzolillo DL (1986): Factors affecting intertroop transfer by adult male *Papio anubis*. In Else JG, Lee PC (eds): "Primate Ontogeny, Cognition and Social Behaviour." Cambridge: Cambridge University Press, pp 371–380.
Martin P (1984): The time and energy costs of play behaviour in the cat. Z Tierpsychol 64:298–312.
Oates JF (1977): The guereza and its food. In Clutton-Brock TH (ed): "Primate Ecology." London: Academic Press, pp 276–323.
Oates JF, Waterman PG, Choo GM (1980): Food selection by south Indian leaf monkeys *Presbytis johnii* in relation to leaf chemistry. Oecologia 45:45–56.
Ohsawa H, Dunbar RIM (1984): Variation in the demographic structure and dynamics of gelada baboon populations. Behav Ecol Sociobiol 15:231–240.

Packer CR (1979): Inter-troop transfer and inbreeding avoidance in *Papio anubis*. Anim Behav 27:1–36.
Popp JL (1983): Ecological determinism in the life histories of baboons. Primates 24:198–210.
Richard AF (1974): Intra-specific variation in the social organization and ecology of *Propithecus verreauxi*. Folia Primatol 22:178–207.
Richard AF (1977): The feeding behavior of *Propithecus verreauxi*. In Clutton-Brock TH (ed): "Primate Ecology." London: Academic Press, pp 72–96.
Richard AF (1985): "Primates in Nature." New York: WH Freeman.
Rowell TE (1966): Forest living baboons in Uganda. J Zool Lond 159:344–364.
Rowell TE (1967): Variability in the social organization of primates. In Morris D (ed): "Primate Ethology." London; Weidenfeld and Nicholson, pp 219–235.
Rowell TE (1974): The concept of social dominance. Behav Biol 11:131–154.
Rowell TE, Olson DK (1983): Alternative mechanisms of social organization in monkeys. Behaviour 86:31–54.
Sackett GP, Holm RA, Fahrenbruch CE (1979): Ponderal growth in colony and nursery-reared pigtail macaques *(Macaca nemestrina)*. In Ruppenthal GC (ed): "Nursery Care of Nonhuman Primates." New York: Plenum Press, pp 187–201.
Sailer LD, Gaulin SJC, Boster JS, Kurland JA (1985): Measuring the relationship between dietary quality and body size in primates. Primates 26:14–27.
Seyfarth RM (1980): The distribution of grooming and related behaviours among adult female vervet monkeys. Anim Behav 28:798–813.
Seyfarth RM, Cheney DL, Hinde RA (1978): Some principles relating social interactions and social structure among primates. In Chivers DJ, Herbert J (eds): "Recent Advances in Primatology I." London: Academic Press, pp 39–51.
Southwick CH (1967): An experimental study of intra-group agonistic behavior in rhesus monkeys *(Macaca mulatta)*. Behaviour 28:182–209.
Sugiyama Y, Ohsawa H (1982): Population dynamics of Japanese monkeys with special reference to the effect of artificial feeding. Folia Primatol 39:238–263.
Terborgh J, Goldizen AW (1985): On the mating system of the cooperatively breeding saddle-backed tamarins *(Saguinus fuscicollis)*. Behav Ecol Sociobiol 16:293–300.
de Waal FBM (1979): Exploitation and familiarity dependent support strategies in a colony of semi-free ranging chimpanzees. Behaviour 66:268–312.
de Waal FBM (1982): "Chimpanzee Politics." London: Jonathan Cape.
Wrangham RW (1977): Feeding behaviour of chimpanzees in Gombe National Park, Tanzania, In Clutton-Brock TH (ed): "Primate Ecology." London: Academic Press, pp 504–538.
Wrangham RW (1980): An ecological model of female-bonded primate groups. Behaviour 75:262–300.
Wrangham RW (1981): Drinking competition among vervet monkeys. Anim Behav 29:904–910.
Wrangham RW (1983): Ultimate factors determining social structure. In Hinde RA (ed): "Primate Social Relationships." Oxford: Blackwells, pp 255–262.
Waser PM, Homewood K (1979): Cost-benefit approaches to territoriality: A test with forest primates. Behav Ecol Sociobiol 6:115–119.
Zuckerman S (1932): "The Social Life of Monkeys and Apes." New York: Harcourt, Brace.

Ecology and Behavior of Food-Enhanced Primate Groups, pages 313–346

15

Studies of Food–Enhanced Primate Groups: Current and Potential Areas of Contribution to Primate Social Ecology

Dennis R. Rasmussen

Animal Behavior Research Institute, and Department of Behavioral Endocrinology, Wisconsin Regional Primate Research Center, Madison, Wisconsin 53715

INTRODUCTION

Social ecology is the study of covariation in environment and social organization (Crook, 1970a). All studies that relate differences in food enhancement to variation in social organization are therefore social ecological studies. This chapter is focused on how data collected from food-enhanced primate groups can contribute to social ecology. In order to delimit accurately the role of studies on food-enhanced primate groups in primate social ecology, the field is partitioned into three areas: historical, functional, and proximate social ecology.

Tinbergen's conceptual division of questions in biology into categories similar to those used here (1951, 1963, 1965, 1969) has had an important and increasing impact on our abilities to focus questions precisely. The conceptual division of social ecology used here is therefore hoped to encourage a similar clarity of focus in social ecology. By recognizing differences in questions concerning historical, proximate, and functional social ecology (Rasmussen, 1981a), the student of behavior may understand why some field researchers justifiably do not value studies of food-enhanced primate groups as highly as those conducted on undisturbed, natural groups. Some researchers who study food-enhanced groups may accept this view and attempt to minimize the description of the degree of enhancement on their study group to the point where it is difficult or impossible to compare the amount of food provided across studies (this volume, chapter 8). This conceptual division also indicates that studies of food-enhanced groups may, at times, be preferable to studies of nonprovisioned groups. Studies of food-enhanced groups have an important role in the study of social ecology, a role that can only be realized with a strong understanding of the levels of focus in social ecology.

Before turning to the conceptual division of social ecology, food enhancement will first be shown to be a pervasive phenomenon and then the likelihood of an increasing number of studies conducted on food-enhanced groups will be discussed.

Food Enhancement: A Continuum

Food enhancement varies along a continuum (Fa, 1986) from total supply of all food to indirect and subtle augmentation of food sources as a result of human presence. Food may be directly enhanced by provisioning captive or free-ranging groups. It may also occur indirectly, by manipulation of the habitat of primates so there is more food available than would normally be found in a given area (this volume, chapter 2). Enhancement of food may persist over many generations: the major source of food of Guatemalan howler *(Alouatta villosa pigra)* and spider monkeys *(Ateles geoffroyi)* living at Tikal during at least 2 months of each year are *Brosimum alicastrum* trees originally planted by Mayans (Coelho et al., 1976, 1977a, 1977b, 1979). Enhancement may occur in very subtle ways: the presence of observers may permit a study group to supplant neighboring groups from food sources more frequently. When less habituated neighboring groups avoid observers within a study group they are necessarily supplanted by that group (Rasmussen, 1979); thus mere presence of observers in an otherwise completely natural setting can actually augment the observed individuals' access to food as well as other resources.

Increasing Studies on Food-Enhanced Groups

Reports on the behavior of food-enhanced groups seem likely to increase. Human impact on the habitats of nearly all nonhuman primates is increasing, so subtle forms of food enhancement will increasingly permeate our data sources. Habituation of nonhuman primates can take months of tedious dawn to dusk effort; provisioning of food is sometimes used to reduce this period and extend observation time (Goodall, 1971; Wrangham, 1974). Providing food to primate groups can also lead to particularly high population densities; these high population densities may sometimes be valuable when 1) the species is endangered, 2) surplus animals are needed for medical research, and 3) when a particularly large population of animals is desired for behavioral observation.

Groups of free-ranging nonhuman primates can be kept in unnatural habitats by providing food often at a considerable reduction of cost compared with laboratory or colony housing. Food-enhanced nonhuman primate groups seem likely to become an increasingly cost-effective means of producing subjects for research. Costs of keeping nonhuman primates in laboratory and colony housing facilities may increase due to activities of animal rights groups; these activities can necessitate hiring of additional security, animal health personnel, and animal care committee members as well as add to construction costs (Holden, 1986). Provisioned groups of nonhuman primates kept in seminatural habitats are not so closely monitored by animal rights groups, because injuries, diseases, or fatalities are attributed to the natural history of the species rather than to inadequate care. For example, the death of a monkey in a provisioned but otherwise free-ranging group resulting from wounds inflicted by a predator or from injuries in intrasexual agonistic encounters tends to be properly viewed by the public as an unpleasant result of natural selection; the unplanned death of a monkey in captivity from any cause other than old age immediately suggests that health care and/or housing was inadequate (see Huntingford, 1984 for an implicit recognition of this perceptual difference).

HISTORICAL SOCIAL ECOLOGY

Definition

Historical social ecology[1] uses inferential methods to estimate past selection pressures that have shaped genetic bases of differences between social organizations (Rasmussen, 1981a).

The Comparative Method in Historical Social Ecology

Several articles appeared in the late 1960s and early 1970s that focused attention of primatologists on how past selection pressures may have given rise to genetic bases of currently observed differences in species' social organizations (Hall, 1965; Crook and Gartlan, 1966; Kummer, 1971a; Eisenberg et al., 1972; Altmann, 1974). Patterns of primate social organization were categorized and the resultant categories then matched to presumed major differences in the selective pressures exerted in the species' ancestral habitats. For example, Crook and Gartlan (1966) felt one-male groups were due to ". . . the interaction of selection pressures from food shortage, predation and habitat topography. . . ." Altmann (1974) stated: "Predation selects for large groups and for groups containing at least one male." Past selective forces are the "ultimate factors" responsible for current genetic bases of differences in social organizations of compared species.

As in ethology (Tinbergen, 1959, 1965; Crook, 1965), the use of the

[1]I have changed to the term historical social ecology from evolutionary social ecology used in Rasmussen (1981a) since I feel it conveys a more accurate description of this level of social ecological enquiry and since both historical social ecology and functional social ecology are based upon evolutionary theory.

comparative method in social ecological studies is most likely to yield valid results when species compared are closely related and recently speciated from the same ancestral stock. Genetic similarities will therefore most likely be homologues (Altmann, 1974). If the compared species have only recently diverged, then differences in the current natural habitats in which they are observed are more likely to be those which selected the genetic bases of observed differences. If the species are recently diverged, however, particular care must be used to determine if differences attributed to selection actually have a genetic basis (Rasmussen, 1981a).

Historical social ecology is a historical science where past selection pressures are determined through reconstruction and through use of current natural environments. Therefore direct experimental confirmation of postulated selection pressures is not possible. Experiments may be used to support historical reconstructions, such as those described in the section on functional social ecology (see section on functional social ecology, this chapter).

Importance of the Ancestral Habitat in Historical Social Ecology

Use of the current habitat to model past selection pressures. In comparative analyses, aspects of current habitats are related to differences in current social organization. The current habitat is thus used as a model of the habitat which actually exerted selection pressures on ancestors of observed populations. In order to accurately model past selective forces, observations must therefore be collected in natural habitats similar to those in which their recent ancestors were subject to selection. Natural habitats of this type are referred to as "ancestral habitats" throughout the rest of this paper.[2] Field researchers often go to considerable logistic effort to collect data in ancestral habitats because they are interested in historical social ecology and hence in past selection pressures.

There is another, less obvious, reason for conducting historical social ecological studies in ancestral habitats. If the habitat deviates from the

[2]There are at least two meanings used for "natural habitats." Natural habitats may be referred to as those in which free-ranging populations exist even though these may be very different from those in which the recent ancestors of those populations existed. For example, studies on wild baboons *(Papio anubis & P. ursinus)* have been conducted in areas devoid of predators (Harding, 1973; Anderson, 1982; Hamilton, 1985). Food-enhanced groups may also be found in a natural environment such as the Koshima troop of Japanese macaques, *Macaca fuscata* (Mori, 1979). Similarly, populations may be found on remnant "islands" of once continuous tropical forests (Neyman, 1978) and these may be referred to as natural habitats. Alternatively, the term "natural habitat" may be used to refer to habitats believed to be similar to those in which the progenitors of present populations were subject to natural selection (Bernstein, 1967; Rasmussen, 1981a). Unfortunately a dwindling number of nonhuman primates are found in natural habitats of the latter type, therefore it was considered useful to adopt the term "ancestral" environment when referring specifically to an environment believed very similar to that in which the species existed during its recent evolutionary history.

ancestral habitat, then we have no assurance, without functional analyses (see section on functional social ecology, this chapter), that modal patterns of behavior observed are in fact adaptive patterns. When the study population is in an ancestral environment that it has occupied for many generations a possible initial assumption for many patterns of behavior is that a modal pattern may be most adaptive; the form of selection here is most likely to be stabilizing selection (e.g., Lack, 1954; Patterson, 1965; Tinbergen, 1965). This initial assumption may be used to generate testable hypotheses. The assumption that a modal pattern of behavior is adaptive must then be tested in subsequent functional analyses (see section on functional social ecology, this chapter).

The assumption that a modal pattern is adaptive may be applied only with knowledge of the behavioral pattern and possible proximate correlates of fitness. For example, it might not be applicable to modal agonistic rank since agonistic rank may be positively related with reproductive success (Hausfater, 1975; Smith, 1981). Hence individuals of higher than average agonistic rank throughout their reproductive careers could have greater Darwinian fitness (Duvall et al., 1976).

A change from the ancestral environment implies that genetic bases of behavior that were specifically adapted to that past environment may change (Rensch, 1959). Directional selection would therefore be more likely to be found in a population after a change from the ancestral environment.[3] If the environment is considerably different from the ancestral habitat, previously adaptive patterns of behavior may no longer be adaptive and may even become pathological (Calhoun, 1962). In studies of food-enhanced groups the assumption that the modal pattern of behavior is linked with the highest fitness would thus be more difficult to justify than in studies conducted on groups in ancestral environments. If ancestral and altered environments are not differentiated it is not possible to make valid initial assumptions about whether a trait is or is not adaptive on the basis of its frequency of occurrence (Rowell, 1979).

Need to account for recent changes in habitats. Increasing human impact on nonhuman primate habitats has made the current environments in which many wild populations are found unlike those of their ancestors. Ancestral habitats are therefore rapidly disappearing. If we are to understand

[3]Stabilizing selection may sometimes persist even after rather large changes from ancestral environments: for example, Karn and Penrose (1951) found children of modal body weight born at a London obstetric hospital had a greater likelihood of survival than those of lesser or greater weight. Similarly, Drickamer (1974, Figure 2, Table XII) and, more recently, Small and Smith (1986) report data indicating a negative relationship between likelihood of infant survival and deviation of birth date from the birth peak of rhesus macaques *(M. mulatta).*

past ultimate factors, current human impact must be taken into account in historical social ecology.

For example, large felid and canid predators have been drastically reduced or completely eliminated in many areas of the world. Sometimes natural predators may have been eliminated and then replaced with those introduced by humans, such as feral dogs (Dittus, 1979). Predators may avoid people so observers may protect groups by virtue of their presence (Rasmussen, 1979). Because predator pressure on nonhuman primate groups has therefore often been reduced from that in their recent ancestral habitats, it is difficult to assess the impact of predators on patterns of social organization observed in nonhuman primates.

Discussions of the selective influence of predators may therefore be explicitly excluded because of insufficient information (Martin, 1981) or this selective influence may be disregarded. An example of this myopic focus on present environments in historical discussions may be found in some recent literature on the positive correlation between degree of sexual dimorphism and body size of nonhuman primates (Clutton-Brock et al., 1977; Leutenegger, 1978; Gaulin and Sailer, 1984): Although influences on predators are mentioned (e.g., Alexander et al., 1979; Leutenegger, 1982), a serious attempt has not been made to estimate the relationship between size and form of predator defense. Sexual dimorphism may be greater in larger species because large size may permit more effective defense against predators and increase difficulty of avoiding predators by hiding. Absence of discussion of the effects of predators on sexual dimorphism seems particularly odd as sexual dimorphism increases more relative to size in terrestrial species (Figure 2 in Clutton-Brock et al., 1977; Leutenegger, 1978). There are strong reasons to believe predators may have been responsible for selection of many characteristics of nonhuman primate social organization (Washburn and DeVore, 1961; Schaller, 1963; Kruuk and Turner, 1967; Turnbull-Kemp, 1967; Altmann and Altmann, 1970; Saayman, 1971; Busse, 1977, 1980; Sigg, 1980; Rasmussen, 1981b; Collins, 1984; Anderson, 1986). A careful historical analysis of factors that have shaped a present population must take into account any modern changes in numbers and amount of predators (e.g., MacKinnon, 1974).

Predators may have been either eliminated or reduced in most modern habitats relative to ancestral environments. It is not difficult to imagine a near future when there will be an equivalent impact of food enhancement (e.g., this volume, chapter 2 and chapter 6); most habitats are influenced in the subtle ways mentioned at the beginning of this chapter. There is thus a pressing need for more studies in remaining ancestral habitats and the protection of these habitats if we are to reconstruct selective pressures responsible for the genetic bases of the diversity of current primate social organizations.

Food enhancement and comparative analyses. Nonhuman primate species which have been sympatric with humans during their historical past have an ancestral environment which includes humans. Historical social ecological studies of such species should attempt to use environments in which contact with humans is as similar as is possible to that believed to have existed during their recent evolutionary history.

There are species of nonhuman primates whose recent evolutionary history has included considerable contact with humans and there is evidence which suggests this contact has exerted selective pressures on these species. For example, the rhesus macaques in India appear to have lived in contact with humans for many generations (Southwick et al., 1961, 1965; Siddiqi and Southwick, this volume, chapter 6). This contact has included food enhancement and may therefore have also increased population density. Another example is provided by the Panamanian tamarin *(Saguinus geoffroyi)*. This species is most abundant in areas of secondary neotropical growth (Moynihan, 1970; Dawson, 1978; Eisenberg, 1979); particularly in growth that has been allowed to regenerate for 20 to 40 years after being cleared for agriculture (Skinner, 1985; personal observation). Thus human agricultural activities during the past several hundreds of years may actually have fostered the spread of this species throughout Panama. Humans may exert selection pressures other than through food enhancement; for example, many current and past human populations have hunted nonhuman primates for food and thus exerted specialized predator pressures on their social organizations (e.g., MacKinnon, 1974; Jolly, 1985).

Unless food enhancement is believed to make an environment more similar to the ancestral habitat than currently extant habitats, food enhancement definitely detracts from comparative historical social ecological studies that use current environments for models of past selection pressures.

Genetically Based Differences in Social Organizations

Historical social ecology is focused on differences in past selection pressures and hence necessarily focused on the genetic bases of differences between social organizations (Rasmussen, 1981a). An attempt is made to match past selection pressures with current genetic differences in social organizations. Articles may be very carefully worded so the term ''genetic'' is never used. If, however, differences in social organizations are attributed to selection pressures, this necessarily implies that the selected differences have a genetic basis. Such genetic differences may exist at multiple levels: for example, in tendencies toward intrasexual agonism, food preferences, birth rate, longevity, size and therefore the types of predators, habitat selection, disease resistance, and learning abilities.

Comparative historical social ecology must identify genetic differences between social organizations; this type of social ecological study therefore can disregard neither the sources of variation as suggested by Crook (1970b) nor be unconcerned with identification of the genetic bases of differences as suggested by Grafen (1984). This is particularly true with primates, since they have exceptional powers of learning and considerable flexibility in patterns of social organization (Crook, 1970a; Kummer, 1971a; Eisenberg, 1979; Jolly, 1985). Analyses of genetic bases of differences are exceedingly difficult and time consuming (e.g., Wecker, 1963), yet if we do not attempt these analyses, discussion of selection pressures can become reduced to academic game playing. It seems rather unsatisfying to leave analyses at the point where we assume differences between social organizations are due to genetic differences in order to estimate the selective forces responsible for those assumed differences.

What if we make a leap of faith and assume differences have a genetic basis when in fact they do not? We would then have gone to considerable effort to describe differences in selection pressures for a difference that does not exist. More seriously, we may have been misled and missed the important evolutionary question of why compared species have the shared flexibility permitting adaptive modifications[4] to different habitats (Rasmussen, 1981a; Gottlieb, 1984).

One of the hallmarks of primates is their ability to modify adaptively their social organization and behavior as a function of their habitat. Plasticity that makes genetic change unnecessary is a characteristic of anagenesis (Rensch, 1959; Gottlieb, 1984). Adaptive flexibility of primates may have resulted from exposure of populations to variable environments; individuals capable of most rapid adaptive modification to change therefore may have been those who made proportionately greater contributions of offspring to subsequent generations. This may be especially true for more terrestrial species, since they tend to have particularly wide geographic distributions (DeVore, 1963).

If we ignore sources of variation it may be impossible to understand the selective forces that gave rise to some of the most unique aspects of primates such as their learning abilities. For example, advanced learning abilities may have been selected so that behavior will be altered in adaptive ways to

[4]"Adaptive modification" is a term used to indicate proximally induced adaptive differences among individuals or populations that are not due to a genetic difference (Lorenz, 1965; Kummer, 1971a). For example, I found juvenile yellow baboons *(P. cynocephalus)* to sit closer to adult males when cover was more dense and predator detection more difficult (Rasmussen, 1983). It seems likely that this is an adaptive modification in the behavior of juveniles since the alteration in spacing may decrease the likelihood that they will be killed by predators.

environmental changes. If so, then no single environment is representational of the ultimate factors responsible for these abilities, rather, changes in environments are the ultimate factors. Learning abilities would increase adaptation to variable environments because the reinforcers for learning would be directly associated with fitness (Rasmussen, 1981a). Food, water, access to mates, and avoidance of injury have all been used successfully as reinforcers and they are also clearly associated with fitness. The genetic bases of reinforcers as well as learning processes seem likely to be subject to natural selection.

Initial Analyses of the Genetic Bases of Differences in Social Organizations

One way of increasing our confidence that a difference we observe has a genetic basis is through analysis of the flexibility of the difference to environmental variation.[5] If two species are both observed across a wide range of environments, including some common to both species (Hall, 1965; Crook and Aldrich-Blake, 1968), and they always have a characteristic difference, then this evidence suggests the difference may have a genetic basis. The difference then becomes a potential focus for further developmental studies (e.g., Kaufman, 1975). Intra-specific characteristics that do not alter across a wide range of environmental variation are likely to be controlled by processes such as genetic canalization (Waddington, 1957). Kummer (1971a) provided a finely reasoned example of this technique in his comparison of differences in the social organization of hamadryas *(P. hamadryas)* and anubis *(P. anubis)* baboons in Ethiopia. Nagel (1971, 1973) further supported Kummer's results with a study of the hybrid zone of these two species.

Studies of food-enhanced groups may help us to learn if search for differences in selection pressures in historical analyses are warranted, even though they may not permit accurate models of those selection pressures. Provisioning may profoundly alter the distribution of food from that of the ancestral environment of a species. If a difference in the social organization of two species persists when they are provisioned or kept in captivity (Kummer and Kurt, 1965; Stammbach, 1978), then there is

[5]Analyses of heritability are not necessarily suited to questions focused on past selection pressures, since characteristics that have been subjected to intense selection may have had their genetic variability markedly reduced and hence have low current heritability. Thus aspects of social organization that seem to characterize a species since they do not covary with environmental factors are precisely those we might expect on an a priori basis to have low heritability. Daly and Wilson (1983) provide a discussion on limitations of the measure of heritability.

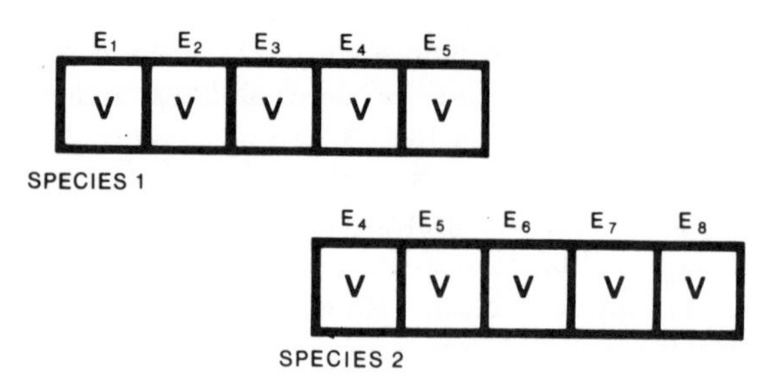

Fig. 1. Diagrammatic illustration of a comparison used for initial investigation of genetic differences in aspects of social organization. A variable (V) is compared across a range of 8 different environments (E_1 to E_8) 2 of which are the same for both species (E_4 and E_5). If there is little variation across environments within species but a considerable difference between species, then this evidence suggests the difference may have a genetic basis. Subsequent developmental studies are necessary to confirm this evidence. Confidence that the difference has a genetic basis is further increased if the behavioral difference persists even in environments markedly different from ancestral environments. For example, if differences persist during food enhancement or captivity (e.g., Kummer and Kurt, 1965; Loy, this volume, chapter 8). Like the deprivation experiment (Lorenz, 1965), the expression of a behavior in an environment much different from the ancestral environment suggests the behavior may be controlled by a process such as genetic cannalization (Waddington, 1957). Lorenz (1965) did not, however, explicitly indicate the necessity of comparison with other species. This comparison is necessary since it is meaningful to speak only of genetic differences between species: All behavioral characteristics must have a genetic basis since all are dependent upon the genetic characteristics of the individual for phenotypic expression.

If there is considerable variability across environments and little between species, this may indicate the compared species have the same anagenetic ability to modify the variable in an adaptive manner to environmental context.

stronger reason to believe the difference may have a genetic basis (Fig. 1). For example, Johnson and Southwick (1984) found spatial and developmental relationships of mother-infant pairs to have similar patterns in structurally diverse environments; Loy (this volume, chapter 8) summarizes evidence that rhesus have a higher birth rate than Japanese macaques across a wide range of amount of food provided; Siddiqi and Southwick (this volume, chapter 6) report a diurnal cycle of activity that does not qualitatively change with amount of provisioning or group size.

Knowledge of the degree of covariation between an aspect of social organization and environmental factors is of use in functional analyses as well as in historical reconstructions of past selection pressures. Traits which do not vary with the environment are those which we would expect to require genetic alteration if there were a strong alteration of the environment from the ancestral habitat (Rensch, 1959).

FUNCTIONAL SOCIAL ECOLOGY

Definition

Functional social ecology is the study of how the current environment influences Darwinian fitness or inclusive fitness (Hamilton, 1964) of individuals through their social organization and social interactions (Rasmussen, 1981a). Since this is not a historical science, experimental data can be collected to prove current environmental influences on social organization and, ultimately, fitness.

Functional Studies and Natural Habitats

Most functional analyses have been conducted in natural habitats because of simultaneous interest in historical and functional questions (e.g., Tinbergen, 1965). As Crook and Gartlan (1966) noted, analyses of function provide understanding of current selection pressures and, if conducted in ancestral environments, these analyses provide evidence of past selection pressures. As in Darwin (1859) selective processes observed today are hypothesized to be similar to those that occurred in preceding generations.

A more recent rationale for conducting functional studies in an ancestral environment is that the coefficient of relatedness, r, necessary in analyses of inclusive fitness, may only be estimated accurately if selection is weak (Charlesworth, 1980). As is indicated below, when nonhuman primates are in habitats considerably different from those which selected the genetic bases of their behavior they may be subject to strong selection. Hence, functional studies in altered environments, including food-enhanced environments (Kurland, 1977), may offer particular difficulties for analyses of inclusive fitness.

Functional Analyses in Food-Enhanced Groups

While there are reasons for conducting functional studies in natural environments, they need not be conducted in such environments. Many essential questions concerning the functions of social organization may best be addressed by experimental manipulation of the social and physical environment (Rasmussen, 1981c, 1984). Indeed, functional social ecological studies must be conducted on free-ranging and food-enhanced groups as well as on groups kept in laboratory and zoo colonies in order to determine how such environmental changes alter selection pressures. Absence of such information will prevent us from understanding possible differences in the genetic bases of behavior of subjects in an increasingly greater proportion of the literature on nonhuman primates. We are in the process of domesticating several species of nonhuman primates for life in captivity. Failure to document the selective forces at work will prevent us from having an accurate understanding of the eventual genetic differences that will exist between wild and domesticated populations. Failure to study such selective forces would also represent waste of exceptional research opportunities. Very subtle selective differences may exist: for example, nonhuman primates who are both more responsive and more capable of interacting in human nonverbal communication may be more likely to be kept as tractable subjects in colonies. Hence any genetic bases for such communicative traits may currently be increasing in captive populations.

One method of conducting functional social ecological studies is to relate environmental variation to changes in social behavior and to measures of reproductive success (Patterson, 1965; Rasmussen, 1981a). Comparison of groups during periods with and without food enhancement provide the best example of this approach so far available in studies of nonhuman primates. These comparisons indicate the considerable power of studies involving manipulation of food enhancement for functional analyses.

Data collected on troops of Japanese macaques on Mt. Ryozen during periods with and without provisioning (Sugiyama and Ohsawa, 1982) indicate food enhancement may result in a greater differential between numbers of young surviving per female in high social classes and those in lower social classes (Table I): infants 2 years of age or younger whose mothers were of low social class experienced a higher mortality rate; this disparity was reduced when provisioning stopped. These data suggest social class is related to the Darwinian fitness of female Japanese macaques and that selection pressures associated with social class are increased by food enhancement. Food enhancement at Mt. Ryozen may thus have led to stronger selection pressure for genetically based characteristics associated with high social rank of females.

At Mt. Ryozen the greater ability of females of high agonistic rank to gain access to food provided in an area of about 3 × 20 m may well be responsible for the increased disparity in reproductive success of females as a function of social class. Several other studies of food-enhanced macaques have indicated females of high agonistic rank have higher reproductive rates and increased surviving infants (Drickamer, 1974; Sade et al., 1976; Dittus, 1979; Mori, 1979; Wilson et al., 1978). Fairbanks and McGuire (1984) have found high-ranking female vervet monkeys *(Cercopithecus aethiops sabaeus)* also to have shorter interbirth intervals, more births, and more surviving infants. Differential reproductive success of females in these groups is consistent with observations that agonistic behavior becomes more frequent (Kummer and Kurt, 1965; Southwick et al., 1965; Fa, 1986) and rank orders more clearly defined when groups are food enhanced (Rowell, 1967). The most clearly defined pattern to emerge from Lyles and Dobson's review of empirical differences between unprovisioned and provisioned

TABLE I. Number of Infants Surviving Per Female Japanese Macaque by Social Class During Provisioned (1969–1973) and Nonprovisioned Periods (1974–1980) on Mt. Ryozen

Social class[a]	No. of females [b]	Provisioned: Surviving infants per female	No. of females[b]	Not provisioned: Surviving infants per female
A	15	.8	40	.33
B	28	.5	20	.20
C	23	.26	41	.27
D	15	.6[c]	33	.18
Coefficient of variation		.42		.28

Data from Tables II and IV, Sugiyama and Ohsawa (1982).

[a]Social class was determined by distance from the male of highest agonistic rank using the central-peripheral classification favored by Japanese primatologists (Imanishi, 1960). Individuals in a higher social class were generally of higher agonistic rank although quantitative data were not collected on all dyads.

[b]Number of females in the "high reproductive ages."

[c]It is possible that females of the lowest social class had the second highest number of surviving infants because they were so peripheral to the provisioned food that they obtained a greater proportion of natural foods (this volume, chapter 4) and did not get drawn into the competition associated with access to the provisioned food (e.g., Calhoun, 1962).

primates (this volume, chapter 9) was the increased survival rates of infants when food was enhanced. This result, and the references cited above, suggest the pattern of greater proportionate reproductive success amongst females of high agonistic rank found at Mt. Ryozen may be found in other species of nonhuman primates.

Increased food supply has been found correlated with increased population growth rates (Dittus, 1977a, 1977b; Mori, 1979; Sugiyama and Ohsawa, 1982; Malik et al., 1984; Berman, this volume, chapter 13; Lyles and Dobson, this volume, chapter 9; Paul and Kuester, this volume, chapter 10) and decreased food supplies with negative growth rates (Galat and Galat-Luong, 1973; Struhsaker, 1973, 1976; Hausfater, 1975; Dittus, 1977a). If food enhancement changes both the intensity of selection for characteristics such as high social class and causes an increase in population growth, then there will be an increased number of individuals emigrating from the enhanced groups who are the product of the stronger selection for high social class. Food enhancement may therefore not only alter the genetic composition of troops provisioned, but also that of neighboring and nonprovisioned troops.

Potential Areas for Functional Analyses in Food-Enhanced Groups

Fu:· onal social ecological studies are potentially our most important source of information on how selection actually operates on nonhuman primate groups. Attention is restricted here to a few topics of particular interest in food-enhanced groups.

Studies of food enhancement must be expanded to include influences of the quantity of food provided, its spatial distribution, and its quality (see section on proximate social ecology, this chapter). Current social ecological theory tends to attribute exceptional importance to the distribution, abundance, and quality of food as factors influencing group size, particularly to the number of females found in groups and their distribution in space (Wrangham, 1980). Analyses of shifts in selection pressures as a function of alterations in food abundance, distribution, and quality would provide a means of testing this theory.

Studies of provisioned, free-ranging groups have so far mostly provided information on how abundant food distributed in a small area influences social interactions and measures of reproductive success. This limited study of the alteration of food supply is the result of using food enhancement as a means of concentrating individuals for study and habituation. Many scientists have avoided the study of factors influenced by food enhancement and instead concentrated on variables believed to provide data as similar as possible to that obtained from groups in ancestral environments. Food enhancement has therefore not been used widely as an intentional experimental manipulation of food distribution, abundance, and quality. As indicated in Table I, the distribution pattern of food in provisioned groups may increase differential reproductive success as a function of social class. In addition, this distribution pattern of food seems directly responsible for the central peripheral structure used to determine social class (Rasmussen and Rasmussen, 1979; Fa, 1986; Wada and Matsuzwa, 1986).

There is a strong need for functional analyses of the social ecology of captive groups. Food enhancement may be found to have rather different influences on measures of fitness in these groups. In contrast to food-enhanced and free-ranging groups, captive groups often have food distributed on an ad libitum basis 24 hours a day. Many of these groups are maintained for production of subjects in medical research, so special care is taken to assure all group members obtain plenty of food and are maintained in excellent health. The availability of food to all group members and complete protection from predators may be anticipated to result in altered ecological influences on social organization and rank-related differences in reproductive success. More offspring born into such groups survive, so covariation in rank and survival of infants diminish (e.g., Nieuwenhuijsen et al., 1985). Since more infants tend to survive to reproductive age (e.g., this volume, chapter 10) intrasexual and intersexual competition for mates would be anticipated to have a greater proportional selective influence than differential survival of offspring. Kummer and Kurt's (1965) comparative study of captive and wild hamadryas baboons *(P. hamadryas)* and numerous qualitative comparisons (e.g., Rowell, 1967) suggest agonistic competition is greater in captive groups and thus support this hypothesis. Food-enhanced captive groups are therefore ideal for functional analyses concerned with intrasexual and intersexual competition (Rasmussen, 1984). Even in groups where food is continuously available, infant mortality may remain associated with the agonistic rank of mothers. Here mortalities may be the direct result of social behavior such as "kidnapping" (this volume, chapter 10).

Functional analyses also need to be conducted on groups at the opposite end of continua of food quantity and predator presence. Deductive reasoning parallel to that in the previous paragraph indicates groups with scarce food and abundant predators are those in which intrasexual and intersexual selection would be expected to have proportionately the weakest selective influence. Some evidence exists to support this line of reasoning: intrasexual competition diminishes when food is very limited. Chacma baboons were observed to become lethargic and the troop dispersed when food was limited due to their entrapment on an island (Hall, 1963). Similar observations were made on rhesus released on an island off the coast of Puerto Rico (Morrison and Menzel, 1972). In an important experimental study, Southwick (1967) found rhesus macaques to engage in less agonistic behavior during starvation, an observation he found so paradoxical that he replicated his results. Predators seem more likely to kill juveniles than adults, and particularly juvenile males in primate societies where males transfer between groups more frequently than do females (Dittus, 1980; Rasmussen, 1981b). Predators would thus tend to decrease the number of individuals reaching the age at which they become capable of competing for mates and particularly of killing males when they transfer between groups; in polygynous societies males have a greater variance in reproductive success (Daly and Wilson, 1983) and therefore engage in more intrasexual competition. Abundant predators would therefore be expected to decrease the number of individuals competing for mates. Indeed, some evidence suggests larger groups of long-tailed macaques *(M. fascicularis)* engage in more agonistic interactions (Van Schaik et al., 1983).

Functional analyses need to be conducted on more subtle changes in selective factors that may be anticipated as a result of food enhancement. For example, group members particularly good at finding food and who have exceptional abilities in memory of locations where food may be found (Eisenberg and Wilson, 1978; Clutton-Brock and Harvey, 1980; Martin, 1981) would not be favored by an increase in their own access to resources or their attractiveness to other group members based on these abilities. Berman's research on rhesus (this volume, chapter 13) provides an example of rather subtle shifts in maternal care as a function of the rapid increase in growth of groups on Cayo Santiago due to food enhancement; she suggests the functional significance of these shifts and initial means of verifying her hypotheses.

PROXIMATE SOCIAL ECOLOGY

Definition

Proximate social ecology is the study of influences of the current environment on phenotypic aspects of social organization. Proximate analyses do not necessarily focus on aspects of behavior related to fitness. Proximate social ecological studies therefore do not necessarily focus on the behavior of individuals as must functional studies; the level of focus may, for example, be the group as a whole. Two major methods are in use. First, a comparative method similar to that used in historical social ecology: species or different populations of the same species are observed in different environments and variations in social organization are attributed to differences in environments. Second, a population of primates may be observed for several months and analyses made of covariation in environmental factors and aspects of social organization.

The Comparative Method

Many comparative analyses of variation observed in social organizations of primate species have attributed all or part of the variation to current habitats (DeVore, 1963; Hall, 1965; Rowell, 1967; Denham, 1971; Kummer, 1971b; Eisenberg et al., 1972; Altmann, 1974; Mori, 1979; Martin, 1981). In these analyses an aspect of social organization, such as number of individuals in groups, is related to an aspect of the environment, such as the percentage of foliage in the diet (Clutton-Brock and Harvey, 1979). Comparative analyses may also focus on the relationship between how groups use their environments and aspects of their social behavior. For example, the relationship between group size and area of home range across several species (Clutton-Brock and Harvey, 1977). Appropriate interpretations of such comparative analyses attribute discovered systematic relationships neither to selective pressures nor to current environmental differences; only empirical relationships are described.

If the goal of a comparative analysis is to understand how differences in species-typical environments may be responsible for phenotypic differences in patterns of social organizations, then these comparative analyses must draw on studies conducted in ancestral environments. Because species are genetically different by definition, and the environments in which they are observed are also different if they are allopatric, this type of comparative analysis confounds environmental and genetic sources of variation. The results of these analyses will, like those described in the section on historical social ecology, this chapter, require subsequent studies of sources of variation. For example, Kummer (1971a) found use of cliffs or sleeping trees to vary more directly with environment than with species differences in hamadryas baboons, anubis baboons, and hybrids of these two species.

If a comparative analysis is concerned with the influence of variation in a specific environmental factor such as food distribution, on an aspect of social organization such as group size, then data may be drawn from studies in ancestral environments and environments with varying degrees of provisioning (e.g., Mori, 1979). In such an analysis, it would be preferable to restrict attention to populations with as homogenous genetic makeup as possible. Such restriction decreases the confounding between genetic differences and influences of the immediate environment. In practice, comparisons may be restricted to a primate family (e.g., Clutton-Brock and Harvey, 1979), a genus (e.g., Nagel, 1973), or a species (e.g., this volume, chapter 8).

Covariation in Environment and Social Organization of Groups Over Time

The second method used in proximate social ecology is the analysis of covariation in environmental factors and social organization of a population. Since this type of analysis takes place during a very short period relative to that which would lead to major changes in the genetic structure of a population, the method does not confound environmental and genetic sources of variation as much as does the comparative method.

Many studies of functional social ecology must include this type of proximate social ecological analysis in order to link current environmental variation with variation in social behavior (e.g., Table I). Proximate analyses do not, however, necessarily include measures of fitness. As a result, studies of proximate social ecology need not focus on the behavior of individuals as must functional studies. The level of focus may be at the level of the group or even of several groups and never touch on analyses of individual behavior necessary for determination of measures of fitness in functional analyses. Unfortunately, the absence of the necessity of focus on measures of fitness has sometimes resulted in investigators ignoring the importance and potential predictive power added to proximal studies by knowledge of these measures so intimately associated with the significance of social behavior (Rasmussen, 1984).

Analyses at the level of the group might be made for several applied reasons. For example, it might be desirable to determine the environmental factors and group composition that leads to least aggression to avoid injury to display animals at zoos or to medical research subjects (Alexander and

Roth, 1971; Erwin and Erwin, 1976). Since increasing demand has been placed on captive colonies for production of nonhuman primates, studies may also be conducted to determine the group housing and food distribution arrangements that produce the maximum number of healthy surplus animals for least cost. On a more theoretical level, it may be of interest to determine how the distribution and abundance of resources influence patterns of social interaction and behavior of group members (Southwick, 1967; Rasmussen and Rasmussen, 1979).

Field studies. Most field studies of primate social ecology are proximate analyses of covariation in environmental factors and aspects of social organization, even though many of these have been conducted, in part, because of interest in historical social ecology (e.g., Altmann and Altmann, 1970; Chivers, 1974; Struhsaker, 1974; Clutton-Brock, 1975; Hladik, 1975; Oates, 1977; Wrangham, 1977; Rasmussen, 1979, 1983; Marsh, 1981; Harrison, 1983; Isbell, 1983).

Many of these studies have focused on environmental factors influencing patterns of range use. These field studies indicate that resources essential for survival and limited in abundance attract nonhuman primates. Hazardous locations are avoided; these locations include areas of dense cover where predators are difficult to detect or areas where predators are likely to be encountered (Rasmussen, 1983), and locations where encounters with groups potentially capable of injuring group members (Wrangham, 1981). As a result, patterns of use of the area in which animals live tend to reflect a trade-off of benefits and costs of using each area (Altmann and Altmann, 1970). This is an exciting potential area of contact between experimental analyses of learning, extinction, and social ecology (Altmann, 1974). There is a growing recognition of the need to consider environmental factors in addition to food on patterns of range use (Rasmussen, 1980; Post, 1984). Theoretical analyses in early research sometimes disregarded the importance of estrous females and intergroup interactions on patterns of group movements, although these have since been shown to be strong correlates of the way in which groups use their range (Struhsaker, 1974; Rasmussen, 1979; Isbell, 1983).

A social ecological study conducted on yellow baboons *(P. cynocephalus)* in an ancestral environment at Mikumi National Park, Tanzania, provides an example of how variation in environmental factors may be quantitatively linked with aspects of social organization (Rasmussen, 1979, 1981b, 1983). The greatest number of females in the troop were in estrous and the most males emigrated and immigrated when 1) the troop's use of its range was most likely to bring it into contact with neighboring troops, 2) visibility was least obscured by vegetation, 3) juveniles were farthest from adult males, and 4) the least amount of time was spent gathering food. When cover was most

Fig. 2. Muroto, a high-ranking adult male playing with and surrounded by a group of juveniles during the period when his troop's range was small in area and the troop was late in leaving its sleeping sites (May to July). During this time of year cover was much more dense and visibility more restricted so predators were more difficult to detect. Juveniles may therefore have been closer to adult males during this portion of the annual cycle, since they were less likely to be killed by the abundant feline and canine predators which shared this troop's home range.

dense and detection of the abundant predators most difficult, juveniles stayed closer to adult males (Fig. 2), the troop's range was smallest in area, and the baboons stayed latest in their sleeping trees. This study thus revealed an adaptive synchronization to seasonal environmental variation in reproductive biology of the troop and patterns of space use.

Studies of food-enhanced and captive groups. Studies conducted on provisioned groups have often focused on the pattern of spatial distribution of these groups on provisioning grounds (Imanishi, 1960; Yamada, 1966; Alexander and Bowers, 1969; Stephenson, 1974; Casey and Clark, 1976; Fedigan, 1976; Fa, 1986). Although several authors have discussed the importance of environmental design on patterns of social interactions, only a few have quantitatively investigated these relationships (Southwick, 1967; Alexander and Roth, 1971; Casey and Clark, 1976; Rasmussen and Rasmussen, 1979; Mori, 1979; Sugiyama and Ohsawa, 1982; Fa, 1986).

Levels of Analysis in Proximate Social Ecology

Covariation in several types of variables may be analyzed to determine quantitative relationships between environmental design and patterns of social organization (Fig. 4). Examples of how these relationships may be analyzed and their use are discussed below. This is by no means an exhaustive inventory of methods. Rather, the methods are provided as examples of useful means for unraveling the relationships between the environment in which a group is located and aspects of its social organization.

Relationship between actual resources and use of space. Many field studies have attempted to link the manner in which primates use the area in which they live with the actual distribution of resources in those areas. Specifically, these studies have attempted to link aspects of the patterns of range use of groups with the distribution and abundance of food (Rasmussen, 1980).

Some field studies have correlated occupational density of quadrats with the distribution of ecological factors within those quadrats following the approach introduced into primatology by Rodman (1973) and Clutton-Brock (1975). A map of the distribution of a resource is compared with a map of the occupational density of quadrats. Thus if a large amount of time is spent using a resource area a strong relationship can be graphically and quantitatively depicted. This approach is therefore useful for showing how the amount of time spent in an area is related to resources in those areas (see, for example, Figs. 4 and 5, this volume, chapter 2.) A limitation of this analytic approach, which I have called a static analysis elsewhere (Rasmussen, 1980), is that the degree of correlation found is necessarily contingent upon the proportion of time devoted to the use of a given resource (Fig. 3). If members of a group spend 80% of their time gathering and eating food, then we would anticipate a reasonably strong relationship between the occupational density of a given area and a measure of the quality of food in that area. On the other hand, if a group spends a small proportion of time gathering and eating food and the rest of its time is involved in activities associated with use of other resources (mates, shelter areas, water sources, sleeping trees, etc.) and avoidance of hazards (competitors for mates, predators, areas providing no shelter from harsh climatic conditions, etc.) then we would expect a rather poor correlation between patterns of range use and the distribution of food.

A more revealing method than static analysis is the dynamic analysis of covariation in aspects of resource distribution and patterns of range use (Chivers, 1974; Struhsaker, 1974; Armitage, 1975; Oates, 1977; Rasmussen, 1979, 1983; Rasmussen and Rasmussen, 1979; Isbell, 1983; Harrison, 1983). Since changes in environmental variables are quantitatively related to changes in patterns of range use, even a resource that is only used for a few minutes a day, such as water, potentially can be shown to be a major correlate of the manner in which a group uses the area in which it lives. This method is also more consistent with goals of proximate social ecology, since variation in environmental factors can be correlated with aspects of behavior

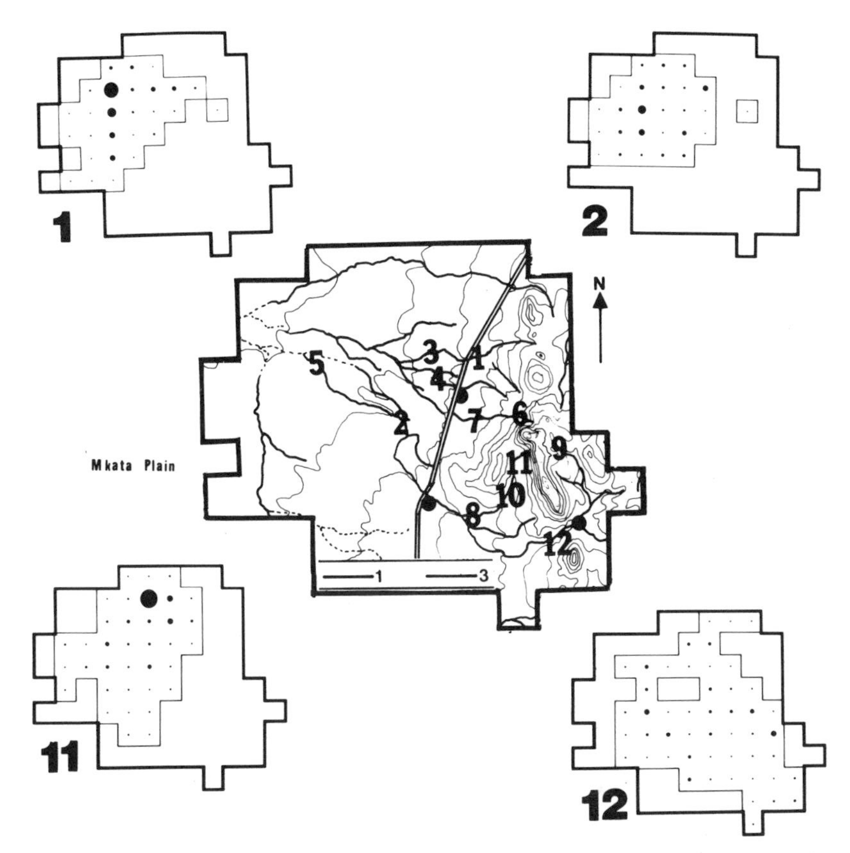

Fig. 3. Comparison of logic used static analyses of range use and dynamic analyses of covariation in patterns of range use and environmental factors. **Center** is a topographical map of the range of a troop of 130 yellow baboons in Mikumi National Park, Tanzania. Large numbers are sleeping sites, circles are water holes, double line is highway, heavy dendritic lines are flash flood channels, fine lines are topographical contours. The scale is in kilometers. Small diagrams indicate patterns of use of range during approximately month-long data blocks **(1,2,11,12).** Diameter of the circles are proportional to the time spent in quadrats.

In a static analysis the amount of time spent in quadrats during a single data block would be correlated with the distribution of resources in those quadrats. Presence or absence of a significant correlation in static analyses is dependent upon the amount of time the troop uses a resource. In a dynamic analysis, measures of variation across patterns of range use, such as area or clumping of occupational density, are correlated with variation in environmental factors, such as food distribution or cover density. In a dynamic analysis, presence or absence of a correlation is dependent upon the degree to which the environmental factors and patterns of range use covary; for example, if cover density increases as area of the range decreases.

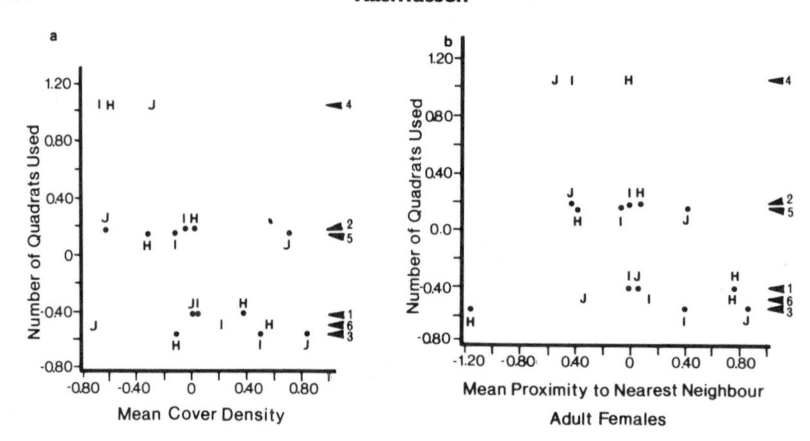

Fig. 4. An example of an analysis of covariation in environmental factors, use of the environment and social behavior. The scattergram to the **left** indicates that when "A" troop, a troop of olive baboons at the Gombe National Park, was using the smallest area during 6 month-long data blocks, the 3 adult males in the troop (H,I,J,) were in least dense cover ($P<0.01$). Cover was less dense because the troop was gathering the remains of fish dried on beaches along the shore of Lake Tanganyika. The scatter diagram to the **right** indicates males also tended to be closer to adult females when their troop's area was smaller in area and more time was spent using the unintentionally provisioned food ($P<0.05$). Arrows indicate value on y-axis for each of the 6, approximately month long, data blocks. The statistical test used was a nonparametric trend analysis for ranked and correlated data (Ferguson, 1965). Values graphed are residuals after statistical control of variation in number of days in data blocks, variation in number of samples, and time since initiation of data collection (Rasmussen, in prep.).

such as intergroup interactions (Struhsaker, 1974; Isbell, 1983), intergroup transfers of individuals and changes in interpersonal spacing (Rasmussen, 1979, 1983). An example of this method is illustrated in Figure 4. Static analyses, as implied by their name, do not allow analysis of this covariation when used in a simple manner.

Relationship between actual resources and used resources. Patterns of range use are not maps of "ranging behavior," an inexact term. Rather, they are patterns made by charting the location of individuals or groups without respect to behavior. Occupational density of space may be analyzed as a function of the major resource or resources being used by a group or individual (Rasmussen and Rasmussen, 1979; Rasmussen, 1980). Division of data on geographical location by resource use provides maps of used resource areas.

Division of data on patterns of range use by used resource areas highlights the necessary relationship between use of space and amount of time spent using a resource area. For example, in the Oregon troop of Japanese macaques (Rasmussen and Rasmussen, 1979), the 20 adult males in the troop tended to rest in the areas that contained shelter and objects of distinctive figure-ground contrast (Menzel, 1966; Estrada and Estrada, 1976), the resources associated with resting; they tended to move on trails, along walls, and on horizontal logs, resources associated with moving (Estrada and Estrada, 1976); and they tended to eat on the food pad, the location where food was provided. Food was the most clumped resource in the corral (1.76% of its area) followed by shelter and areas of figure-ground contrast (18.75% of the corral's area) and by trails and paths (19.92% of the corral's area). Analyses indicated males' patterns of range use were most clumped while feeding, less so when resting, and least while moving. There was thus a strong correspondence between the actual distribution of resources associated with feeding, resting, and moving and the way the males used geographical space when feeding, resting, or moving.

If a correlation coefficient would have been calculated between occupational density of all quadrats and use of the food pad in a static analysis, we would not have found a significant correlation between the distribution of food and males' use of space because the mean amount of time males spent feeding was only 5.59% (SD=2.22). Here, the location where food was provided had to be visited and hence it was an essential resource area in the corral where the troop is housed. Yet static analyses would have failed to indicate a relationship between distribution of food and the pattern with which male Japanese macaques used the area in which they lived.

Analyses of used resource areas may reveal specific aspects of actual resources most attractive to primates. For example, if we find that two areas contain nearly equal quantities of a given resource, say a type of food, and one of the two areas is much preferred over the others; we can then examine differences in the resource areas to determine why one is preferred. Differences might exist in qualities of the food source, distance from other food sources, and distances from resources such as sleeping sites and potentially hazardous areas (Rasmussen and Rasmussen, 1979; Wrangham, 1981).

Relationships between actual and used resource distribution and aspects of social behavior. There is another reason why analyses of used resource distributions are useful in proximate analyses: the distribution of used resource areas contains most of the variation in resource distribution that may be related to aspects of social organization. For example, the abundance and distribution of used food will determine the maximum size of a group of primates (Wrangham, 1980). Resources that are not used will only indirectly influence social behavior.[6]

[6]Food sources that are not used may indirectly influence aspects of social behavior. For example, a high-quality food resource may be avoided because it is often used by a group of higher agonistic rank. Although the food source is not used, the study group might alter its

There is a practical reason for the analysis of used, rather than actual, resource areas: collection of data on used resources takes much less time away from behavioral observations. Charting actual locations of resources can be exceedingly time consuming and this task must usually be performed when not observing primates. In contrast, used resource distributions may be collected during routine behavioral observations (Rasmussen, 1983). If the social ecologist is primarily interested in social behavior and has limited time, it is possible to more than double the amount of behavioral data by focusing on used, rather than actual, resources.

Used resource distributions have been found to be environmental variables strongly related to aspects of social behavior: males of higher agonistic rank in the Oregon troop of Japanese macaques were found to have more clumped distributions of areas used for feeding, resting, and movement (Fig. 5). Males of higher agonistic rank were also found to have a more consistent pattern of use of movement areas because they did not make detours to avoid higher ranking males or females. Older males were found to have more clumped distributions of resting areas. Analyses indicated this latter relationship was not due to decreased movement of older males; agonistic rank and age were not correlated in these males (Eaton, 1976; Rasmussen and Rasmussen, 1979). With advancing age, the males seemed to form greater attachment to specific areas for resting.

The actual distribution of food has also been correlated with aspects of behavior. For example, Southwick (1967) found an increase in aggression in an experimental study of captive rhesus macaques when all food was placed in one basket instead of on eight feeding boards. Chalmers (1968) observed mangabeys *(Cercocebus albigena)* engaged in more aggressive interactions when feeding on clumped and large fruit than when feeding on more dispersed small fruit. Wrangham (1974) found the clumped distribution of provisioned food was linked to frequencies of attacks between chimpanzees *(Pan troglodytes)* and baboons *(P. anubis)*.

Individual differences in use of preferred resource areas. Analyses of differential use of preferred resource areas allow determination of how social relationships between individuals give rise to individual differences in use of geographical space and to characteristic patterns of group spatial structure. For example, in our analyses of the social ecology of adult males in the Oregon troop of Japanese macaques we found males of highest agonistic rank had a much greater frequency of use of the preferred areas for resting and movement. A central tower in the corral where the troop is housed was one

range use to monitor the presence or absence of the higher ranking group. This type of indirect influence could be determined by analysis of the distribution of resource areas used by neighboring groups.

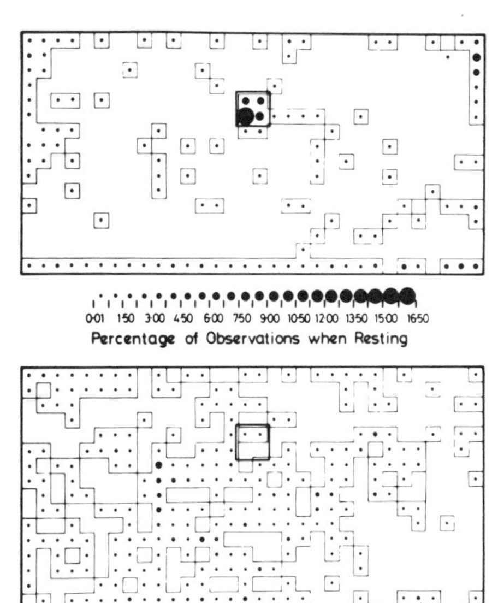

Fig. 5. Patterns of use of resting areas by the male of highest (**top**) and lowest (**bottom**) agonistic rank in the 2-acre corral in which the Oregon troop of Japanese macaques were housed. The size of the circles is proportional to the percentage of time the males spent in each 4 × 4 m quadrat. The 4 quadrats outlined in the center of the corral is the location of the tower.

of the most preferred resting areas. The tower and logs were favorite locations for displays linked to the frequency of male ejaculation (Modahl and Eaton, 1977). There was a strong correlation between the agonistic rank of males and the percentage of time they were observed to rest on this tower (Fig. 6, $r_s = .81$, $p < .001$). Further analyses indicated that significant correlations between agonistic rank of the males and the degree to which their use of resting and moving areas were clumped were due to the limited number of preferred resting and movement areas. Males of high agonistic rank tended to have a more clumped spatial distribution when resting and moving because they used the most preferred and least abundant resource areas.

There was no correlation between agonistic rank and frequency of use of the pad where food was distributed, although males of higher agonistic rank had a more clumped pattern of use of space while feeding. High-ranking males ate earlier from more dense piles of food (Alexander and Bowers, 1969; Casey and Clark, 1976). Higher ranking males also had access to the centers of the piles of food so that a central-peripheral structure formed on piles with highest ranking animals in the center (Alexander and Bowers, 1969). Lower ranking males had a less clumped pattern of use of feeding

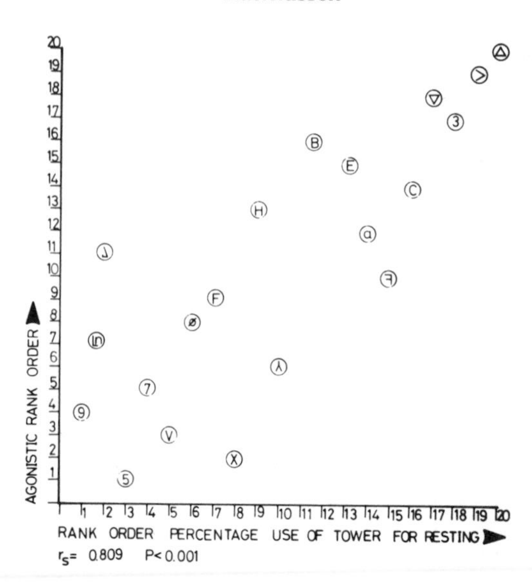

Fig. 6. Scattergram of the relationship between increasing agonistic rank and the rank percentage of time the 20 adult males in the Oregon troop of Japanese macaques were observed to rest on the central tower.

areas since they had to shift position more frequently as they depleted their less abundant food supplies.

The interaction between the distribution of resources in the corral and agonistic rank therefore resulted in rank-related patterns of space use (Rasmussen and Rasmussen, 1979). These results suggest that males' use of space can be indirectly manipulated by altering the distribution and abundance of resources associated with feeding, resting, and movement. These results also suggest that the central-peripheral structure so characteristic of food-enhanced, free-ranging Japanese macaques (Imanishi, 1960; Sugiyama and Ohsawa, 1982) is due to the distribution of provisioned food. Other spatial structures could be created, with possible attendant alterations in social behavior, by manipulating food distribution.

In a functional study, Wrangham (1981) analyzed relationships between agonistic rank and use of preferred resource areas. He found high-ranking female vervet monkeys *(Cercopithecus aethiops)* used drinking areas closer to sleeping areas than did lower ranking females. A result which suggests low-ranking females had to travel farther because they were at a competitive disadvantage. Wrangham found mortality due to poor physical condition associated with limited food and water related to the females' agonistic rank.

CONCLUSION

Studies of food-enhanced groups may have the most ancillary role in historical social ecology. Only if the ancestral habitat includes food enhancement should historical studies be conducted on populations with food enhancement. Studies of food-enhanced groups may help us to understand how current environments may alter aspects of social organization away from that which would exist in an ancestral environment. Studies of food-enhanced groups may also help to determine whether differences between groups have a genetic basis, an essential part of historical studies.

Functional analyses may be conducted with as great facility on food-enhanced groups as on nonprovisioned groups. Studies of food-enhanced groups are less valuable in functional studies conducted to support historical reconstructions of past selective pressures. However, functional studies of food-enhanced groups are essential if we are to understand selective factors currently influencing many of the current populations of nonhuman primates. If these studies are not conducted, in the future we may not be able to understand the reasons for genetic differences that will exist between populations of nonhuman primates kept in zoos, provisioned in the wild, in colonies, and in wild groups. Nonhuman primates are currently being domesticated and functional analyses are essential for understanding the genetic consequences of this domestication. Functional studies of provisioned groups may be useful for models of the manner in which our own very modified environment may influence the genetic structure of our species.

Like functional analyses, proximate studies of primate social ecology can be conducted on both wild and food-enhanced groups. Studies involving the systematic manipulation of food hold the potential for understanding how the distribution, quality, and abundance of food can influence phenotypic patterns of social organization. These studies offer the exciting possibility of fertile merger of theory and methods from the experimental psychology of learning and social ecology. Application of knowledge gained from proximate social ecological studies will aid in successful management of nonhuman primates in captivity. Eventually proximate social ecological studies on nonhuman primates may provide models and theory for sciences concerned with the effects of our environment on human behavior.

ACKNOWLEDGMENTS

I thank Drs. Berman, Dobson, Fa, Forthman Quick, Iwamoto, Kuester, Loy, Lyles, Paul, and Southwick for generously sharing preliminary manuscripts of their chapters. The editors and Dr. Forthman Quick made several useful comments. My field and colony studies have been aided by far too many to list, to all I give my sincere thanks. Research reported in this article has been supported by grants from the L.S.B. Leakey Foundation, NIMH Grant R03-MH24921-01, NIH Grants HD-05969, RR00169-PHS/ DHEW, RR-00163, NSF Grants BMS74-17531, BMS75-05732, and University of California-Riverside Biomedical Science Grants 5-S05-RR07010-0851 & 09. My current research on Panamanian tamarins is supported in part by the School for Field Studies, the L.S.B. Leakey Foundation and the World Wildlife Fund. The writing of this chapter was partially supported by NIMH National Research Senior Service Award 1 F32 MH09419-01 and NIH grant RR00167.

REFERENCES

Alexander BK, Bowers JM (1969): Social organization of a troop of Japanese macaques in a two-acre enclosure. Folia Primatol 10:230–242.

Alexander BK, Roth EM (1971): The effects of acute crowding on aggressive behavior of Japanese monkeys. Behaviour 39:73–90.

Alexander RD, Hoogland JL, Howard RD, Noonan KM, Sherman PW (1979): Sexual dimorphisms and breeding systems in pinnipeds, ungulats, primates, and humans. In Chagnon NA, Irons W (eds): "Evolutionary Biology and Human Social Behavior: An Anthropological Perspective." Duxbury, North Scituate, Massachusetts: pp 402–435.

Altmann SA (1974): Baboons, space, time, and energy. Am Zool 14:221–248.

Altmann SA, Altmann J (1970): "Baboon Ecology: African Field Research." Chicago: University of Chicago.

Anderson CM (1982): Baboons below the tropic of capricorn. J Human Evol 11:205–217.

Anderson CM (1986): Predation and primate evolution. Primates 27:15–39.

Armitage KB (1975): Social behavior and population dynamics of marmots. Oikos 26:341–354.

Bernstein IS (1967): Defining the natural habitat. In Starck D, Scheider R, Kuhn HJ (eds): "Progress in Primatology." Gustav Stuttgart, Fischer Verlag, pp 177–179.

Busse C (1977): Chimpanzee predation as a possible factor in the evolution of red colobus monkey social organization. Evolution 31:907–911.

Busse C (1980): Leopard and lion predation upon chacma baboons living in the Moremi Wildlife Reserve. Botswana Notes Rec 12:15–21.

Calhoun JB (1962): A "behavioral sink." In Bliss EL (ed): "Roots of Behavior." New York: Harper & Brothers, pp 295–315.

Casey DE, Clark TW (1976): Some spacing relations among the central males of a transplanted troop of Japanese macaques (Arashiyama West). Primates 17:433–450.

Charlesworth B (1980): Models of kin selection. In Markl H (ed): "Evolution of Social Behaviour: Hypotheses and Empirical Tests." Weinheim: Verlag Chemie, pp 11–16.

Chalmers NR (1968): Group composition, ecology, and daily activities of free-living mangabeys in Uganda. Folia Primatol 8:247–262.

Chivers DJ (1974): The siamang in Malaya. Contrib Primatol 4:1–335.

Clutton-Brock TH (1975): Ranging behaviour of red colobus *(Colobus badius terphrosceles)* in the Gombe National Park. Anim Behav 23:706–722.

Clutton-Brock TH, Harvey PH (1977): Primate ecology and social organization. J Zool 183:1–39.

Clutton-Brock TH, Harvey PH (1979): Home range size, population density and phylogeny in primates. In Bernstein IS, Smith EO (eds): "Primate Ecology and Human Origins." New York: Garland, pp 210–214.
Clutton-Brock TH, Harvey PH (1980): Primates, brains and ecology. J Zool, Lond 190:309–321.
Clutton-Brock TH, Harvey PH, Rudder B (1977): Sexual dimorphism, socionomic sex ratio and body weight in primates. Nature 269:797–800.
Coelho AM Jr, Bramblett CA, Quick LB, Bramblett SS (1976): Resource availability and population density in primates: A socio-bioenergetic analysis of the energy budgets of Guatemalan howler and spider monkeys. Primates 17:63–80.
Coelho AM Jr, Bramblett CA, Quick LB (1977a): Ecology, population characteristics, and sympatric association in primates: A socio-bioenergetic analysis of howler and spider monkeys in Tikal, Guatemala. Yearbook Phys Anthropol 20:96–135.
Coelho AM, Bramblett CA, Quick LB (1977b): Social organization and food resource availability in primates: A socio-bioenergetic analysis of diet and disease hypotheses. Am J Phys Anthropol 46:253–264.
Coelho AM Jr, Bramblett CA, Quick LB (1979): Activity patterns in howler and spider monkeys: An application of socio-bioenergetic methods. In Bernstein IS, Smith EO (eds): "Primate Ecology and Human Origins: Ecological Influences on Social Organization." New York: Garland, pp 175–199.
Collins DA (1984): Spatial pattern in a troop of yellow baboons *(Papio cynocephalus)* in Tanzania. Anim Behav 32:536–553.
Crook JH (1965): The adaptive significance of avian social organizations. Symp Zool Soc Lond 14:181–218.
Crook JH (1970a): The socio-ecology of primates. In Crook JH (ed): "Social Behaviour in Birds and Mammals." London: Academic Press, pp 103–166.
Crook JH (1970b): Social organization and the environment: Aspects of contemporary social ethology. Anim Behav 18:197–209.
Crook JH, Aldrich-Blake P (1968): Ecological and behavioural contrasts between sympatric ground dwelling primates in Ethiopia. Folia Primatol 8:192–227.
Crook JH, Gartlan JS (1966): Evolution of primate societies. Nature 210:1200–1203.
Daly M, Wilson M (1983): "Sex, Evolution and Behavior." Boston: Willard Grant.
Darwin C (1859): On the origin of species. (edition cited here is a facsimile of the first) New York: Atheneum, 1967.
Dawson GA (1978): Composition and stability of social groups of the tamarin, *Saguinus oedipus geoffroyi* in Panama: Ecological and behavioral implications. In Kleiman DG (ed): "The Biology and Conservation of the Callitrichidae." Washington, DC: Smithsonian Institution Press, pp 23–37.
Denham WW (1971): Energy relations and some basic properties of primate social organization. Am Anthropol 73:77–95.
DeVore I (1963): A comparison of the ecology and behavior of monkeys and apes. Vik Fund Publ Anthropol 37:301–319.
Dittus WP (1977a): The social regulation of population density and age-sex distribution in the toque monkey. Behaviour 63:281–322.
Dittus WPJ (1977b): The socioecological basis for the conservation of the toque monkey *(Macaca sinica)* of Sri Lanka (Ceylon). In Rainier III, Prince, Bourne GH (eds): "Primate Conservation." New York: Academic Press, pp 237–265.
Dittus WP (1979): The evolution of behaviors regulating density and age-specific sex ratios in a primate population. Behaviour 69:265–302.
Dittus WP (1980): The social regulation of primate populations: A synthesis. In Lindburg DG

(ed): "The Macaques: Studies in Ecology, Behavior, and Evolution." New York: Van Nostrand-Reinhold, pp 263–286.

Drickamer LC (1974): A ten year summary of reproductive data for free-ranging *Macaca mulatta*. Folia Primatol 21:61–80.

Duvall SW, Bernstein IS, Gordon TP (1976): Paternity and status in a rhesus monkey group. J Reprod Fertil 47:25–31.

Eaton GG (1976): The social order of Japanese macaques. Sci Am 235:96–106.

Eisenberg JF (1979): Habitat, economy, and society: Some correlations and hypotheses for the neotropical primates. In Bernstein IS, Smith EO (eds): "Primate Ecology and Human Origins." New York: Garland, pp 215–262.

Eisenberg JF, Muckenhirn NA, Rudran R (1972): The relation between ecology and social structure in primates. Science 176:863–874.

Eisenberg JF, Wilson D (1978): Relative brain size and feeding strategies in the chiroptera. Evolution 32:740–751.

Erwin N, Erwin J (1976): Social density and aggression in captive groups of pigtail monkeys *(Macaca nemestrina)*. Appl Anim Ethol 2:265–269.

Estrada A, Estrada R (1976): Establishment of a free-ranging colony of stumptail macaques *(Macaca mulatta)*: Relations to the ecology I. Primates 17:337–355.

Fa JE (1986): Use of time and resources by provisioned troops of monkeys: Social behaviour, time and energy in the Barbary Macaque *(Macaca sylvanus* L.) at Gibraltar. Contrib Primatol 23:1–377.

Fairbanks LA, McGuire MT (1984): Determinants of fecundity and reproductive success in captive vervet monkeys. Am J Primatol 7:27–38.

Fedigan LM (1976): A study of roles in the Arashiyama west troop of Japanese monkeys *(Macaca fuscata)*. Contrib Primatol 9:1–95.

Ferguson GA (1965): "Nonparametric Trend Analysis: A Practical Guide for Research Workers." Montreal: McGill University.

Galat G, Galat-Luong A (1973): Demographie et regime alimentaire d'une troupe de *Cercopithecus aethiops* sabaeus en habitat marginal au Nord Senegal. Terre et Vie 3:557–577.

Gaulin SJC, Sailer LD (1984): Sexual dimorphism in weight among primates: The relative impact of allometry and sexual selection. Int J Primatol 5:515–535.

Goodall J van Lawick (1971): "In the Shadow of Man." Boston: Houghton Mifflin.

Gottlieb G (1984): Evolutionary trends and evolutionary origins: Relevance to theory in comparative psychology. Psychol Rev 91:448–456.

Grafen A (1984): Natural selection, kin selection and group selection. In Krebs JR, Davies NB (eds): "Behavioural Ecology: An Evolutionary Approach." Sunderland, MA: Sinauer, pp 62–84.

Hall KRL (1963): Variations in the ecology of the chacma baboon, *Papio ursinus*. Symp Zool Soc Lond 10:1–28.

Hall KRL (1965): Ecology and behavior of baboons, patas, and vervet monkeys in Uganda. In Vagtborg H (ed): "The Baboon in Medical Research." Austin: University of Texas, pp 43–61.

Hamilton WD (1964): The genetical evolution of social behavior I. & II. J Theoret Biol 7:1–52.

Hamilton WJ III (1985): Demographic consequences of a food and water shortage to desert Chacma baboons, *Papio ursinus*. Int J Primatol 6:451–462.

Harding RSO (1973): Predation by a troop of olive baboons *(Papio anubis)*. Am J Phys Anthropol 38:587–591.

Harrison MJS (1983): Patterns of range use by the green monkey, Cercopithecus sabaeus, at Mt. Assirik, Senegal. Folia Primatol 41: 157–179.

Hausfater G (1975): Dominance and reproduction in baboons *(Papio cynocephalus):* A quantitative analysis. Contrib Primatol 7:1–150.

Hladik CM (1975): Ecology, diet, and social patterning of old and new world primates. In Tuttle RH (ed): "Socioecology and Psychology of Primates." The Hague: Mouton, pp 3–36.

Holden C (1986): A pivotal year for lab animal welfare. Science 232:147–150.

Huntingford FA (1984): Some ethical issues raised by studies of predation and aggression. Anim Behav 32:210–215.

Imanishi K (1960): Social organization of subhuman primates in their natural habitat. Curr Anthropol 1:393–407.

Isbell LA (1983): Daily ranging behavior of red colobus *(Colobus badius tephrosceles)* in Kibale forest, Uganda. Folia Primatol 41:34–48.

Johnson RL, Southwick CH (1984): Structural diversity and mother-infant relations among rhesus monkeys in India and Nepal. Folia Primatol 43:198–215.

Jolly A (1985): "The Evolution of Primate Behavior." New York: Macmillan.

Karn MN, Penrose LS (1951): Birth weight and gestation time in relation to maternal age, parity, and infant survival. Ann Eugenet 161:147–164.

Kaufman IC (1975): Learning what comes naturally: The role of life experience in the establishment of species typical behavior. Ethos 3: 131–142.

Kruuk H, Turner M (1967): Comparative notes on predation by lion, leopard, cheetah and wild dog in the Serengeti area, East Africa. Mammalia 31:1–27.

Kummer H (1971a): "Primate Societies: Group Techniques of Ecological Adaptation." Chicago: Aldine-Atherton.

Kummer H (1971b): Immediate causes of primate social structures. Proc 3rd Int Congr Primatol 3:1–11.

Kummer H, Kurt F (1965): A comparison of social behavior in captive and wild hamadryas baboons. In Vagtborg H (ed): "The Baboon in Medical Research." Austin: University of Texas, pp 65–80.

Kurland JA (1977): Kin selection in the Japanese monkey. Contrib Primatol 12:1–145.

Lack D (1954): "The Natural Regulation of Animal Numbers." Oxford: Clarendon Pres.

Lorenz K (1965): "Evolution and the Modification of Behavior." Chicago: University of Chicago.

Leutenegger W (1978): Scaling of sexual dimorphism in body size and breeding system in primates. Nature 272:610–611.

Leutenegger W (1982): Sexual dimorphism in nonhuman primates. In Hall RL (ed): "Sexual Dimorphism in *Homo sapiens,* A Question of Size." New York: Praeger, pp 11–36.

MacKinnon J (1974): The behaviour and ecology of wild orang-utans *(Pongo pygmaeus).* Anim Behav 22:3–74.

Malik I, Seth PK, Southwick CH (1984): Population growth of free-ranging rhesus monkeys at Tughlaqabad, northern India. Am J Primatol 7:311–321.

Marsh CW (1981): Ranging behaviour and its relation to diet selection in Tana River red colobus *(Colobus badius rufomiteatus).* J Zool, Lond 195:473–492.

Martin RD (1981): Field studies of primate behavior. Symp Zool Soc Lond 46:287–336.

Menzel EW (1966): Responsiveness to objects in free-ranging Japanese monkeys. Behaviour 26:130–150.

Modahl KB, Eaton GG (1977): Display behaviour in a confined troop of Japanese macaques *(Macaca mulatta).* Anim Behav 25:525–535.

Mori A (1979): Analysis of population changes by measurement of body weight in the Koshima troop of Japanese macaques. Primates 20:371–398.

Morrison JA, Menzel EW Jr (1972): Adaptation of a free-ranging rhesus monkey group to division and transplantation. Wildl Monogr 31:5–78.

Moynihan M (1970): Some behavior patterns of Platyrrhine monkeys. II. *Saguinus geoffroyi* and some other tamarins. Smith Contrib Zool 28:1–60.

Nagel U (1971): Social organization in a baboon hybrid zone. Proc 3rd Int Congr Primatol 3:48–57.

Nagel U (1973): A comparison of anubis baboons, hamadryas baboons and their hybrids at a species border in Ethiopia. Folia Primatol 19:104–165.

Neyman PF (1978): Aspects of the ecology and social organization of free-ranging cotton-top tamarins *(Saguinus oedipus)* and the conservation status of the species. In Kleiman DG (ed): "The Biology and Conservation of the Callirichidae." Washington, DC: Smithsonian Institution, pp 39–71.

Nieuwenhuijsen K, Lammers AJJC, de Neef KJ, Slob AK (1985): Reproduction and social rank in female stumptail macaques *(Macaca arctoides)*. Int J Primatol 6:77–99.

Oates JF (1977): The guereza and its food. In Clutton-Brock TH (ed): "Primate Ecology: Studies of Feeding and Ranging Behaviour in Lemurs, Monkeys and Apes." London: Academic Press, pp 275–321.

Patterson IJ (1965): Timing and spacing of broods in the black-headed gull *(Larus ridibundus* L.). Ibis 107:433–460.

Post DG (1984): Is optimization the optimal approach to primate foraging? In Rodman PS, Cant JGH (eds): "Adaptations for Foraging in Nonhuman Primates." New York: Columbia University Press, pp 280–303.

Rasmussen DR (1979): Correlates of patterns of range use of a troop of yellow baboons *(Papio cynocephalus)*. I. Sleeping sites, impregnable females, births, and male emigrations and immigrations. Anim Behav 27:1098–1112.

Rasmussen DR (1980): Clumping and consistency of primates' patterns of range use: Definitions, sampling, assessments, and applications. Folia Primatol 34:111–139.

Rasmussen DR (1981a): Evolutionary, proximate and functional primate social ecology. In Bateson PPG, Klopfer PR (eds): "Perspectives in Ethology. Vol. 4." New York: Plenum Press, pp 75–103.

Rasmussen DR (1981b): Communities of baboon troops *(Papio cynocephalus)* in Mikumi National Park, Tanzania: A preliminary report. Folia Primatol 36:232–242.

Rasmussen DR (1981c): Pair-bond strength and stability and reproductive success. Psychol Rev 88:274–290.

Rasmussen DR (1983): Correlates of patterns of range use of a troop of yellow baboons *(Papio cynocephalus)*. II. Spatial structure, cover density, food gathering, and individual behavior patterns. Anim Behav 31:834–856.

Rasmussen DR (1984): Functional alterations in the social organization of bonnet macaques *(Macaca radiata)* induced by ovariectomy: An experimental analysis. Psychoneuroendocrinology 9:343–374.

Rasmussen DR (in prep.): Seasonal covariation in a troop of olive baboon's *(Papio anubis)* range use, behavior and environment at the Gombe National Park, Tanzania.

Rasmussen DR, Rasmussen KL (1979): Social ecology of adult males in a confined troop of Japanese macaques *(Macaca fuscata)*. Anim Behav 27:434–445.

Rensch B (1959): "Evolution Above the Species Level." New York: Columbia University Press.

Rodman PS (1973): Synecology of Bornean primates. I. A test for interspecific interactions in spatial distribution of five species. Am J Phys Anthropol 38:655–660.

Rowell TE (1967): Variability in the social organization of primates. In Morris D (ed): "Primate Ethology." Garden City, New York: Anchor Books, pp 283–305.

Rowell TE (1979): How would we know if social organization were not adaptive? In Bernstein IS, Smith EO (eds): "Primate Ecology and Human Origins: Ecological Influences on Social Organization." New York: Garland STPM, pp 1–22.

Saayman GS (1971): Baboons' responses to predators. Afr Wild Life, 25:46–49.

Sade DS, Cushing K, Cushing P, Dunai J, Figueroa A, Kaplan JR, Lauer S, Rhodes D, Schneider J (1976): Population dynamics in relation to social structure on Cayo Santiago. Yearbook Phys Anthropol 20:253–262.

Schaller GB (1963): "The Mountain Gorilla: Ecology and Behavior." Chicago: Chicago University Press.

Sigg H (1980): Differentiation of female positions in hamadryas one-male-units. Z Tierpsychol 53:265–302.

Skinner C (1985): A field study of Geoffroy's tamarin *(Saguinus geoffroyi)* in Panama. Am J Primatol 9:15–25.

Small MF, Smith DG (1986): The influence of birth timing upon infant growth and survival in captive rhesus macaques. Int J Primatol 7:289–304.

Smith DG (1981): The association between rank and reproductive success of male rhesus monkeys. Am J Primatol 1:83–90.

Southwick CH (1967): An experimental study of intragroup agonistic behavior in rhesus monkeys *(Macaca mulatta)*. Behaviour 28:182–209.

Southwick CH, Beg MA, Siddiqi, MR (1961): A population survey of rhesus monkeys in Northern India. II. Transportation routes and forest areas. Ecology 42:699–710.

Southwick CH, Beg MA, Siddiqi MR (1965): Rhesus monkeys in North India. In DeVore I (ed): "Primate Behavior: Field Studies of Monkeys and Apes." New York: Holt, Rinehart & Winston, pp 11–159.

Stammbach E (1978): On social differentiation in groups of captive female hamadryas baboons. Behaviour 67:322–338.

Stephenson GR (1974): Social structure of mating activity in Japanese macaques. Symp 5th Cong Int Primatol Soc. pp. 63–115.

Struhsaker TT (1973): A recensus of vervet monkeys in the Masai-Amboseli game reserve, Kenya. Ecology 54:930–932.

Struhsaker TT (1974): Correlates of ranging behavior in a group of red colobus monkeys *(Colobus badius tephrosceles)*. Am Zool 14:177–184.

Struhsaker TT (1976): A further decline in numbers of Amboseli vervet monkeys. Biotrop 8:211–214.

Sugiyama Y, Ohsawa H (1982): Population dynamics of Japanese monkeys with special reference to the effect of artificial feeding. Folia Primatol 39:238–263.

Tinbergen N (1951): "The Study of Instinct." London: Clarendon.

Tinbergen N (1959): Comparative studies of the behaviour of gulls (Laridae): A progress report. Behaviour 15:1–70.

Tinbergen N (1963): On aims and methods of ethology. Z für Tierpsychol 20:410–433.

Tinbergen N (1965): Behaviour and natural selection. In Moore JA (ed): "Ideas in Modern Biology." New York: Natural History Press, pp 521–542.

Tinbergen N (1969): Ethology. In Harre R (Ed): "Scientific Thought." Oxford: Clarendon Press, pp 238–268.

Turnbull-Kemp P (1967): "The Leopard." Cape Town, Timmins.

Van Schaik CP, Van Noordwijk MA, deBoer R, den Tonkelaar I (1983): The effect of group size on time budgets and social behaviour in wild long-tailed macaques *(Macaca fascicularis)*. Behav Ecol Sociobiol 13:173–181.

Wada K, Matsuzawa T (1986): A new approach to evaluating troop deployment in wild Japanese monkeys. Int J Primatol 7:1–16.

Waddington CH (1957): "The Strategy of Genes." London: Allen & Unwin.
Washburn SL, DeVore I (1961): Social behavior of baboons and early man. Vik Fund Publ Anthropol 31:91–104.
Wecker SC (1963): The role of early experience in habitat selection by the prairie deer mouse, *Peromyscus maniculatus bairdi*. Ecol Monogr 33:307–325.
Wilson ME, Gordon TP, Bernstein IS (1978): Timing of births and reproductive success in rhesus monkey social groups. J Med Primatol 7:202–212.
Wrangham RW (1974): Artificial feeding of chimpanzees and baboons in their natural habitat. Anim Behav 22:83–93.
Wrangham RW (1977): Feeding behaviour of chimpanzees in Gombe National Park, Tanzania. In Clutton-Brock TH (ed): "Primate Ecology: Studies of Feeding and Ranging Behaviour in Lemurs, Monkeys and Apes." London: Academic Press, pp 503–538.
Wrangham RW (1980): An ecological model of female-bonded primate groups. Behaviour 75:262–300.
Wrangham RW (1981): Drinking competition in vervet monkeys. Anim Behav 29:904–910.
Yamada M (1966): Five natural troops of Japanese monkeys on Shodoshima Island. I. Distribution of social organization. Primates 7:315–362.

Index